ACS SYMPOSIUM SERIES **630**

Herbicide Metabolites in Surface Water and Groundwater

M. T. Meyer, EDITOR
U.S. Geological Survey

E. M. Thurman, EDITOR
U.S. Geological Survey

Developed from a symposium sponsored
by the Division of Agrochemicals and
the Division of Environmental Chemistry, Inc.,
at the 209th National Meeting
of the American Chemical Society,
Anaheim, California,
April 2–7, 1995

American Chemical Society, Washington, DC

Library of Congress Cataloging-in-Publication Data

Herbicide metabolites in surface water and groundwater / M. T. Meyer, editor, E. M. Thurman, editor.

    p.   cm.—(ACS symposium series, ISSN 0097–6156; 630)

"Developed from a symposium sponsored by the Division of Agrochemicals and the Division of Environmental Chemistry, Inc., at the 209th National Meeting of the American Chemical Society, Anaheim, California, April 2–6, 1995."

Includes bibliographical references and indexes.

ISBN 0–8412–3405–1

1. Herbicides—Environmental aspects. 2. Metabolites—Environmental aspects. 3. Water—Pollution.

I. Meyer, M. T., 1952– . II. Thurman, E. M. (Earl Michael), 1946– . III. American Chemical Society. Division of Agrochemicals. IV. American Chemical Society. Division of Environmental Chemistry. V. American Chemical Society. Meeting (209th: 1995: Anaheim, Calif.) VI. Series.

TD427.H46H47   1996
628.1'.6842—dc20   96–20170
                                                        CIP

This book is printed on acid-free, recycled paper.

Advisory Board

ACS Symposium Series

Foreword

THE ACS SYMPOSIUM SERIES was first published in 1974 to provide a mechanism for publishing symposia quickly in book form. The purpose of this series is to publish comprehensive books developed from symposia, which are usually "snapshots in time" of the current research being done on a topic, plus some review material on the topic. For this reason, it is necessary that the papers be published as quickly as possible.

Before a symposium-based book is put under contract, the proposed table of contents is reviewed for appropriateness to the topic and for comprehensiveness of the collection. Some papers are excluded at this point, and others are added to round out the scope of the volume. In addition, a draft of each paper is peer-reviewed prior to final acceptance or rejection. This anonymous review process is supervised by the organizer(s) of the symposium, who become the editor(s) of the book. The authors then revise their papers according to the recommendations of both the reviewers and the editors, prepare camera-ready copy, and submit the final papers to the editors, who check that all necessary revisions have been made.

As a rule, only original research papers and original review papers are included in the volumes. Verbatim reproductions of previously published papers are not accepted.

ACS BOOKS DEPARTMENT

Contents

Preface ix

1. **Herbicide Metabolites in Surface Water and Groundwater: Introduction and Overview** 1
   E. M. Thurman and M. T. Meyer

ANALYTICAL METHODS

2. **Coordinating Supercritical Fluid and Solid-Phase Extraction with Chromatographic and Immunoassay Analysis of Herbicides** 18
   Martha J. M. Wells and G. Kim Stearman

3. **A High-Performance Liquid Chromatography–Based Screening Method for the Analysis of Atrazine, Alachlor, and Ten of Their Transformation Products** 34
   Blaine R. Schroyer and Paul D. Capel

4. **Factors Influencing the Specificity and Sensitivity of Triazine Immunoassays** 43
   Timothy S. Lawruk, Charles S. Hottenstein,
   James R. Fleeker, Fernando M. Rubio,
   and David P. Herzog

5. **Standardization of Immunoassays for Water and Soil Analysis** 53
   Bertold Hock, P.-D. Hansen, A. Krotzky, L. Meitzler,
   E. Meulenberg, G. Müller, U. Obst, F. Spener,
   U. Strotmann, L. Weil, and C. Wittmann

6. **In Situ Derivatization–Supercritical Fluid Extraction Method for the Determination of Chlorophenoxy Acid Herbicides in Soil Samples** 63
   Viorica Lopez-Avila, Janet Benedicto,
   and Werner F. Beckert

7. Application of In Vivo Fluorometry To Determine Soil
   Mobility and Soil Adsorptivity of Photosynthesis-Inhibiting
   Herbicides 77
   Daisuke Yanase, Misako Chiba, Katsura Yagi,
   Mitsuyasu Kawata, and Yasushi Takagi

FATE AND TRANSPORT

8. Interactions Between Atrazine and Smectite Surfaces 86
   David A. Laird

9. Estimation of the Potential for Atrazine Transport in a Silt
   Loam Soil 101
   David A. V. Eckhardt and R. J. Wagenet

10. The Effect of Ammonia on Atrazine Sorption and Transport 117
    S. A. Clay, D. E. Clay, Z. Liu, and S. S. Harper

11. Fate of a Symmetric and an Asymmetric Triazine Herbicide
    in Silt Loam Soils 125
    W. C. Koskinen, J. S. Conn, and B. A. Sorenson

12. Fate of Atrazine and Atrazine Degradates in Soils of Iowa 140
    Ellen L. Kruger and Joel R. Coats

13. Transport of Nutrients and Postemergence-Applied Herbicides
    in Runoff from Corrugation Irrigation of Wheat 151
    A. J. Cessna, J. A. Elliott, K. B. Best, R. Grover,
    and W. Nicholaichuk

14. Potential Movement of Certain Pesticides Following
    Application to Golf Courses 165
    A. E. Smith and D. C. Bridges

15. Relation of Landscape Position and Irrigation
    to Concentrations of Alachlor, Atrazine, and Selected
    Degradates in Regolith in Northeastern Nebraska 178
    Ingrid M. Verstraeten, D. T. Lewis, Dennis L. McCallister,
    Anne Parkhurst, and E. M. Thurman

WATER QUALITY STUDIES

16. The Environmental Impact of Pesticide Degradates
    in Groundwater 200
    Michael R. Barrett

17. Herbicide Mobility and Variation in Agricultural Runoff
    in the Beaver Creek Watershed in Nebraska.................................. 226
    Li Ma and Roy F. Spalding

18. Monitoring Pesticides and Metabolites in Surface Water
    and Groundwater in Spain... 237
    D. Barceló, S. Chiron, A. Fernandez-Alba, A. Valverde,
    and M. F. Alpendurada

19. Hydroxylated Atrazine Degradation Products in a Small
    Missouri Stream.. 254
    Robert N. Lerch, William W. Donald, Yong-Xi Li,
    and Eugene E. Alberts

20. Assessment of Herbicide Transport and Persistence
    in Groundwater: A Review.. 271
    S. K. Widmer and Roy F. Spalding

21. Cyanazine, Atrazine, and Their Metabolites as Geochemical
    Indicators of Contaminant Transport in the Mississippi
    River 288
    M. T. Meyer, E. M. Thurman, and D. A. Goolsby

Author Index 305

Affiliation Index 305

Subject Index.. 306

Preface

Since the introduction of the herbicide 2,4-D in 1945, many thousands of journal articles and several volumes have been published on herbicides. The majority of publications have explored herbicide characteristics related to weed control and crop tolerance. These characteristics include sorption, bonding, leaching, solubility, vapor pressure, half-life, and degradation pathways. In addition, many studies have been conducted on herbicides in surface water, and these studies have demonstrated that herbicides are common contaminants in surface water. Much less has been written on herbicide transport and occurrence in groundwater. However, this is changing.

As a result of mounting interest in the state of our groundwater resources in the 1980s and the improved ability over the last decade to detect compounds by gas chromatography–mass spectroscopy and high-performance liquid chromatography, research on the occurrence and transport of herbicides into groundwater has expanded. This research has found that herbicides are commonly detected in groundwater, but at much lower concentrations than in surface water. Because research has shown that herbicides are common contaminants in surface water and groundwater, a logical hypothesis to test was that herbicide metabolites are also present in water. Several recent studies have shown this to be the case. The term metabolite in this volume refers to direct and indirect biotic and abiotic degradation products of a parent herbicide.

The study of herbicide metabolites has become an important research area in agricultural and environmental chemistry. As a result, this symposium was designed to give an overview of the environmental chemistry of herbicide metabolites and to provide specific examples of research being conducted in the broad categories of analytical methods, fate and transport, and water-quality surveys. It is our hope that this symposium will lead to future research dedicated to the fate of herbicide metabolites in surface water and groundwater and provide a better understanding of the environmental chemistry of pesticides. This volume was prepared to benefit environmental and agrochemical researchers, residue chemists, environmental regulators, and those in positions to recommend areas of study or establish policy at agrochemical companies, universities, and federal, state, and local government agencies.

Acknowledgments

We acknowledge the financial support provided by Ciba Geigy, DuPont, Monsanto, and the Agrochemicals Division and Environmental Chemistry, Inc., Division of the American Chemical Society, which helped to make this symposium a successful endeavor.

M. T. MEYER
U.S. Geological Survey
3816 Sunset Ridge Road
Raleigh, NC 27607

E. M. THURMAN
U.S. Geological Survey
4821 Quail Crest Place
Lawrence, KS 66049

March, 1996

Chapter 1

Herbicide Metabolites in Surface Water and Groundwater: Introduction and Overview

E. M. Thurman[1] and M. T. Meyer[2]

[1]U.S. Geological Survey, 4821 Quail Crest Place, Lawrence, KS 66049
[2]U.S. Geological Survey, 3916 Sunset Ridge Road, Raleigh, NC 27607

Several future research topics for herbicide metabolites in surface and ground water are outlined in this chapter. They are herbicide usage, chemical analysis of metabolites, and fate and transport of metabolites in surface and ground water. These three ideas follow the themes in this book, which are the summary of a symposium of the American Chemical Society on herbicide metabolites in surface and ground water. First, geographic information systems allow the spatial distribution of herbicide-use data to be combined with geochemical information on fate and transport of herbicides. Next these two types of information are useful in predicting the kinds of metabolites present and their probable distribution in surface and ground water. Finally, methods development efforts may be focused on these specific target analytes. This chapter discusses these three concepts and provides an introduction to this book on the analysis, chemistry, and fate and transport of herbicide metabolites in surface and ground water.

Organic contaminants are a major concern for most of the rivers, streams, and ground water of agricultural areas. Each year millions of dollars are spent by private, government, and university laboratories analyzing water samples for organic compounds. In the United States, there have been numerous studies completed by the U.S. Environmental Protection Agency (*1*), the U.S. Department of Agriculture (*2*), the U.S. Geological Survey (*3-5*), and by many manufacturers of pesticides (*6*). Numerous surveys of ground water, both at the State level (*7*) and at the Federal level (*8-9*) have been conducted. The political outcome of these efforts has resulted in three major changes in pesticide usage in the Corn Belt. First is a label change for atrazine to reduce its usage in sensitive areas and a full re-evaluation of atrazine by EPA, which is still underway. Second is a shift from alachlor to acetochlor as a major chloracetanilide herbicide in the Corn Belt. Third is the removal of cyanazine from the market by the year 2000. Thus, the numerous studies have had a significant effect on usage of parent compounds.

However, there has been much less study on the metabolites of these compounds. In the European Community, regulations for both herbicides and their metabolites in ground water are at 0.1 µg/L per compound, and a total of all parents and metabolites of 0.5 µg/L. However, in the United States, health advisories have not been established for metabolites, and the possibility of summing parent and metabolites to meet the U.S. Environmental Protection Agency's MCL (Maximum Contaminant Level) is still being considered. For example, the state of Wisconsin has set a limit of 3.0 µg/L for the sum of atrazine, deethylatrazine, and deisopropylatrazine concentrations in ground water under the Wisconsin Ground-Water Act 410 (1983), Rule under the law, Enforcement Standard, Chapter NR 140, Wisconsin ADM CODE (1991). Other states may follow their example.

At this time (1995), there is a critical environmental concern about herbicides and their metabolites in the environment. One example is the soil metabolite of alachlor, 2-[(2',6'-diethylphenyl)(methoxymethyl)-amino]-2-oxoethanesulfonate, or ESA. A recent survey of ~8,000 samples of ground water for alachlor by immunoassay resulted in a high frequency of false positives, which have been attributed to the ESA metabolite (10). Apparently the ESA metabolite of alachlor leaches much more rapidly through the soil than does the parent compound and makes an important contribution to the total organic contaminant load of ground water in the central United States, whereas alachlor does not (10). Likewise, ESA has been reported as an important metabolite of alachlor in surface water and reservoirs of the central United States, often occurring at higher concentrations than parent compound (11).

Another example is deethylatrazine (DEA) and deisopropylatrazine (DIA), which are two dealkylated metabolites of four parent triazines, atrazine, cyanazine, propazine, and simazine (12). Both DEA and DIA are found frequently in surface water of the central United States (13). Their concentrations generally vary from 10 to 50% of parent concentrations. Because both metabolites are chlorinated they still retain some herbicidal activity (12). Recent work shows that these metabolites also may be used as indicators of surface- and ground-water interaction (14). These two examples show that metabolites of herbicides are important contributors to the pool of organic contaminants that enter surface and ground water, and they are an area of water chemistry that deserves considerably more attention and study.

A final consideration is the toxicity of herbicide metabolites in the environment. Although toxicity studies are not required by U.S. Environmental Protection Agency for minor herbicide metabolites (less than 10% of the C-14 metabolites of parent compound), these minor metabolites often become more important in the environment because of their persistence in soil and water.

Furthermore, halogenated metabolites of parent compounds often have herbicidal activity and the possibility exists for toxic effects on aquatic life.

This book, "*Herbicide Metabolites in Surface and Ground Water*", deals with these topic in a series of chapters looking at a variety of herbicides in surface and ground water from both Europe (Spain) and the United States. There are chapters on runoff, sorption, and transport in both surface and ground water. There is work on fate in soil using C-14 atrazine, as well as chapters on chemical analysis of herbicides using enzyme linked immunosorbent assay (ELISA), supercritical fluid extraction (SFE), solid phase extraction (SPE), and high performance liquid chromatography (HPLC). There are also chapters dealing with conceptual models and landscape geochemistry. Thus, this monograph, which is the result of the American Chemical Society Symposium on "Herbicide Metabolites in Surface and Ground Water", examines the chemistry, fate, and transport of herbicides and their metabolites.

This introductory chapter will be divided into three sections to introduce the topics of the book. The first section will deal with herbicide usage and potential sources of metabolites. The second section will discuss chemical analysis of metabolites, and the third section will discuss fate and transport of herbicide metabolites in surface and ground water.

Herbicide Usage

Usage for the top 10 herbicides applied in the United States for 1992 (current data) is shown in Figure 1. The major herbicides in order of usage are atrazine, metolachlor, alachlor, 2,4-D, cyanazine, trifluralin, pendimethalin, glyphosate, EPTC, and butylate. Of these top 10 compounds, only atrazine, alachlor, metolachlor, and cyanazine have had field studies completed on metabolites in surface and ground water (*3-9*). Although there have been studies of the degradation of most of these herbicides in laboratory and field experiments, there have been few detailed studies of the occurrence of these metabolites in the environment.

One reason for hesitancy in studying herbicide metabolites is the difficulty and cost associated with the analyses of numerous compounds. For example, the degradation pathway of cyanazine, a major triazine herbicide that is used across the Corn Belt and in the southern United States on cotton, contains at least nine soil metabolites. First is hydroxycyanazine and the de-alkylated metabolites of hydroxycyanazine (Figure 2). These compounds are not soluble and are tightly bound to soil, where they undergo degradation (*15-16*). The chlorinated de-alkylated metabolites are more soluble than parent compound because of dealkyation and may be quite mobile in surface water (17). Meyer (*17*) found that cyanazine amide is a major metabolite of cyanazine and often has concentrations in surface water that are nearly equal to cyanazine, especially later in the growing season after degradation in the soil. Furthermore, when the halogen atom is still present on the metabolite, then there often may be herbicidal activity and sometimes toxicity. Thus, the first problem is correct selection of metabolites for methods development and water-quality studies. This question will be addressed later in this chapter in the "Fate and Transport" section.

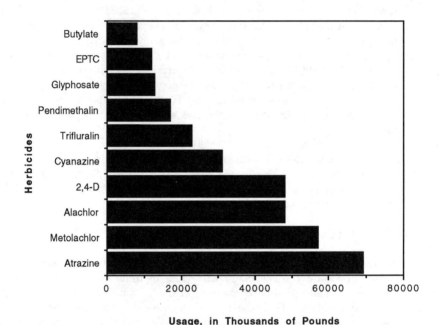

Figure 1. Top ten herbicides used in the United States, 1992, data from reference (*31*).

A second question is where are the areas of the country that are of concern for herbicide metabolites in surface and ground water? One approach to this question may be addressed by a geographic information system (GIS). A GIS map of herbicide use could give insight into the areas of the country where herbicide studies of both parent compound and metabolites might be valuable. For example, Figure 3 shows the use of cyanazine and trifluralin in the United States by county. Cyanazine is used in the Midwestern United States on corn (Minnesota, Iowa, Nebraska, Wisconsin, and Illinois) and in the south and west on cotton (Mississippi, Louisiana, Arkansas, and California). Thus, the analysis of key metabolites shown in Figure 2 could be focused on these areas of the country. Trifluralin (used on soybeans and wheat) shows a similar distribution in the Midwest and extends even farther north to North Dakota. It too is used in the south and west on cotton. Little has been reported on either trifluralin or its metabolites in the United States, in spite of its heavy use.

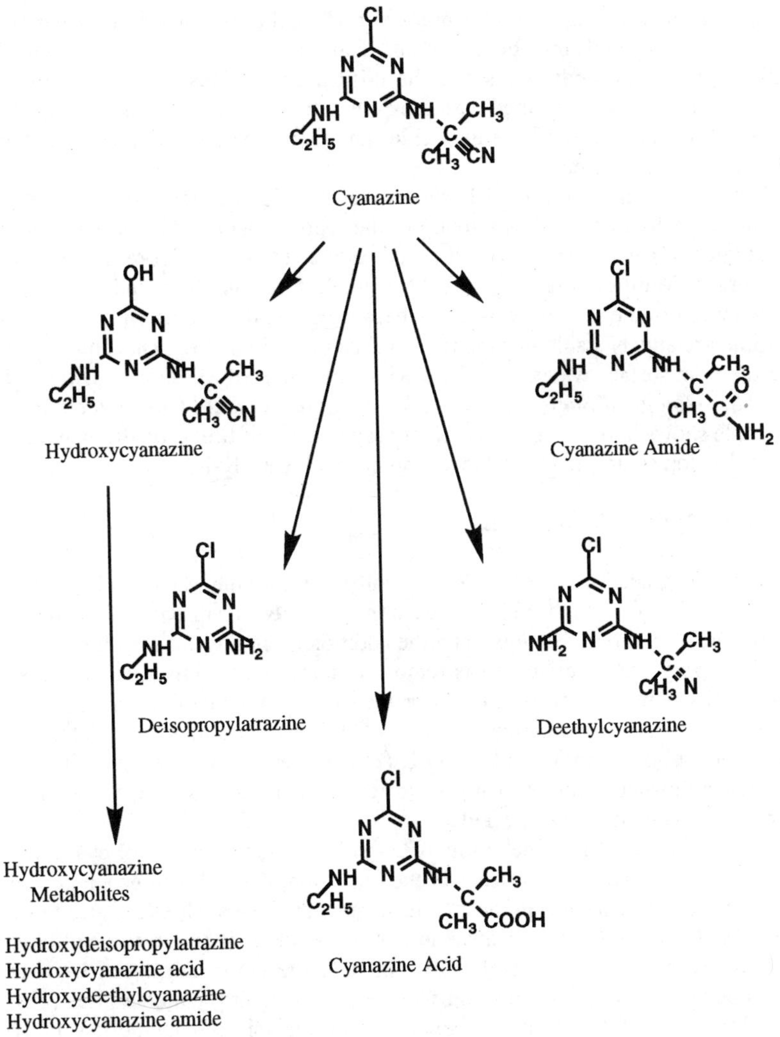

Figure 2. Degradation pathway for the corn and cotton herbicide, cyanazine (*15-17*).

The distribution of the rice herbicides, molinate and propanil, is shown by county in the United States in Figure 4. These compounds are not among the top 10 herbicides shown in Figure 1, but because of the focused use of these rice herbicides in a small area of the country (the Delta Region in Mississippi, Louisiana, and

Texas, and the rice-growing area of central California), the concentrations of parent and metabolite compounds may be important. Thus, a GIS approach is quite useful for outlining potential study areas and the critical compounds that are applied. When this information is combined with previous studies of fate and transport, then areas of focus for new research on herbicides and their metabolites in surface and ground water become obvious.

The propanil map (Figure 4B) shows that this rice herbicide is commonly used in the Delta Region of Mississippi, but the reported half life of propanil is 2 days in studies of aerobic and anaerobic rice paddies (*18*). Therefore, in spite of intensive use (7.5 million pounds in a focused area), it may not be found in surface and ground water of the rice-growing areas, because of rapid degradation. However, it does form a relatively stable metabolite, 3,4-dichloroaniline (18). This compound is being found in surface waters that drain the rice paddies of Mississippi (*Richard Coupe, U.S. Geological Survey, Jackson, Mississippi, Personal Communication*). This example shows how the combination of usage data, half life, and GIS maps can be used to develop strategies for herbicide and metabolite analysis.

Chemical Analysis of Metabolites

The process of degradation of herbicides generally increases the water solubility and the polarity of the compound. The increase in solubility is caused by the loss of carbon, the incorporation of oxygen, and the addition of carboxylic-acid functional groups. For every carbon atom that is removed, the water solubility will increase from two to three times. For example, atrazine's maximum solubility is 33 mg/L, deethylatrazine (loss of two carbon atoms) is 670 mg/L, and deisopropylatrazine (loss of three carbon atoms) is 3200 mg/L (*19*). Increases in water solubility is dramatic when the metabolite is a carboxylic acid, which often happens when there is an alkyl side chain that can be degraded.

The example of cyanazine in Figure 2 shows that the solubility of cyanazine acid will increase substantially because of the formation of a carboxyl group and an anion. Likewise the cyanazine amide and the de-alkylated metabolites are also more soluble (17). Even the hydroxycyanazine is more soluble, although the hydroxyl metabolites are not mobile in soil (*12*). In fact, the entire family of triazine herbicides behaves in a similar fashion with respect to hydroxy metabolites (12). The increase in polarity makes the isolation and analysis of metabolites more difficult than the analysis of the parent compounds. Thus, the analysis of herbicide metabolites almost always begins with the problem of the isolation of metabolites from the water sample.

Recent advances in solid phase extraction (SPE) have greatly increased the ability to isolate herbicide metabolites from water. For example, there are a number of new methods for the isolation of polar organic metabolites that are considerably more efficient than the typical liquid extraction. Recent innovations in SPE include: graphitized carbon (*20*), C-18 and styrene-divinylbenzene extraction disks (*21*), and anion-exchange extraction disks. The graphitized carbon has been shown to isolate water soluble compounds through the high surface area of the carbon and possibly

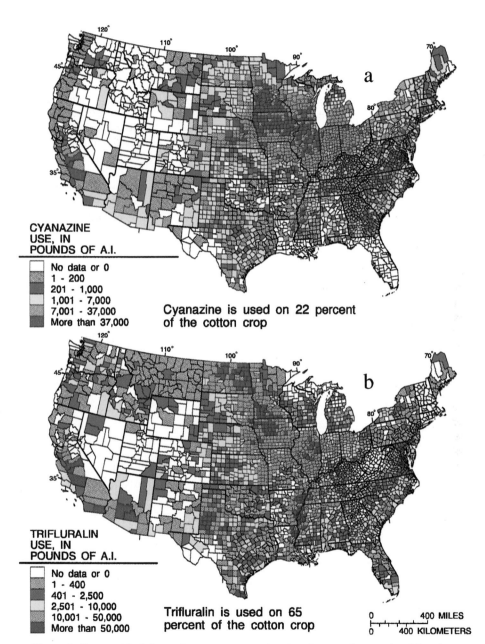

Figure 3. Geographic-information-system map showing the use of cyanazine (a) and trifluralin (b) in the United States *(32)*, as pounds of active ingredient (A.I.).

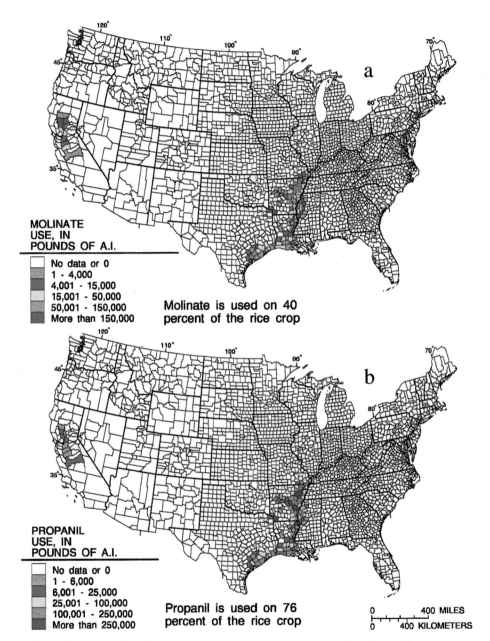

Figure 4. Geographic-information-system map showing the use of molinate (a) and propanil (b) in the United States (32), as pounds of active ingredient (A.I.).

due to the incorporation of oxygen into the graphitized carbon and the creation of positively charged sites, which give a certain degree of anion-exchange capacity to the packing material (20). A recent paper outlines procedures for use of carbon for recovery of herbicides and their metabolites (*20*). See also chapters by Wells and Stearman and by Papilloud and Haerdi on solid phase extraction and supercritical fluid extraction methods in this book.

The Empore disk is another innovation in SPE that has been used in environmental chemistry for the isolation of herbicides and metabolites. The disk consists of a Teflon web that has 8 micron particles of packing material (C-18, for example) that are embedded in a Teflon matrix. This is a smaller particle size than is conventionally used in SPE (~40 μm is conventional), and the smaller particles give the disks excellent mass transfer during sorption, which means that the rate of uptake is fast. Consequently, the water sample may be quickly passed through the disk with good recovery of both parent compound and metabolites. The retention characteristics of the packing material are generally equivalent to the larger particle size, but the rates of sorption are considerably faster with the disks.

There are a variety of packing materials that have been used in the disks, including styrene divinylbenzene (SDB) and anion-exchange packing. The SDB disks have greater surface area than the C-18 disks often resulting in higher capacities for metabolites than C-18, but usually less than the graphitized carbon. Anion exchange has not been used frequently for contaminant analysis because recoveries may be poor as a result of the slow kinetics of sorption. Thus, the faster rates of mass transfer in the disks may be useful for anion exchange and the disks appear to work well for metabolite isolation when carboxyl groups are involved (*22*).

Yet another recent innovation is the derivatization of metabolites on the anion-exchange disks before elution. Field and Monihan (*22*) used this method to look at metabolites of dacthal (DCPA) in ground water. The analytes contain as many as two carboxylic acid groups, which makes the molecule quite soluble. Nonetheless, they isolate effectively and were derivatized on the anion-exchange disk, eluted in the sample vial, and then directly injected onto the gas chromatograph. The result is that mass losses during derivatization are greatly reduced and the method is much more efficient (*22*).

Following isolation of the metabolites the compounds are identified usually by a chromatographic method, such as gas chromatography/mass spectrometry (GC/MS). When GC/MS is used, it is first necessary to obtain authentic standards for identification. Many times the spectra of the metabolites are not present in the mass spectrometric data bases. Metabolite standards may be obtained from the U.S. Environmental Protection Agency repository of pesticides (Columbus, Ohio) or from the chemical manufacturer of the parent compound. Generally, the manufacturer has been required by EPA to carry out metabolism studies, and the relevant standards have been identified and synthesized. This work is first done in the laboratory using C-14 labeled compounds; afterwards, an extensive effort is expended to synthesize the metabolites. Most of the manufacturers are generous, not only with metabolites but also with C-14 labeled parent compound and obtaining standards is possible.

If a metabolite standard is not available, synthesis may be required. The effort and cost is sometimes great, and this fact has discouraged studies of metabolites in surface and ground water. The literature is generally rich in information on the types of metabolites found in field dissipation studies. It would seem worthwhile to have an agency, such as EPA, consolidate a complete library of metabolite compounds as part of their registration efforts for new pesticides, which could be quite easily done in the early stages of compound registration.

Another method commonly used for metabolite identification is high performance liquid chromatography (HPLC). It often will happen that the compound is not volatile in the GC because of an ionic functional group or because of a hydroxyl group that has been added to the molecule. In this case, the compound is analyzed most easily by HPLC. Two common problems with the use of HPLC are that the sensitivity is less than with selected-ion monitoring GC/MS, and that other compounds occur in the chromatogram, such as humic substances, which interfere with identification. Diode array detection has helped with purity of peaks and with the identification of metabolites, but sometimes the spectrum is not definitive. Thus, the use of HPLC/MS has opened new avenues for metabolite identification. See chapters by Schroyer and Capel and by Barcelo and others that deal with analysis by HPLC and HPLC/MS.

In the past, HPLC/MS has had low sensitivity and poor ionization, such as with thermospray. However, the use of particle-beam ionization and electrospray promises to provide powerful new methods for analysis of triazine herbicides and their hydroxy metabolites (23). The latest gas chromatography/mass spectrometry/mass spectrometry (GC/MS/MS) systems offer the opportunity to identify the hydroxy metabolites with increased sensitivity and great selectivity, even in ground water (24). Unfortunately, the systems are expensive and commonly are used on research problems; thus, they are not yet routine analytical instruments for metabolite analysis.

However, the proliferation of bench-top GC/MS has made it a common tool for metabolite analysis. The combination of derivatization and bench-top GC/MS is an appealing alternative to HPLC/MS. The major analytical problem with this alternative is quantitation because of incomplete derivatization or because of hydrolysis after derivatization; therefore, the use of a deuterated standard may be needed, which creates a new problem of obtaining a deuterated internal standard of the metabolite.

Another method that recently has gained popularity for herbicide analysis is enzyme-linked immunosorbent assay or ELISA. During the last 5 years, a number of ELISA methods for herbicides and their metabolites have reached the market (25). For example, there is an ELISA for deethylatrazine and deisopropylatrazine (26), which may be used to analyze for both metabolites. Sometimes the parent-compound ELISA will also cross react with metabolites, which has been viewed as an analytical problem. However, cross reactivity may be used for methods development, as has been the case for alachlor metabolite, ESA. Aga and others (27) found that SPE could be used to separate parent and metabolite and then the ELISA could be used to identify both compounds. Thus, the analysis of a difficult metabolite, ESA, was simplified. See also chapters in this book by Lawruk and others, Hock and others, and Yanase and others on ELISA and fluorescence methods for herbicide metabolites.

For all these reasons, the analysis of herbicide metabolites in surface and ground water has not proceeded as quickly as the identification of parent compounds. Furthermore, the problem of which metabolite to identify has complicated the question of chemical analysis. For example, if a parent compound consists of 10 to 20 metabolites (C-14 studies of the manufacturers often are quite extensive), knowing which of these compounds may be important in surface and ground water is a difficult question and requires a different type of understanding-- one that considers fate and transport.

Metabolite Fate and Transport

There are several considerations when examining fate and transport of herbicide metabolites in surface and ground water. They include: aqueous solubility, sorption capacity and partition coefficient (Koc), vapor pressure or volatility, importance of the metabolite in laboratory degradation studies, and stability in the environment. These factors should be considered in order to predict metabolite mobility and occurrence in surface water and ground water. For example, in the consideration of cyanazine in Figure 2, there are nine metabolites that have been identified; which are most likely to occur in surface water and ground water?

An examination of the literature shows that cyanazine is commonly found in surface water (*3-5*) but is rarely detected in ground water (*6-9*). The nondetections in ground water have been attributed to its sorption to soil followed by rapid degradation, the half life for cyanazine is ~14 days (*15*) . Thus, the study of cyanazine metabolites is a relevant question. Although there have been no studies of hydroxy metabolites of cyanazine in the environment, the likelihood of finding them is small, based on past studies of hydroxyatrazine (24). Recently, there have been reports of the transport of hydroxyatrazine into surface water (*23 and chapter by Lerch and others this book)*, but hydroxyatrazine has been found not to be an important metabolite in ground water (*24*). Thus, the five hydroxy metabolites of cyanazine are probably less important for a first look at surface water and ground water.

Cyanazine acid has not been reported, but based on solubility, it should be a mobile compound in both surface and ground water. Cyanazine amide, which is a further metabolite of cyanazine acid (*15*), and deisopropylatrazine have been found frequently in surface water (*17*) and in ground water (*9*). Thus, the solubility and the sorption capacity of the metabolite may be used to predict transport.

The fate of herbicide metabolites may be estimated from half-life calculations, when available. Commonly, there have been C-14 studies of the parent compound, especially in conjunction with registration of the parent herbicide. These data may be quite valuable in predicting the fate of metabolites in soil and water. Two useful references are the Farm Chemicals Handbook '95 (29) and the Herbicide Handbook (18). Both contain information on fate, transport, solubility, partition coefficient (Koc), half life, and references on the parent compound and are a good starting point for metabolite studies.

An example of how one might use Koc, vapor pressure, and half life to predict transport is shown for propanil. Propanil is a major rice herbicide used in the Delta Region of Mississippi, Louisiana, and Arkansas (Figure 4). Its structure and two reported metabolites (18) are shown in Figure 5. The solubility of parent compound is high (at about 500 mg/L), vapor pressure is high (4 x 10^{-5} mm Hg), and the half life in soil is rapid (~2 days). The metabolites include two compounds-- propionic acid and 3,4-dichloroaniline.

The vapor pressure of the propionic acid metabolite would be considerably less than the parent compound in the dissociated state at the pH of most water samples and would not be volatile. The solubility would be at least 50 times greater than the parent compound. The half life also may be short, similar to the parent compound, due to the ease of removing the propyl group. The other metabolite, 3,4-dichloroaniline, has a vapor pressure less than the parent compound, is more soluble by at least three times, and is thought to be relatively stable, with a half life of 60 days or longer (30). Thus, the information that may be deduced from this information would be the following.

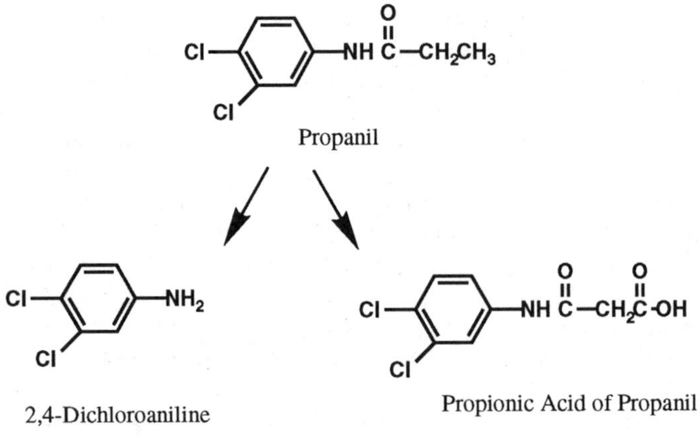

Figure 5. Metabolites of the rice herbicide, propanil (18).

Propanil is added to rice at rates of 3.6 to 5.6 kg/ha (18), where rapid degradation occurs. Rates of degradation in both aerobic and anaerobic rice paddies varied from 1 to 3 days (18). Because of the relatively high volatility of propanil, there may be escape of propanil to the atmosphere, and in fact, it has been found in air in the rice areas of Mississippi (*Coupe, Richard, U.S. Geological Survey, Jackson, Mississippi, personal communication*). The conclusion is reached that the rapid degradation of propanil results in production of propionic acid and 3,4-dichloroaniline. Analysis of water samples from Mississippi as part of the U.S. Geological Survey's National Water Quality Assessment shows that indeed 3,4-dichloroaniline is a common metabolite of propanil that frequently occurs, even in the absence of parent compound (*Coupe, Richard, U.S. Geological Survey, personal communication*).

Thus, the fate of herbicide metabolites in surface and ground water may be linked with the basic physical parameters of solubility, vapor pressure, and half life. These factors may be used in conjunction with published studies of metabolism to develop a plan of action for field and monitoring studies of herbicides and their metabolites in soil and water. See the chapters in this book by Cessna and others, Clay and others, Eckhardt and Wagenet, Smith, Spalding and Ma, Verstraeten and others, and Widmer and Spalding on the transport of metabolites in surface and ground water. See chapters by Kruger and Coats, Koskinen and others, and Laird on the fate of metabolites in soil and water.

In conclusion, the study of herbicides and their metabolites in surface and ground water is an important topic for environmental chemistry because of the frequency of occurrence, possible herbicidal activity of chlorinated metabolites, and concentrations approaching that of the parent compound. Currently, there are nearly 100 herbicides that are commonly used in the United States with an annual use of more than 300 million pounds (*31*). In general, these compounds must degrade to CO_2 through a series of metabolites, which if summed would be nearly equal to the parent load of herbicide in the environment. Early in the season after application of herbicides, the concentrations of parent compounds are of greatest concern (*4,14*). Later in the season after degradation as much as 50% or more of the load of an herbicide is metabolite (*13*), and these compounds also enter surface water and ground water. Much less is known about the fate and transport of these compounds in the environment than the parent compound. This symposium volume addresses this important topic.

Acknowledgments. Use of brand names is for identification purposes only and does not imply endorsement by the U.S. Geological Survey.

Literature Cited

1. U.S. Environmental Protection Agency, *Another Look-National Survey of Pesticides in Drinking Water Wells, Phase 2 Report:*, EPA/579/09-91/020, U.S. Environmental Protection Agency, Washington, D.C, 1992.

2. Leonard, R.A, In *Environmental Chemistry of Herbicides;* Grover, R. Ed., CRC: Boca Raton FL, 1988, pp. 45-88.

3. Pereira, W.E; Rostad, C.E. *Environ. Sci. Technol.*, **1990**, *24,* 1400-1406.

4. Thurman, E.M.; Goolsby, D.A.; Meyer, M.T.; Kolpin, D.W. *Environ. Sci. Technol.*, **1991**, *25,* 1794-1796.

5. Scribner, E.A; Goolsby, D.A.; Thurman, E.M.; Meyer, M.T.; Pomes, M.L., *Concentrations of Selected Herbicides, Two Triazine Metabolites, and Nutrients in Storm Runoff from Nine Stream Basins in the Midwestern United States*, U.S. Geological Survey Open-File Report 94-396 (1994), 144p.

6. Holden, L.R.; Graham, J.A.; Whitmore, R.W.; Alexander, W.J.; Pratt, R.W.; Liddle, S.K.; Piper, L.L. *Environ. Sci. Technol.*, **1992**, *26,* 935-943.

7. Hallberg, G.R. *Agric. Ecosys. Envir.*, **1989**, *26,* 299-367.

8. Burkart, M.R.; Kolpin, D.W. *J. Environ. Qual.*, **1993**, *22,* 646-656.

9. Kolpin, D.W.; Burkart, M.R.; Thurman, E.M., *Herbicide and Nitrate in Near-surface Aquifers in the Midcontinental United States, 1991*; U.S. Geological Survey Water-Supply Paper 2413, Denver, CO, 1994.

10. Baker, D.B.; Bushway, R.J.; Adams, S.A.; Macomber, C. *Environ. Sci. Technol.* **1993**, *27,* 562-564.

11. Thurman, E.M.; Goolsby, D.A.; Aga, D.S.; Pomes, M.L.; Meyer, M.T. *Environ. Sci. Technol.* **1996**, in press.

12. Gunther, F.A.; Gunther, J.D. *Residue Reviews Volume 32 The Triazine Herbicides* ; Springer-Verlag, New York; 1970, 413p.

13. Thurman, E.M.; Meyer, M.T.; Mills, M.S.; Zimmerman, L.R.; Perry, C.A.; Goolsby, D.A. *Environ. Sci. Technol.*, **1994**, *28,* 2267-2277.

14. Thurman, E.M.; Goolsby, D.A.; Meyer, M.T.; Mills, M.S.; Pomes, M.L.; Kolpin, D.W. *Environ. Sci. Technol.*, **1992**, *26,* 2440-2447.

15. Benyon, K.I.; Stoydin, G.; Wright, A.N. *Pest. Sci.* **1972**, *3,* 293-305.

16. Sirons, G.J.; Frank, R.; Sawyer, T. J. *Agri. Food Chem.*, **1973**, *21,* 1016-1020.

17. Meyer, M.T., *Geochemistry of Cyanazine Metabolites*, Ph.D. Thesis, University of Kansas; Lawrence, KS, 1994, 362 p.

18. *Herbicide Handbook*; Ahrens, W.H., Ed.; Weed Science Society of America: Champaign, Illinois, 1994; 7th Edition.

19. Mills, M.S.; Thurman, E.M. *Environ. Sci. Technol.*, **1994**, *28,* 73-79.

20. Di Corcia, A.; Samperi, R. *Anal. Chem.*, 1990, *28*, 1490-1494.

21. Markell, C.; Hagen, D.F.; Bunnelle, V.A.; *LC/GC*, **1991**, *9*. 332-338.

22. Field, J.A.; Monihan, K., *Anal. Chem.*, **1995**, *67*, 3357-3362.

23. Lerch, R.N.; Donald, W.W.; Li, Y.X.; Alberts, E.E. *Environ. Sci. Technol.*, **1995**, *29*, in press.

24. Zongwei, C.; Ramanujam, V.M.S.; Gross, M.L.; Monson, S.J.; Cassada, D.A.; Spalding, R.F. *Anal. Chem.*, **1994**, *66*, 4202-4209.

25. *Immunochemical Methods for Environmental Analysis*; Van Emon, J.M.; Mumma, R.O., Eds.; ACS Symposium Series 442, American Chemical Society; Washington, D.C., 1990.

26. Wittmann, C.; Hock, B. *J. Agric. Food Chem.*, **1991**, *39*, 1194-1200.

27. Aga, D.S.; Thurman, E.M.; Pomes, M.L. *Anal. Chem.*, **1994**, *66*, 1495-1499.

29. *Farm Chemicals Handbook '95*; Meister, R.T., Ed.; Meister Publishing Company; Willoughby, Ohio, 1995.

30. Correa, I.E.; Steen, W.C. *Chemosphere*, **1995**, *30*. 103-110.

31. Gianessi, L.P.; Anderson, J.E. *Pesticide Use in U.S. Crop Production, National Data Report (1992)*, National Center for Dood and Agricultural Policy: Washington, D.C., 1995.

32. Battaglin, W.A.; Goolsby, D.A. *Spatial Data in Geographic Information System Format on Agricultural Chemical Usage, Land Use, and Cropping Practices in the United States*, U.S. Geological Survey Water Resources Investigation 94-4176, (1995) 87p.

ANALYTICAL METHODS

Chapter 2

Coordinating Supercritical Fluid and Solid-Phase Extraction with Chromatographic and Immunoassay Analysis of Herbicides

Martha J. M. Wells[1,2] and G. Kim Stearman[1,3]

[1]Center for the Management, Utilization and Protection of Water Resources, and Departments of [2]Chemistry and [3]Plant and Soil Science, Tennessee Technological University, Box 5033, Cookeville, TN 38505

Protocol development for the determination of herbicide residues and their metabolites from environmental matrices can be tedious and expensive. Very few environmental samples can be analyzed directly without sample pretreatment, but it is possible to do so in some cases. Concentration and purification of the sample are usually required. Extraction and cleanup can require multiple processing steps. Coordinating various processes of sample preparation with methods of final determination is the subject of this discussion. An integrated look at strategies that take advantage of inherent solute properties and optimize sample matrix control is presented.

In today's environmental analytical laboratories, techniques for isolating analytes from the sample matrix such as liquid-solid extraction (LSE), supercritical fluid extraction (SFE), and solid-phase extraction (SPE), and methods for final determination including enzyme immunoassay analysis (EIA), high-performance liquid chromatography (HPLC), and gas chromatography (GC) are applied to the analysis of herbicides and their metabolites. Individual analytical procedures are often linked together to process a sample by what have become known as hyphenated techniques, such as SFE-EIA or LSE-SPE-HPLC. Coordinating the sample flow from step-to-step between procedures can require additional manipulation due to solvent incompatibility or background matrix effects. With a continuing emphasis on the effect of nonpoint source pollution, techniques that meet the analytical goals of (a) maximizing detector response to the components-of-interest while minimizing the response of interferences, and (b) controlling analysis time and labor costs are needed. The strategies that will keep pace with increasing analytical demands exploit inherent solute properties and optimize sample matrix control.

0097–6156/96/0630–0018$15.00/0

Hyphenated Techniques for Sample Preparation and Final Determination

Depending upon herbicide concentrations, some samples, particularly those collected soon after field treatment (anticipated to have the highest concentrations), may be analyzed directly with little sample preparation. We have successfully taken extracts from LSE and SFE directly to final determination by EIA and HPLC, and extracts from SFE to GC for herbicides in soil matrices, in instances in which concentrations were high and when interferences could be controlled. Aqueous agricultural runoff samples, and soil and sediment LSE extracts were monitored for simazine and 2,4-D without sample cleanup by EIA and by direct injection HPLC using a diode array detector (Sutherland, D.J.; Stearman, G.K.; Wells, M.J.M. Tennessee Technological University, in preparation). SFE extracts of trifluralin in soil were collected in acetone through C_{18} traps and analyzed directly by GC (Garimella, U.; Stearman, G.K.; Wells, M.J.M. Tennessee Technological University, in preparation). An SFE method coupled with EIA was developed (*1*) for extraction and analysis of the soil herbicides 2,4-D, simazine, atrazine, and alachlor that required no additional sample cleanup. The SFE-EIA method is a quick, alternative procedure for soil herbicide analyses if cross-reacting components and interfering matrices are not present. Chromatographic alternatives often require time-consuming purification procedures. Although not as specific for the analyte as HPLC or GC, EIA can be very useful as a screening procedure or in instances in which the nature of the analyte present is well-defined.

As these examples indicate, some environmental samples can be analyzed directly with minimal sample pretreatment. However, a solvent system optimized to produce maximum extraction efficiency may be inappropriate for proceeding directly to chromatographic or immunoassay analysis, or the extracted sample volume may require preconcentration before final determination, or the background interferences may simply be too great. In the analysis of herbicides from soils and other complex environmental matrices, it can be the role of SPE to concentrate, purify, and/or fractionate an LSE or SFE extract prior to final determination by EIA, HPLC, or GC (Figure 1). Current trends are to automate these procedures, such as by SPE-HPLC (*2-4*) and SPE-GC (*5*).

Much of the analytical literature discusses chemical procedures segregated by each technique. Here, the approach is to examine techniques in terms of their ability to exploit solute properties or influence sample matrix effects.

Solute Properties

Hydrophobicity/Polarity/Ionogenicity. Most herbicide metabolites are more polar than the parent compounds. A corollary to this statement is that most herbicide metabolites are less toxic than the parent compounds. There are notable exceptions to each of these statements, but it is usually nature's way to attempt to detoxify contaminants and render them excretable by conversion into more polar, water soluble compounds. What this means to the analyst is that herbicide metabolites may be less amenable to analysis by GC than their parent compounds. Hydroxyatrazine is an example of a metabolite more polar than the parent compound that is successfully analyzed only by HPLC (*6,7*). On the other hand, the elution of metribuzin DA after the parent compound, metribuzin,

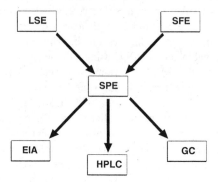

Figure 1.   Coordinating techniques for sample preparation with final
             determination.

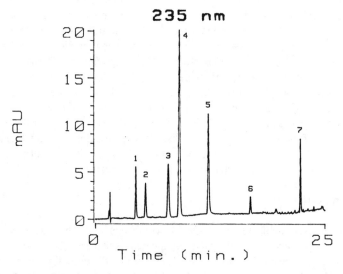

Figure 2.   Reversed-phase liquid chromatographic elution profile of the
             herbicides and metabolites (1) metribuzin DK, (2) metribuzin
             DADK, (3) metribuzin, (4) metribuzin DA, (5) atrazine, (6)
             metolachlor, and (7) esfenvalerate.

during reversed-phase HPLC (RP-HPLC) analysis (Figure 2) indicates that it is more hydrophobic (less polar) than the parent compound (Riemer, D.D.; Wells, M.J.M. Tennessee Technological University, unpublished results). Likewise, RP-HPLC indicates that α,α,α-trifluoro-4,6-dinitro-N,N-di-(n-propyl)-o-toluidine is more hydrophobic than its parent compound, trifluralin (Garimella, U.; Stearman, G.K.; Wells, M.J.M. Tennessee Technological University, in preparation). The analyst must be aware that the parent compound and its metabolites may differ widely in hydrophobicity, polarity, and ionogenicity.

Supercritical Fluid Extraction. Successful extraction of herbicides by supercritical fluids requires knowledge of the properties of the solutes in order to select the appropriate extracting phase. Advances in SFE of analytes from environmental samples were recently reviewed (*8,9*). Analyte solubility in the supercritical phase can be manipulated by controlling temperature and pressure. Carbon dioxide (CO_2) is the most commonly used supercritical fluid because it is readily available and achieves supercritical state at relatively low pressure (78 atmospheres) and temperature (31°C). The addition of an organic modifying solvent such as acetone or methanol to CO_2 can also improve recoveries. In fact, SFE extraction of herbicides from soil often requires addition of polar organic modifiers to the supercritical CO_2. SFE extraction of soil atrazine was improved by adding modifiers or by increasing the temperature of extraction to as much as 140°C (*10*) or 300°C (Hawthorne, S.B., University of North Dakota, personal communication, 1994).

Solubility of organic modifiers in CO_2 varies. For example, acetonitrile is only soluble at 4% in supercritical fluid CO_2, while acetone and methanol are soluble up to 20%, thereby limiting the use of acetonitrile as a modifier. The amount of organic modifier needed for complete extraction recovery varies depending on the soil and the pesticide. Robertson and Lester (*10*) showed improved recovery using 20% rather than 10% acetone modifier to the CO_2 for SFE extraction of triazines. Various modifiers, including water, need to be examined to determine their effects on different soil matrices (*11*). The purpose of the modifier is to increase the solubility of the analyte and/or increase the surface area of the soil by swelling, or to competitively adsorb with the analyte to the soil. Extraction temperature must be increased as the modifier percentage increases so it will continue to exist as a one-phase system in supercritical CO_2. In some cases, improved recovery results are achieved by adding solvents as entrainers, i.e. directly to the extraction cell as compared to adding them as a modifier with the CO_2 (*12-14*).

The significance of acid-base modifiers that interact with the binding or solubilization of herbicides is important. The polar modifiers are thought to weaken the bonds between the analyte and the soil matrix, and also to solubilize the analytes by raising the pH, although the mechanism by which this enhanced extraction recovery occurs is unknown. Alzaga et al. (*15*) found the most effective extraction fluid for SFE extraction of pirimicarb from soil was carbon dioxide modified with basic compounds, pyridine or triethylamine. In our laboratory, the addition of triethylamine to an acetone-water modifier significantly improved recoveries for atrazine, simazine, and 2,4-D, while alachlor was extracted equally well with pure CO_2(*1*). Other important controllable SFE parameters include flow rates, sample size and packing, and moisture content of the sample.

SFE flow is controlled by using stainless steel or fused silicon tubing restrictors. Recently variable flow restrictors have been used (16). Restrictors plug when water freezes at the tip or when co-extracted organic molecules accumulate in the opening. Restrictor plugging resulting in erratic flow has been an ongoing problem with SFE technology. Restrictors are commonly heated to 150°C to prevent or reduce plugging. We have found that plugged restrictors were a problem when methanol was used as a modifier, but not with acetone (1).

The collection of the analyte is important. Extraction recoveries may be 100% while collection recoveries are less than 50% because of the formation of aerosols or volatilization of analyte from the collection trap. We use liquid collection or solid-phase traps (C_{18}) to collect the extracted analyte. The C_{18} traps improved mean recoveries in SFE by almost 25% (1). An extraction kinetics study (1) conducted for a 20-minute interval showed that the majority of the herbicides atrazine, alachlor, simazine, and 2,4-D were extracted in the first 5 minutes. After 15 minutes, about 80-90% of the herbicide had passed through the C_{18} trap; that is, the herbicide was in the modifier solution that collected in the collection vial. The collection vials are cooled to 4°C to inhibit volatilization and improve collection efficiency. Langenfeld, et al., (17) found that greater than 90% trapping of 66 test analytes could be attained by controlling the solvent temperature at 5°C.

Packing the extraction cells is also important. If cells are packed too loosely the CO_2 will pass through without contact with the soil and recoveries will be low. Unless glass wool or sand are added to the sample, the frits in the extraction cells can clog, preventing flow.

Multistage Solid-phase Extraction (SPE) Procedures. Single-residue methodology can be developed into multiresidue procedures by taking advantage of differences in solute hydrophobicity, polarity, and ionogenicity. The improved selectivity produced by multistaging SPE is very useful for the preparation of multicomponent or multiresidue samples for analysis. Initially, the generation of multiresidue procedures requires more laboratory time for development, but in the long run saves time and money because fewer field samples are necessary, requiring less laboratory processing overall. Multistage processes lead to multiple extracts or fractions that may improve individual analyses. Three multistage SPE procedures used in herbicide analysis are selective adsorption, selective desorption, and chromatographic mode sequencing.

Selective adsorption in SPE can be accomplished by controlling the sample matrix or the sorbent. The type and quantity of sorbent, hydrophobicity and ionizability of the components, and sample volume and pH interactively determine the breakthrough volume. The extraction of hexazinone, its metabolites, and picloram (18) uses selective adsorption by pH control of the sample matrix (Figure 3). At a sample pH of 4.5, hexazinone and its metabolites are retained on an octadecylsilica stationary phase, while picloram, ionized at this pH, passes through the column and remains in the aqueous fraction. Elution with methanol removes hexazinone and five metabolites into a single fraction. When the aqueous fraction is adjusted to pH 2.0 and passed through an octadecylsilica sorbent, picloram is retained and subsequently eluted by 25% acetic acid.

Selective desorption of the herbicides picloram and 2,4-dichlorophenoxyacetic acid (2,4-D) is accomplished by utilizing differences in the eluotropic strength of elution solvents (Figure 4). When picloram and 2,4-D are simultaneously adsorbed onto an octydecyl silica column (*18*), a two-step serial desorption, first by 25% acetic acid followed by methanol, separates these two compounds into distinct fractions. The fractionation is produced by the difference in eluotropic strength between 25% acetic acid and methanol.

Chromatographic mode sequencing is the serial use of different chromatographic sorbents for SPE. Chromatographic mode sequencing with both strong cation exchange (SCX) and hydrophobic (C_{18}) sorbents was used (Sutherland, D.J.; Stearman, G.K.; Wells, M.J.M. Tennessee Technological University, in preparation) to fractionate and purify simazine and 2,4-D residues from a common sample matrix (Figure 5). An SCX column coupled in tandem above a C_{18} column was used to separately adsorb simazine on the ion exchange column and 2,4-D on the C_{18} column. Following sample loading, the two columns were separated and 2,4-D was eluted from the C_{18} column.

Detectability

Chromophores. The structure of a molecule determines the kind of radiation that will be absorbed. The chromophore of a molecule that results in absorption of energy in the ultraviolet-visible range is due to the presence of functional groups having valence electrons with low excitation energies. Many herbicides have unique ultraviolet-visible spectra. The four chromatograms depicted in Figure 6 were obtained by monitoring a single injection at four wavelengths (204, 210, 221, and 298 nm) using a diode-array detector (*19*). Metribuzin can be monitored at high wavelengths in the ultraviolet range (298 nm) where few interferences are noted (Figure 6). The variation in absorbance of a component at different wavelengths is indicative of a particular compound. Trifluralin has an absorbance in the visible range (Figure 7) that can be exploited for very specific detection (Garimella, U.; Stearman, G.K.; Wells, M.J.M. Tennessee Technological University, in preparation). The application of diode-array detectors for simultaneous analysis of herbicides and their metabolites at multiple wavelengths is useful. Absorbance ratioing can be used to investigate the identities of components. The unique absorbance patterns of metribuzin and trifluralin facilitate analysis by direct injection HPLC with little or no pretreatment.

Electrophores. Analogous to chromophores, it is the electrophore in a molecule that renders it susceptible to electrochemical detection. Detection of metribuzin by GC-ECD is fifty times more sensitive than detection of atrazine (*19*). Both compounds are triazines, but metribuzin is an unsymmetrical triazine, while atrazine is a symmetrical triazine. The structural difference appears to alter the electrophore.

Chemical Derivatization. Chemical derivatization can improve detectability by "tagging" solutes-of-interest with a chromophore, fluorophore, or electrophore to

EXTRACTION SCHEME
HEXAZINONE/HEXAZINONE METABOLITES/PICLORAM

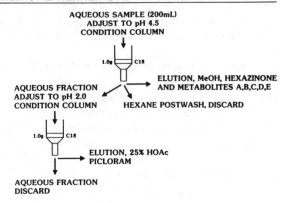

Figure 3.   Solid-phase extraction scheme for the selective adsorption of hexazinone, hexazinone metabolites, and picloram. (Reproduced with permission from ref. 18. Copyright 1986 Varian Sample Preparation Products.)

EXTRACTION SCHEME, PICLORAM/2,4-D

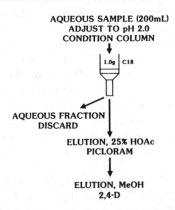

Figure 4.   Solid-phase extraction scheme for the selective desorption of picloram and 2,4-D. (Reproduced with permission from ref. 18. Copyright 1986 Varian Sample Preparation Products.)

EXTRACTION SCHEME, SIMAZINE/2,4-D

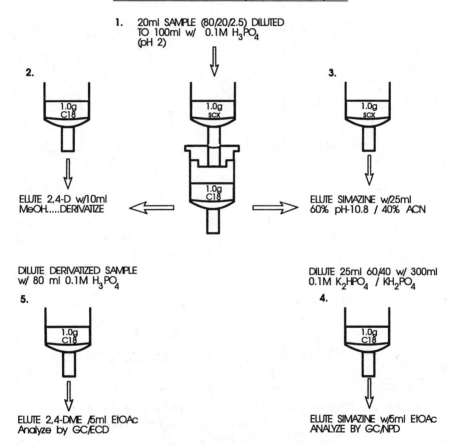

1. 20ml SAMPLE (80/20/2.5) DILUTED TO 100ml w/ 0.1M H₃PO₄ (pH 2)

2.

1.0g C18

ELUTE 2,4-D w/10ml MeOH.....DERIVATIZE

DILUTE DERIVATIZED SAMPLE w/ 80 ml 0.1M H₃PO₄

5.

1.0g C18

ELUTE 2,4-DME /5ml EtOAc Analyze by GC/ECD

1.0g SCX

1.0g C18

3.

1.0g SCX

ELUTE SIMAZINE w/25ml 60% pH-10.8 / 40% ACN

DILUTE 25ml 60/40 w/ 300ml 0.1M K₂HPO₄ / KH₂PO₄

4.

1.0g C18

ELUTE SIMAZINE w/5ml EtOAc ANALYZE BY GC/NPD

Figure 5.   Solid-phase extraction scheme illustrating chromatographic mode sequencing for the fractionation of simazine and 2,4-D.

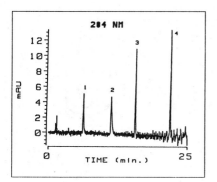

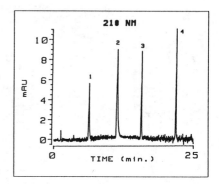

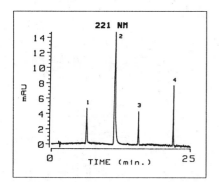

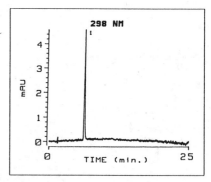

Figure 6.   Reversed-phase liquid chromatographic elution profile of (1) metribuzin, (2) atrazine, (3) metolachlor and (4) esfenvalerate at 204, 210, 221, and 298 nm.

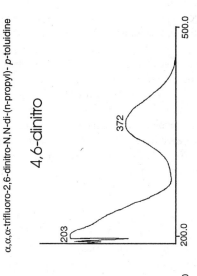

trifluralin

205
272.2
386

200.0   500.0

α,α,α-trifluoro-2,6-dinitro-N,N-di-(n-propyl)- *p*-toluidine

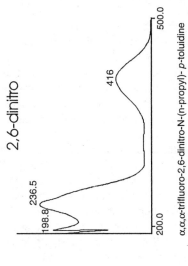

4,6-dinitro

203
372

200.0   500.0

α,α,α-trifluoro-4,6-dinitro-N,N-di-(n-propyl)- *o*-toluidine

TREFLAN™
AND
METABOLITES

™Treflan is the trademark for a brand
of the common herbicide trifluralin.

2,6-dinitro

198.8
236.5
416

200.0   500.0

α,α,α-trifluoro-2,6-dinitro-N-(n-propyl)- *p*-toluidine

Figure 7.   Ultraviolet-visible spectra of trifluralin and metabolites.

enhance ultraviolet, fluorescent, or electrochemical properties. Chemical derivatization is also used to improve solute volatility for gas chromatography. Chemical derivatization following SPE is complicated by the fact that many derivatizing reagents are water-sensitive, and the sample invariably picks up water from the SPE sorbent surface, even with extensive drying. During derivatization of 2,4-D to the methyl ester (2,4-DME) the concentration of the derivatizing reagent (BF_3-methanol) was optimized. (Sutherland, D.J.; Stearman, G.K.; Wells, M.J.M. Tennessee Technological University, in preparation). Because BF_3-methanol is subject to hydrolysis by water, the increased amount of the derivatizing reagent necessary for complete derivatization is surmised to be caused by water that is extracted from the C_{18} sorbent during elution.

Some analysts have combined SFE with derivatization reactions. Coupling derivatization reactions with sample extraction and concentration reduces sample handling and analysis time. Hawthorne et al., (20) reported SFE of polar analytes, including 2,4-D, using trimethylphenylammonium hydroxide as an ion pair and methylating reagent. Soil extraction of chlorophenoxy compounds, including 2,4-D, incorporating a static SFE derivatization technique prior to dynamic extraction, improved recoveries (21). Chatfield, et al., (22) displaced chlorophenoxyacetic acids from ion exchange resins as their methyl esters using supercritical carbon dioxide containing methyl iodide.

Sample Matrix Control

Aqueous Dilution. EIA methods are sensitive to organic extracting solvents, including acetone and acetonitrile, and therefore, must normally be diluted below 5% concentration of these solvents (23).   Final soil extract volumes are diluted 25:1 - 200:1 before pipetting into EIA microtiter wells, depending on herbicide concentrations. EIA microtiter plate kits in a 96-well format are commonly used for simazine, atrazine, alachlor, and 2,4-D (Millipore, Inc., New Bedford, MA). Standards, including a blank, are made in the same matrix as the diluted soil extracts.

EIA responses to atrazine and metolachlor in potentially interfering matrices of acetonitrile, methanol, and humic acid were optimized by dilution (24). We reported matrix interference at humic acid concentrations of 100 mg/L for EIA of atrazine and metolachlor.

Dilution is also an important strategy to alter the matrix of high organic content extracts of herbicides from soil. The LSE of simazine and 2,4-D from soil was optimized in a solvent of acetonitrile/ water/acetic acid, 80/20/2.5 v:v:v. As is, the herbicides will not be well-retained from this solution by SPE due to its high organic content. Dilution with phosphoric acid (0.1M) increases the total sample volume, but reduces the eluotropic strength of the sample matrix, allowing the herbicides to be adsorbed by SPE (Sutherland, D.J.; Stearman, G.K.; Wells, M.J.M. Tennessee Technological University, in preparation).

Solvent Exchange. Solvent exchange of atrazine and alachlor following SFE from soil is accomplished by SPE to render the solvent amenable for analysis by GC with a nitrogen-phosphorus detector (NPD). SPE extracts of atrazine and alachlor were collected in acetone. The sample was diluted with reagent grade water to a final concentration of 2% acetone, sorbed by reversed-phase SPE, and eluted with ethyl acetate effecting a solvent exchange that rendered the sample compatible with GC analysis (*1*).

Surface Tension. The antithesis to diluting with water samples that are water-miscible in order to reduce the percentage of organic solvent present, is to intentionally alter an aqueous sample matrix by the addition of water-miscible organic solvents. When a liquid sample matrix is altered in this way, the fundamental property most affected is the surface tension. The surface tension of pure water is quite large, 72.00 dyne/cm. Increasing the water-miscible, organic component of a solution by the addition of polar organic solvents, such as methanol or acetonitrile, decreases surface tension. Surfactants exert their chemical character by the ability to reduce surface tension. Humic and fulvic acids reduce the surface tension of aqueous solutions. Conversely, certain inorganic solutes such as sodium chloride or sodium nitrate increase surface tension relative to that of pure water.

In a factorial study of the SPE of metribuzin, atrazine, metolachlor, and esfenvalerate (*19*), retention on the sorbent was controlled by variation in the surface tension of the sample matrix. The addition of organic modifier to the sample reduces the retention factor of analytes on a reversed-phase sorbent, potentially improving recovery of highly hydrophobic compounds. Adding solutes, such as sodium chloride, that increase ionic strength of the sample matrix enhances the salting-out effect and may counteract secondary sorbent interferences.

Soil Moisture. Soil matrix moisture content is an important parameter to consider when extracting environmental samples, especially by SFE. Snyder, et al., (*25*) found that sodium sulfate used as a drying agent aided in recovery of pesticides but caused plugging of restrictors during SFE. They stated that Hydromatrix or other desiccating material may be more appropriate with SFE. In some cases the sodium sulfate was displaced into the collection vial (*25*). Burford et al. (*26*) evaluated 21 potential drying agents and found that five successfully prevented restrictor plugging by water during SFE extraction. These included anhydrous and monohydrate magnesium sulfate, molecular sieves 3A and 5A, and Hydromatrix.

Soil Texture. Matrix effects are the major factor that have to be understood for a successful application of SFE to soil analysis (*27*). Extraction protocols do not necessarily achieve high recoveries on all soil types. High extraction efficiencies on one soil do not translate to similar extraction efficiencies on another soil with different properties. Clay content and type (Table I) are important soil properties that influence recovery (*1*). With high organic matter content or high clay containing soils, sample pretreatment with an entrainer may be necessary to achieve high recovery of analyte. Recovery is in many instances dependent upon soil herbicide concentration. There may be large differences in recovery between herbicide spikes of 10 mg/kg versus 50 µg/kg using the same SFE parameters. This may be due to soil matrix effects which have variable bonding energies for different concentrations of pesticides. Also, at lower

Table I. Supercritical fluid extraction recoveries and standard deviations (RSD) of spiked atrazine and alachlor soils.

Soil Texture	Clay %	Cation Exchange Capacity cmol kg⁻¹	Organic Carbon %	Surface Area m²/g	Recovery % ± RSD (500 ng/g)	
					Atrazine	Alachlor
Sandy loam	7.4	16.1	2.3	9.9	90.1 ± 10.5	90.0 ± 10.5
Silt loam	8.6	6.1	0.9	7.5	88.6 ± 8.4	88.0 ± 6.1
Silt loam	12.5	8.0	0.7	21.2	93.5 ± 14.2	97.7 ± 9.1
Silt loam	24.0	17.5	1.3	29.6	85.0 ± 12.2	98.0 ± 6.4
Silty clay	49.3	40.8	1.5	46.1	74.6 ± 8.1	54.1 ± 5.4

Adapted from reference 1.

concentrations a larger percentage of the total herbicide concentration may become less accessible to extracting solvents than at higher herbicide concentrations, especially in high organic or clay soils.

Cross Reactivity. Sample matrix control is also the key to successful EIA. EIA has gained acceptance as a rapid pesticide analytical technique, complementary to more traditional chromatographic methods. Supercritical fluid extracts of atrazine from soil were analyzed by EIA and compared to analysis by GC-NPD preceded by SPE (Figure 8). The line in Figure 8 represents that expected for a perfect 1:1 relationship; statistics for the best-fit line are noted on Figure 8. EIA may be used both as a screening method and as a semiquantitative method under different conditions (*28*). The major problem with using EIA as a quantitative technique is the cross reactivity of similar compounds. The specificity of an immunoassay is the ability to discriminate among different molecular entities and to respond only to those targeted. Antibodies bind with the targeted analyte. Cross-reactivity is the binding of structurally related compounds to the antibody. A low degree of cross-reactivity makes the EIA suitable for single-compound analysis, while a group-specific analysis requires an antibody having a high degree of cross-reactivity with various members of a chemical family (*29*). EIA can be used for soils that have no baseline cross-reactivity with similar compounds. However, with field weathered samples, i.e. those that have had pesticides applied in the field, there is the possibility that the metabolites will be cross-reactive. In fact, we have found that one of the metabolites of the herbicide trifluralin is more sensitive to the EIA than the parent compound (Garimella, U.; Stearman, G.K.; Wells, M.J.M. Tennessee Technological University, in preparation). Despite potential cross-reacting interferences, EIA microtiter

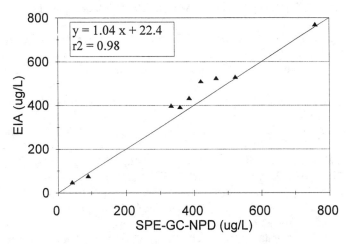

Figure 8.   Comparison of EIA and GC analysis of atrazine soil extracts.

plate techniques are easy to use and allow many samples to be analyzed in a short period of time. In many cases, EIA is also less expensive than traditional GC or HPLC methods (*23*).

Conclusions

The challenges of achieving reduced analysis times, reduced costs and labor, while minimizing organic solvent consumption and disposal, and reducing detection limits, drive the need for development of reliable analytical protocol. Knowledge of the properties of the analyte-of-interest and the potential influence of matrix effects is fundamental to success. Effective integration of a variety of procedures for sample preparation and final determination is necessary to meet the ever increasing demand for the analysis of herbicides and their metabolites in environmental matrices.

Acknowledgments

The participation in this research of Scott Adkisson, Anna Edwards, Uma Garimella, Tadd Ridgill, Daniel Riemer, Binney Stumpf, and Devon Sutherland is gratefully acknowledged.

Literature Cited

(*1*)   Stearman, G.K.; Wells, M.J.M.; Adkisson, S.M.; Ridgill, T.E. *Analyst* **1995,** *120*, 2617-2621.

(*2*)   Chiron, S.; Alba, A.F.; Barcelo, D. *Environ. Sci. Technol.* **1993,** *27*, 2352-2359.

(*3*)   Chiron,S.; Papilloud, S.; Haerdi, W.; Barcelo, D. *Anal. Chem.* **1995,** *67*, 1637-1643.

(*4*)   Nouri, B.; Fouillet, B.; Toussaint, G.; Chambon, P.; Chambon, R. *Analyst* **1995,** *120*, 1133-1136.

(*5*)   Pico, Y.; Louter, A.J.H.; Vreuls, J.J.; Brinkman, U.A.Th. *Analyst* **1994,** *119*, 2025-2031.

(*6*)   Schlaeppi, J.-M.; Fory, W.; Ramsteiner, K. *J. Agric. Food Chem.* **1989,** *37*, 1532-1538.

(*7*)   Thurman, E.M.; Meyer, M.; Pomes, M.; Perry, C.A.; Schwab, A.P. *Anal. Chem.* **1990,** *62*, 2043-2048.

(*8*)   Barnabas, I.J.; Dean, J.R.; Owen, S.P. *Analyst* **1994,** *119*, 2381-2394.

(*9*)   McNally, M.E.P. *Anal. Chem.* **1995,** *67*, 308A-315A.

(*10*)   Robertson, A.M.; Lester, J.N. *Environ. Sci. Technol.* **1994,** *28*, 346-351.

(*11*)   Fahmy, T.M.; Pulaitis, M.E.; Johnson, D.M.; McNally, M.E.P. *Anal. Chem.* **1993,** *65*, 1462-1469.

(*12*)   Liu, M.H.; Kapila, S.; Yanders, A.F.; Clevenger, T.E.; Elseewi, A.A. *Chemosphere* **1991,** *23*, 1085-1095.

(*13*)   Locke, M. *J. Agric. Food Chem.* **1993,** *41*, 1081-1084.

(*14*)   Tomyczk, N.A.; Schneider, J.F. *Proceedings of the 34th ORNL-DOE Conference on Analytical Chemistry in Energy Technology, Gatlinburg, TN,* **1993,** *CONF - 931081*, 52-53.

(*15*) Alzaga, R.; Bayona, J.M.; Barcelo, D. *J. Agric. Food Chem.* **1995**, *43*, 395-400.

(*16*) Levy, J.M. *Proceedings of the 5th International Symposium on Supercritical Fluid Chromatography and Extraction, Baltimore MD,* **1994**, D-4.

(*17*) Langenfeld, J.J.; Burford, M.D.; Hawthorne, S.B.; Miller, D.J. *J. Chromatogr.* **1992**, *594*, 297-307.

(*18*) Wells, M.J.M. *Proceedings of the Third Annual International Symposium, Sample Preparation and Isolation using Bonded Silicas, Cherry Hill, NJ,* **1986**, 117-135.

(*19*) Wells, M.J.M.; Riemer, D.D.; Wells-Knecht, M.C. *J. Chromatogr. A* **1994**, *659*, 337-348.

(*20*) Hawthorne, S.B.; Miller, D.J.; Nivens, D.E.; White, D.C. *Anal. Chem.* **1992**, *64*, 405-412.

(*21*) Lopez-Avila, V.; Dodhiwala, N.S.; Beckert, W.F. *J. Agric. Food Chem.* **1993**, *41*, 2038-2044.

(*22*) Chatfield, S.N.; Croft, M.Y.; Dang, T.; Murby, E.J.; Yu, G.Y.F.; Wells, R.J. *Anal. Chem.* **1995**, *67*, 945-951.

(*23*) Stearman, G.K.; Adams, V.D. *Bull. of Environ. Contam. Toxicol.* **1992**, *48*, 144-151.

(*24*) Stearman, G.K.; Wells, M.J.M. *Bull. of Environ. Contam. Toxicol.* **1993**, *51*, 588-595.

(*25*) Snyder, J.L.; Grob, R.L.; McNally, M.E.P.; Oostdyk, T.S. *J. Environ. Sci. Health,* **1994**, *A29(9)*, 1801-1816.

(*26*) Burford, M.D.; Hawthorne, S.B.; Miller, D.J. *J. Chromatogr.* **1993**, 657(2), 413-427.

(*27*) Wuchner, K.; Ghijsen, R.T.; Brinkman, U.A. Th.; Grob, R.; Mathieu, J. *Analyst* **1993**, *118*, 11-16.

(*28*) Van Emon, J.M.; Lopez-Avila, V. *Anal. Chem.* **1992**, *64*(2), 78A-88A.

(*29*) Meulenberg, E.P.; Mulder, W.H.; Stoks, P.G. *Environ. Sci. Technol.* **1995**, *29*, 553-561.

Chapter 3

A High-Performance Liquid Chromatography—Based Screening Method for the Analysis of Atrazine, Alachlor, and Ten of Their Transformation Products

Blaine R. Schroyer[1] and Paul D. Capel

U.S. Geological Survey, Department of Civil Engineering, University of Minnesota, Minneapolis, MN 55455

A high-performance liquid chromatography (HPLC) method is presented for the for the fast, quantitative analysis of the target analytes in water and in low organic-carbon, sandy soils that are known to be contaminated with the parent herbicides. Speed and ease of sample preparation was prioritized above minimizing detection limits. Soil samples were extracted using 80:20 methanol:water (volume:volume). Water samples (50 μL) were injected directly into the HPLC without prior preparation. Method quantification limits for soil samples (10 g dry weight) and water samples ranged from 20 to 110 ng/g and from 20 to 110 μg/L for atrazine and its transformation products and from 80 to 320 ng/g and from 80 to 320 μg/L for alachlor and its transformation products, respectively.

The use of pesticides in agriculture has become commonplace. Although there are significant beneficial effects of pesticide use, they must be used carefully because they can have adverse effects on the environment. Two of the most commonly used herbicides in the midwestern United States are atrazine and alachlor (*1*). These compounds and their transformation products have been detected in surface and ground waters (*2-7*). Their presence in these waters may or may not represent a hazard to human and ecosystem health.

Several methods have been developed for the analysis of atrazine and alachlor in environmental matrices (*2-6, 8-12*). There are also methods for some of the

---

[1]Current address: Terracon Environmental, 3535 Hoffman Road East, White Bear Lake, MN 55110

transformation products of atrazine and alachlor (*2,3,5,8-12*). Many of these analyses are based on gas chromatography, which cannot be used for the analysis of the most polar transformation products without derivitization (*2-6*). Analysis by high-performance liquid chromatography (HPLC) allows for quantification of both the parent compounds and a broad suite of transformation products without derivitization.

An HPLC method is presented here that has been developed to quantify atrazine and six of its transformation products and alachlor and four of its transformation products. The target analytes are atrazine (ATR) (2-chloro-4-ethylamino-6-isopropylamino-1,3,5-triazine), deisopropyl atrazine (DIA) (2-chloro-4-ethylamino-6-amino-1,3,5-triazine), deethyl atrazine (DEA) (2-chloro-4-amino-6-isopropylamino-1,3,5-triazine), didealkyl atrazine (DDA) (2-chloro-4,6-diamino-1,3,5-triazine), hydroxy atrazine (HA) (2-hydroxy-4-ethylamino-6-isopropylamino-1,3,5-triazine), deisopropyl hydroxy atrazine (DIHA) (2-hydroxy-4-ethylamino-6-amino-1,3,5-triazine), deethyl hydroxy atrazine (DEHA) (2-hydroxy-4-amino-6-isopropylamino-1,3,5-triazine), alachlor (ALAC) (2-chloro-2',6'-diethyl-N-(methoxymethyl) acetanilide), demethoxymethyl alachlor (A-DMM) (2-chloro-2',6'-diethyl-acetanilide), 2,6 diethylaniline (2,6 DiEA), alachlor oxanilic acid (A-OA) (2-[(2',6'diethylphenyl) (methoxymethyl) amino-]oxoacetic acid), and alachlor ethane sulfonic acid (A-ESA) (2-[(2',6'-diethylphenyl)(methoxymethyl)amino-]-2-oxoethane sulfonic acid). Terbutylazine (TERB)(2-tert-butylamino-4-chloro-6-ethylamino-s-triazine) is used as the surrogate because its structure is similar to atrazine, but it has not been used as a pesticide in the United States. This method is designed for the fast, quantitative analysis of these compounds in both water and sandy soils that are known to be contaminated. Therefore, speed and ease of sample preparation were higher priorities than minimizing detection limits, unlike most other HPLC methods reported in the literature (*7-12*).

Methods

Reagents. Atrazine (99.9%), alachlor (99.8%), and deethyl atrazine (98.5%) were obtained from the United States Environmental Protection Agency. Deisopropyl atrazine (97%) and terbutylazine (99%) were obtained from Crescent Chemical Company. Didealkyl atrazine (90%), hydroxy atrazine, deisopropyl atrazine (95%), and deethyl atrazine (97%) were obtained from Ciba-Geigy. 2,6 diethylaniline (99.5%) was obtained from Aldrich Chemicals. Alachlor ethane sulfonic acid (99.5%) and alachlor oxanilic acid were obtained from Monsanto Chemical Company. Pesticide residue grade and HPLC grade methanol was used (Burdick and Jackson). "Organic-free" water was obtained from a Milli-Q system (Millipore) and subsequently passed through a 5 g C-18 solid-phase extraction column (Varian) to further purify the water for use with the HPLC.

HPLC Analysis. The analytes were chromatographed using a 25 cm, 5 μm C-18 column (Sperisorb ODS2) that was maintained at 60°C on a Waters HPLC system (Waters model 600E system controller and pump, model 717+ autosampler, and model 996 photodiode array detector controlled by Millennium 2010, Version 2.00 software). The initial solvent conditions were 20:80 methanol:water (volume:volume) at a flow rate of 1.4 mL/min. The methanol content increased linearly to 100% within 21 minutes. The methanol content remained at 100% until 23 minutes, then the gradient returned to initial conditions in 6 minutes. Data was collected from the photodiode array at a rate of 1.0 spectrum/sec with a resolution of 1.2 nm from 215 to 300 nm. The method takes 31.5 minutes between successive injections. A chromatographic trace is shown in Figure 1.

The wavelength selected for quantification was 216 nm for all compounds. At this wavelength, the best compromise was obtained in the signal-to-noise ratio for all analytes. This allowed for spectral examination of peak purity. A library of each analyte's spectrum was developed from calibration standards and compared with the spectra of peaks in the soil and water samples. If the retention time of each peak and the spectrum of the suspected compound matched, the identity of the compound was considered to be confirmed. Target analytes were quantified by external calibration standards at five concentrations that bracketed the concentrations in the water and soil extracts.

Soil Extraction. Soil samples were homogenized by stirring and two aliquots were obtained. One was for the analysis of herbicides and the other was for determination of moisture content and organic matter content. Herbicides were extracted from soil samples by placing approximately 10 g of moisture soil into a preweighed 50 mL glass centrifuge tube equipped with a screw cap and a Teflon-faced silicon septa. The tube was then weighed again to determine total wet soil weight. A 40 μL injection of terbutylazine from the 999 mg/L stock solution was then added as a surrogate. Following addition of the surrogate, approximately 8 mL of 80:20 methanol:water was added as the extraction solvent. The centrifuge tube was again weighed to determine solvent weight. The tube was placed in a incubator/shaker (New Brunswick) and shaken while heated at 40°C for at least 12 hours. After removing the centrifuge tube from the incubator/shaker, it was weighed again to determine the loss of solvent. The final solvent weight was used to calculate the volume of solvent to determine concentrations. The tube was centrifuged (IEC HN-SII centrifuge) at 1000 rpm for 15 min. Approximately 2 mL of solvent was withdrawn from the tube using a gastight, glass syringe, passed from the syringe through an in-line, 0.2 μm, PTFE membrane filter (Lida, 25 mm diameter), and into an amber autosampler vial. The autosampler vials were stored at 4°C until analysis. Soil moisture content and organic matter content were determined by placing approximately 40 g of the moist, homogenized sample onto a preweighed aluminum pan. The pan and soil were then weighed again to determine the moist soil weight. The sample was then heated in an oven at 105°C for 24 hours and weighed again to determine the dry weight and moisture content. The pans and dried soil were then combusted at 550°C for one hour and weighed to determine the organic matter content (*13*).

Water Analysis. Water samples were placed into amber autosampler vials immediately after collection. These water samples were clear and devoid of solids upon visual inspection. The autosampler vials were stored at 4°C until analysis.

Results and Discussion

HPLC Methodology. The method detection limits were determined by analyzing dilutions of a solution of all the target analytes. These limits were based on triplicate 50 μL injections and assumed no preconcentration of either the water sample or the soil extracting solvent. Table I shows the method detection limits (MDL), the method quantification limits (MQL), and the method reporting limits (MRL) (*14*). The MQL was determined by adding three times the standard deviation of the MDL to the mean MDL. A method reporting limit (MRL), higher than the MQL, also was determined to account for inconsistencies in integration that occurred periodically.

Table I. Method detection limit (MDL), method quantification limit (MQL), and method reporting limit (MRL) for an injection of 50 μL into the HPLC

Analyte	MDL	MQL	MRL
	--------------- μg/L ---------------		
DIHA	20	29	60
DDA	129	22	60
DEHA	22	24	60
DIA	24	27	60
DEA	22	25	60
HA	59	110	110
A-OA	170	270	330
A-ESA	78	120	120
A-DMM	46	84	230
ATR	22	36	60
2,6 DiEA	83	160	160
TERB	26	80	80
ALAC	140	320	320

External calibration, based on standard solutions at five concentrations, was used to quantify the samples. A linear regression of peak area to analyte mass forced through the origin was calculated for each target analyte. A regression line was considered adequate if the correlation coefficients (r^2) were 0.97 or greater, although typically they were greater than 0.99. The five calibration standards were analyzed every 70 samples or when the solvent reservoirs were refilled, whichever came first. A quality control/quality assurance (QA/QC) check sample was injected after every ten samples. This check sample was a mid-concentration calibration standard injected and quantified as a sample. The resulting concentrations from this QA/QC check sample were compared with the known concentrations and plotted over time. If the QA/QC check sample varied from the previously determined mean by more than three standard deviations, the calibration curves were regenerated. Blanks and triplicate injections of samples were also regularly injected. In routine use of this method, the QA/QC samples accounted for about 10% of all samples injected.

Soil Analysis. The procedure for extraction of the analytes from soil was developed for use with sandy soil with a low organic matter content as part of a larger study (*15*). This soil typically had less than 10% clay and silt and less than 3% organic matter (dry weight). In addition, the soil was known to have been contaminated with atrazine and alachlor.   Because of the sandy nature of the soil, there was still a considerable amount of extracting solvent (usually about 60%) remaining in the interstitial space, even after centrifugation. For the HPLC analysis, 2 mL of the supernatant after centrifugation were removed with a syringe and filtered into an autosampler vial, as described above. A 50 µL aliquot was injected onto the HPLC. Recoveries of target analytes that were spiked directly into uncontaminated, moist soil, shaken, and then extracted with the method described here are presented in Table II. Data are presented for two compositions of the solvent mixture injected onto the HPLC (final extract): high methanol content (about 80:20 methanol:water) and low methanol content (about 5:95 methanol:water). For the later eluting compounds, DIA to ALAC with the exception of 2,6 DiEA (Figure 2), the recoveries were generally greater than 70%, regardless of solvent composition. For any of these analytes, the differences between their recoveries in the two solvent compositions only ranged from 5 to 15%. The chromatographic peak shape for all of these target analytes was essentially the same for both solvent compositions (Figures 2 and 3). 2,6 DiEA gave very poor recovery in all extractions. Since it had acceptable chromatography in the standard solutions and water samples, the poor recoveries from soil can be attributed to either its transformation during incubation in the soil or the inability of the extracting solvent to remove 2,6 DiEA from the soil. In either case, 2,6 DiEA should not be considered a viable target analyte in soil with this method. The recoveries were poor for the three early eluting compounds, DIHA, DDA, and DEHA, when the final solvent composition contained the higher percentage of methanol (Table II). Apparently, the higher percentage of methanol in the extraction matrix caused poor chromatography of the most polar compounds with this reversed-phase column and the initial mobile phase composition of 20:80 methanol:water. The resulting peaks were too broad to be correctly integrated. The recoveries of DIHA, DDA, and DEHA were significantly improved by decreasing the methanol content of the final extract (*10*). This was accomplished by placing the sample, in its autosampler vial, under a gentle stream of nitrogen until its volume was decreased by about 80%. The sample was brought back to its exact original volume with addition of water. If analysis of these three polar transformation products of atrazine (DIHA, DDA, and DEHA) is important to a specific investigation, this procedure must be done for each sample.

Water Analysis. The water samples were analyzed directly and without extraction or preconcentration. Quantification of water samples was performed using external standards. This allowed the samples to be injected directly without adding internal standard. The performance of the method for water samples was investigated by

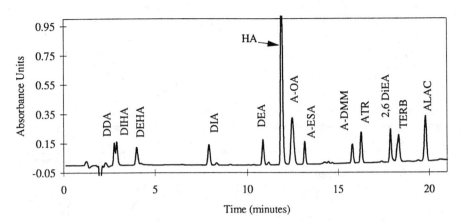

Figure 1. HPLC trace of a 50 μL injection of a standard of atrazine and six of its transformation products, alachlor and four of its transformation products, and terbutylazine, the surrogate. As examples, the concentration of atrazine and alachlor are 4.7 and 29 μg/mL, respectively. Analyte abbreviations are presented in the text.

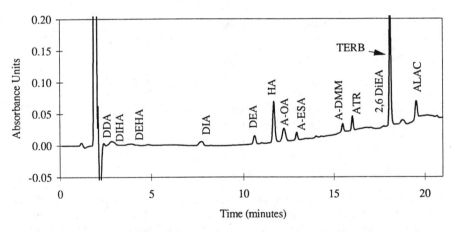

Figure 2. HPLC trace of the target analytes extracted from a soil spike when the final solvent extract composition was about 80:20 methanol:water (volume:volume).

spiking a ground-water sample at two concentrations of target analytes. The spiked
samples were then analyzed in triplicate. The results are shown in Table III.
Recoveries for the high-concentration spike ranged from 95 to 112% with a relative
standard deviation of 0.1 to 1.7%. The recoveries for the low-concentration spike
were similar, the only exceptions being for DIHA and DDA. These two
metabolites are the most polar and elute very early in the chromatogram (Figure 1).
Their anomalously high recoveries in the low-concentration spike were attributed to
integration difficulties.

Table II. Percent recoveries and percent relative standard deviations (%RSD) of
target analytes extracted from spiked soil using high and low methanol content
in final extract

Analyte	Spiked soil concentration mg/kg	High Methanol Content[1]		Low methanol Content[2]
		% Recovery	% RSD	% Recovery
DIHA	0.27	154	29.0	84
DDA	0.37	0	0.0	87
DEHA	0.30	0	0.0	48
DIA	0.33	78	3.4	92
DEA	0.37	90	1.5	96
HA	3.36	49	2.3	52
A-OA	3.12	68	2.8	72
A-ESA	0.85	87	3.4	95
A-DMM	0.69	99	13.4	93
ATR	0.34	97	2.2	102
2,6 DiEA	1.41	6	6.1	0
TERB	0.34	81	12.6	88
ALAC	2.07	87	7.7	92

[1] about 80:20 methanol:water (volume:volume) (n=3)
[2] about 5:95 methanol:water (volume:volume) (n=1)

Conclusions

The analysis of atrazine, alachlor, and nine of their transformation products was
effectively accomplished for sandy soil samples with low organic matter content
and in environmental water samples. The use of HPLC allowed for detection of
the polar metabolites which can not currently be determined by gas
chromatography. This HPLC method has the advantages of small soil quantities (50
g) and solvent extraction volumes (8 mL), relatively fast analyses, and excellent
precision. For soil samples, acceptable recoveries were obtained for all target
analytes, except 2,6 DiEA, when the methanol content of the final extract was
minimized before injection. The direct injection of environmental water samples

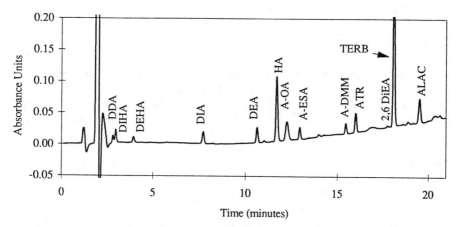

Figure 3. HPLC trace of the target analytes extracted from a soil spike when the final solvent extract composition was about 5:95 methanol:water (volume:volume).

Table III.   Percent recoveries and percent relative standard deviations (%RSD) of target analytes spiked into ground water

Analyte	High concentration spike[1]		Low concentration spike[1]	
	% Recovery	% RSD	% Recovery	% RSD
DIHA	100	1.7	146	0.7
DDA	108	0.8	122	0.5
DEHA	95	0.3	94	1.8
DIA	104	0.1	104	1.7
DEA	105	0.3	106	0.5
HA	106	0.3	109	0.7
A-OA	104	1.1	105	0.8
A-ESA	104	0.3	104	2.1
A-DMM	103	0.3	104	1.5
ATR	110	0.2	111	1.6
2,6 DiEA	112	0.3	111	1.2
TERB	95	0.7	96	2.2
ALAC	104	0.2	104	0.1

[1] There was a five-fold difference between the higher and lower concentration spikes. As examples, these final concentrations for atrazine and alachlor were 0.5 and 2.5 and 3.3 and 15.1 μg/mL, respectively.

provided a precise, yet fast, mode of analysis. This simple and quick HPLCmethod could be a valuable tool for those interested in the fate of atrazine and alachlor, particularly in environments with known use or contamination.

Acknowledgments

This work was funded by the United States Geological Survey's Toxic Substances Hydrology Program as part of the Minnesota Management Systems Evaluation Area study. The use of brand, firm, or trade names in this paper is for identification purposes only and does not constitute endorsement by the U.S. Geological Survey.

Literature Cited

1. Giannessi, L.P.; Puffer, C. *Herbicide Use in the United States*; Resources for the Future: Washington, DC, 1991; 128 p.

2. Thurman, E.M.; Goolsby, D.A.; Meyer, M.T.; Kolpin, D.W. *Environ. Sci.Technol.*, **1991**, *25*, 1794-1796.

3. Schottler, S.P.; Eisenreich, S.J.; Capel, P.D. *Environ. Sci.Technol.*, **1994**, *28*, 1079-1089.

4. Richards, R.P.; Baker, D.B. *Environ. Toxicol. Chem.*, **1993**, *12*, 13-26.

5. Squillace, P.J.; Thurman, E.M.; Furlong, E.T. *Water Resource. Res.*, **1993**, *29*, 1719-1729.

6. Holden, L.R.; Graham. J.A.; Whitmore, R.W.; Alexander, W.J.; Pratt, R.W.; Liddle, S.K.; Piper, L.L., *Environ. Sci.Technol.*, **1992**, *26*, 935-943.

7. Ritter, W.F. *J. Environ. Sci. Health*, **1990**, *B25*, 1-29.

8. Mandelbaum, R.T.; Wackett, L.P.; Allan, D.L. *Appl. Environm. Microbiol.*, **1993**, *59*, 1695-1701.

9. Mandelbaum, R.T.; Wackett, L.P.; Allan, D.L. *Environ. Sci.Technol.*, **1993**, *27*, 1943-1946.

10. Mills, M.S.; Thurman, E.M., *Anal. Chem.*, **1992**, *64*, 1985-1990.

11. Vermeulen, N.M.J.; Apostolides, Z.; Potgieter, D.J.J. *J. Chromat.*, **1982**, *240*, 247-253.

12. Macomber, C.; Bushway, R.J.; Perkins, L.B.; Baker, D.; Fan, T.S.; Fergusen, B.S. *J. Agric. Food Chem.*, **1992**, *40*, 1450-1452.

13. *Standard Methods*; Franson, M.H., Ed.; 15th Edition; Byrd Press: Springfield, VA, 1981; p. 95-96.

14. Stanko, G.H.; Krochta, W.G.; Stanley, A.; Dawson, T.L.; Hillig, K.J.D.; Javik, R.A.; Obryckl, R.; Hughes, B.M.; Saksa, F.I., *Environ. Lab*, **1993**, p.16-20, 60.

15. Delin, G.N.; Landon, M.K.; Lamb, J.A.; Anderson, J.L., *Characterization of the hydrogeology and water quality of the Management System Evaluation Area near Princeton, Minnesota, 1991-1992*, **1994**, U.S. Geological Survey WRI 94-4149, 54p.

Chapter 4

Factors Influencing the Specificity and Sensitivity of Triazine Immunoassays

Timothy S. Lawruk[1], Charles S. Hottenstein[1], James R. Fleeker[2], Fernando M. Rubio[1], and David P. Herzog[1]

[1]Ohmicron Environmental Diagnostics, 375 Pheasant Run, Newtown, PA 18940
[2]Biochemistry Department, North Dakota State University, P.O. Box 55116, Fargo, ND 58105

The use of immunochemical methods to assess the quality of surface and ground water has become a valuable complement to conventional methods such as GC and HPLC. Immunoassays for water analysis have proven to be rapid, sensitive and convenient, allowing a large number of samples to be screened without prior sample preparation. Magnetic particle-based immunoassays for triazines have been developed to offer a wide range of specificities and sensitivities to both parent compounds and metabolites. By varying the design of the triazine immunogen and enzyme conjugate hapten, immunoassay specificity can be manipulated to detect a wide range of triazines or be specific to only the parent compound. The least detectable concentration of each triazine can also be decreased by modification of these components. The development strategies for triazine immunoassays including atrazine, simazine and cyanazine will be discussed.

The worldwide use of the 1,3,5-triazine or *s*-triazine herbicides as pre- and post-emergence herbicides for the control of broad-leaved weeds on corn, sorghum and fruit orchards has necessitated that these compounds and their metabolites be monitored in the environment. For example, atrazine residues are often detected in environmental water samples as a result of spills, spraying, and agricultural run-off (*1-3*). The widespread application, stability and solubility of the triazines in water allow them to leach from soil and be relatively persistent environmental contaminants (*4*).

The use of enzyme immunoassay to screen water samples for the presence of triazine herbicides, in particular atrazine, simazine and cyanazine, has been widely applied (*1-3, 5-11*). Immunoassays provide the analytical chemist with a cost-effective, sensitive, rapid and reliable method for both laboratory and field analyses (*8*). These immunoassay methods have been developed using a variety of

0097–6156/96/0630–0043$15.00/0

solid-phases including antibody adsorbed to microtiter plate wells, polystyrene test tubes and antibody covalently coupled to magnetic particles. The magnetic particle-based triazine immunoassays described herein are used to evaluate the effects of immunogen and enzyme conjugate hapten design on the assay sensitivity and specificity.

Materials and Methods

Antibody Production. The preparation of the immunogen is a critical step in the development of an immunoassay. Since triazines are small molecules (<1000 daltons), they will not elicit an immune response when injected into a rabbit. To obtain an immunogenic response, a triazine analog must first be coupled to a high molecular weight carrier such as bovine serum albumin (BSA) or keyhole limpet hemocyanin (KLH) which will present the triazine as a foreign substance to the animal's immune system and initiate antibody production. The orientation of the triazine analog exposed to the immune system affects the specificity and affinity of the antibody generated.

In this work, the immunogens were prepared by coupling the haptens described to the carrier protein using a mixed anhydride reaction. The immunogens were purified by gel filtration, emulsified with adjuvant and injected into rabbits on a defined schedule. Rabbits were bled every 30 days and antisera collected and screened for activity. Selected antisera was covalently coupled to amine-terminated magnetic particles by glutaraldehyde activation (9).

Triazine-Enzyme Hapten Conjugation. The enzyme conjugates were synthesized by preparing active esters of the carboxylic acid analog of the triazine haptens using N-hydroxysuccinamide. The active esters were coupled to horseradish peroxidase (HRP) by reaction with dicyclohexylcarbodiimide (12). The enzyme conjugates were purified by gel filtration chromatography and titrated to obtain approximately 1.6 absorbance units at 450 nm for the zero triazine concentration under assay conditions.

Immunoassay Format. A competitive assay format was utilized in the development of the triazine immunoassays. In this format, triazine specific antibodies are coupled to magnetic particles. The sample to be tested is added (100-250 μL), along with the triazine enzyme conjugate (250 μL) to a test tube, followed by magnetic particles (500 μL) with antibodies specific for triazines attached. The triazines in the sample and the triazine-labeled enzyme conjugate compete for antibody binding sites on the magnetic particles. After an incubation period of 15-30 minutes, a magnetic field is applied to hold the particles in the tube and the unbound triazines and triazine enzyme conjugate are removed by decanting the reaction tubes. The triazines and enzyme labeled triazine bound to the antibodies on the particles are in proportion to their original concentration. After washing to remove unbound triazine enzyme conjugate, the triazines are detected by adding a color development reagent containing hydrogen peroxide and tetramethylbenzidine. The enzyme labeled triazine bound to the antibody catalyzes the conversion of the color development reagent to a blue-colored product. The color development is stopped and stabilized with the addition of dilute acid. The final concentration of the triazines is determined by measuring the absorbance at 450 nm and comparing the sample absorbances to a linear regression line, using a log/logit transformation, developed from calibrators prepared in the zero standard. Since the enzyme labeled triazine was in competition for antibody binding sites with the unlabeled triazine in the sample, the color developed is inversely proportional to the concentration of triazine in the sample.

Determination of Specificity (Cross-reactivity). Figure 1 illustrates a typical dose response curve for atrazine and deisopropylatrazine, one of its metabolites, in a triazine immunoassay. The least detectable dose (LDD) for each triazine, the lowest concentration that can be distinguished from zero, was defined as 90% B/Bo, the concentration at which the enzyme conjugate is displaced 10% (*13*). B/Bo is the absorbance at 450 nm observed for a sample or standard divided by the absorbance of the zero standard, which produces the maximum absorbance. The 50% B/Bo inhibition concentration (IC_{50}) is the triazine concentration at which the immunoassay is displaced 50%. The percent cross-reactivity is the IC_{50} of the triazine analyte divided by the IC_{50} of each potential cross-reactant (times 100).

The compounds evaluated for cross-reactivity in the various triazine immunoassays are selected *s*-triazines (Table I). These compounds differ at three positions. The 2-position which contains chloro-, methoxy- or methylthio- groups and the 4- and 6-positions which have ethylamino-, isopropylamino-, or tert-butylamino- groups. Cyanazine also has an additional nitrilo substitution on the tert-butyl moiety. The major degradation products of atrazine were also evaluated and included 2-hydroxyatrazine and the dealkylated metabolites.

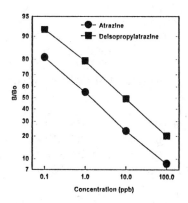

Figure 1. Typical triazine dose response curves used to calculate least detectable dose (LDD) at 90% B/Bo and cross-reactivity at IC_{50} (50% B/Bo), where B/Bo is the absorbance at 450 nm of the standard divided by the absorbance at 450 nm of the zero standard. The logit of B/Bo (*y*) was plotted versus the log of the concentration (*x*).

Table I.   Structures of the *s*-Triazines, Related Metabolites and Haptens.

Compound	R_2	R_4	R_6
Triazines			
Atrazine	Cl	$NHCH(CH_3)_2$	$NHCH_2CH_3$
Cyanazine	Cl	$NHCCN(CH_3)_2$	$NHCH_2CH_3$
Simazine	Cl	$NHCH_2CH_3$	$NHCH_2CH_3$
Propazine	Cl	$NHCH(CH_3)_2$	$NHCH(CH_3)_2$
Terbuthylazine	Cl	$NHC(CH_3)_3$	$NHCH_2CH_3$
Prometryn	SCH_3	$NHCH(CH_3)_2$	$NHCH(CH_3)_2$
Prometon	OCH_3	$NHCH(CH_3)_2$	$NHCH(CH_3)_2$
Ametryn	SCH_3	$NHCH(CH_3)_2$	$NHCH_2CH_3$
Terbutryn	SCH_3	$NHC(CH_3)_3$	$NHCH_2CH_3$
Metabolites			
Desethylatrazine	Cl	$NHCH(CH_3)_2$	NH_2
Didealkylatrazine	Cl	NH_2	NH_2
Deisopropylatrazine	Cl	NH_2	$NHCH_2CH_3$
Hydroxyatrazine	OH	$NHCH(CH_3)_2$	$NHCH_2CH_3$
Cyanazine Amide	Cl	$NHC(CONH_2)(CH_3)_2$	$NHCH_2CH_3$
Cyanazine Chloro Acid	Cl	$NHC(COOH)(CH_3)_2$	$NHCH_2CH_3$
Cyanazine Hydroxy Acid	OH	$NHC(COOH)(CH_3)_2$	$NHCH_2CH_3$
Haptens			
I	$S(CH_2)_3COOH$	$NHCCN(CH_3)_2$	$NHCH_2CH_3$
II	$S(CH_2)_5COOH$	$NHCH(CH_3)_2$	$NHCH_2CH_3$
III	Cl	$NHCH(CH_3)_2$	$NH(CH_2)_5COOH$
IV	Cl	$NHCH_2CH_3$	$NH(CH_2)_3COOH$
V	$S(CH_2)_5COOH$	$NHCH_2CH_3$	$NHCH_2CH_3$

Results and Discussion

Cyanazine Immunoassay Table II illustrates the specificity using Hapten I (Table I) to prepare both the immunogen and enzyme conjugate for the cyanazine immunoassay (7). It is derivatized through the 2-position for conjugation to proteins. As shown by the LDD, 0.035 ppb, and the IC_{50}, 0.44 ppb, the cyanazine assay is very sensitive. This assay is also very specific with less than 0.1% reactivity to both the cyanazine and atrazine metabolites. The cyanazine antibody appears to be most reactive to the cyano- group since the cyanazine amide and chloro acid metabolites which are altered at this position exhibit a substantial decrease in reactivity. The cyanazine immunoassay also demonstrates less than 1% reactivity with related triazines. The greatest reactivity with this method is demonstrated by terbuthylazine (3.2%) and terbutryn (3.1%) both of which have the tert-butyl moiety similar to cyanazine but lack the cyano- group. The cyanazine immunoassay method developed using the hapten structure described provides a sensitive and specific analytical tool.

Atrazine Immunoassay (2-Position Substitution). Table III illustrates the specificity using Hapten II (Table I) to prepare both the immunogen and enzyme conjugate for the atrazine immunoassay, which was derivatized through the 2-position for conjugation to BSA and HRP. The preparation of this hapten has previously been described (*10*). Reactivity to the atrazine metabolites using this hapten design produces significant reactivity to desethylatrazine. It is not surprising

Table II. Specificity of the Cyanazine Immunoassay (Hapten I)

Compound	LDD (ppb)	IC_{50} (ppb)	% Reactivity
Cyanazine	0.035	44	100
Metabolites			
Cyanazine Amide	4.90	419	0.10
Cyanazine Chloro Acid	160	>10,000	<0.1
Cyanazine Hydroxy Acid	1420	>10,000	<0.1
Hydroxyatrazine	3.5	613	0.1
Desethylatrazine	117	2480	<0.1
Deisopropylatrazine	662	>10,000	<0.1
Didealkylatrazine	>10,000	>10,000	<0.1
Related Triazines			
Terbuthylazine	0.05	13.5	3.2
Terbutryn	0.11	14.3	3.1
Ametryn	0.50	80.5	0.55
Simazine	1.60	156	0.28
Propazine	3.5	360	0.12
Prometryn	1.50	527	0.1
Prometon	82.0	2070	<0.1
Atrazine	200	>10,000	<0.1

that although both the isopropylamino and ethylamino groups found on the hapten are present in hydroxyatrazine, cross-reactivity with this metabolite is very low (<1%). Since the hapten design contains a sulfur which closely resembles chlorine in electronic structure and size, production of antibodies which bind the chloro-*s*-triazines and not the hydroxy metabolites was expected (*10*).

Cross-reactivity with related triazines using this hapten design demonstrates the broad specificity of this system. Ametryn contains the same isopropylamino and ethylamino side chains and even a thio substituent in the 2-position, as found on the hapten. As a result, ametryn has the greatest reactivity in the immunoassay (185%). Prometryn and propazine contain isopropylamino groups in both the 4- and 6-positions and have substantial reactivity in the immunoassay, 113% and 97% respectively. Prometryn's slightly greater reactivity is probably due to the thio substituent in the 2-position which again is similar to the hapten. Prometon, which also contain isopropylamino groups at the 4 and 6-positions, exhibits lower reactivity due to the ether linkage at the 2-position which is consistent with the hydroxyatrazine observation above. Simazine, terbutryn, terbuthylazine and cyanazine lack the isopropylamino side chain but do possess the ethylamino side chain and have substantially less reactivity in the assay. The use of this atrazine hapten design provides greater reactivity to the isopropylamino group than the ethylamino group with substantial influence seen from the 2-position substituent. In contrast to the cyanazine immunoassay, this atrazine method has broad specificity to the *s*-triazines, however, it remains relatively sensitive to atrazine with a LDD of 0.05 ppb and a IC_{50} of 0.72 ppb.

Atrazine Immunoassay (6-Position Substitution). The specificity of an alternative approach to the atrazine hapten design is summarized in Table IV. The hapten used to prepare both the immunogen and enzyme conjugate in this study is derivatized at the 6-ethylamino position (Hapten III,Table I). This hapten has previously been described (*10-11*) and found to provide a more sensitive atrazine immunoassay system (*10*). The cross-reactivity to the atrazine metabolites in this method is very similar to the atrazine immunoassay described above, with desethylatrazine having the greatest reactivity (24.3%).

However, when comparing various related triazines, this method is more specific for atrazine. The antibody produced by this hapten design exhibits reactivity to both the isopropylamino and chloro groups. Propazine which has an isopropylamino group at both the 4- and 6-positions, has the greatest reactivity (244%). The propazine structure provides two different molecular orientations of the isopropylamino and chloro groups to the antibody, hence the approximate two-fold increase in reactivity. The influence of the 2-chloro group on reactivity is demonstrated by ametryn (3.6%), which has a structure similar to atrazine except for the substitution of a methylthio group at the 2-chloro position. This atrazine immunoassay design also provides a greater sensitivity to atrazine with an LDD of 0.015 ppb and a IC_{50} of 0.22 ppb. Comparison of the immunoassays developed from the atrazine hapten derivatized at the 6-position (Hapten III) to the previous method derivatized at the 2-position (Hapten II), demonstrates the dramatic effect hapten design can have on assay specificity and sensitivity.

Table III. Specificity of the Atrazine Immunoassay (Hapten II)

Compound	LDD (ppb)	IC_{50} (ppb)	% Reactivity
Atrazine	0.05	0.72	100
Metabolites			
Desethylatrazine	0.06	3.21	22.4
Deisopropylatrazine	0.80	217	0.3
Hydroxyatrazine	1.10	148	0.5
Didealkylatrazine	2000	>10,000	<0.1
Related Triazines			
Ametryn	0.05	0.39	185
Prometryn	0.05	0.64	113
Propazine	0.03	0.74	97
Prometon	0.06	2.22	32.4
Simazine	0.34	4.90	14.7
Terbutryn	0.09	5.50	13.1
Terbuthylazine	0.31	15.5	4.6
Cyanazine	1.00	>10,000	<0.1

Table IV. Specificity of the Atrazine Immunoassay (Hapten III)

Compound	LDD (ppb)	IC_{50} (ppb)	% Reactivity
Atrazine	0.015	0.22	100
Metabolites			
Desethylatrazine	0.027	0.87	25.3
Deisopropylatrazine	0.16	128	0.17
Hydroxyatrazine	0.16	148	0.15
Didealkylatrazine	0.16	981	<0.1
Related Triazines			
Propazine	0.005	0.09	244
Simazine	0.019	2.03	10.8
Prometryn	0.013	2.38	9.2
Prometon	0.015	2.38	9.2
Terbuthylazine	0.019	5.09	4.3
Ametryn	0.019	6.08	3.6
Cyanazine	0.036	50.7	0.4
Terbutryn	0.027	63.5	0.3

Simazine Immunoassay (6-Position Substitution). Table V illustrates the specificity resulting from the use of Hapten IV (Table I), which was derivatized through the 6-position, to prepare both the immunogen and enzyme conjugates for the simazine immunoassay. This hapten design presents the ethylamino and chloro groups distal from the carrier protein of the immunogen. The method exhibits low reactivity to deisopropylatrazine although the chloro and ethylamino groups are present in this compound and the isopropyl position of the hapten was used for protein coupling. Normally it is thought that the hapten position proximal to the protein has little effect on antibody binding, however, this system appears much less sensitive to compounds lacking the second ethylamino group.

The reactivity of this immunoassay to other atrazine metabolites is also relatively low, however, there is substantial reactivity to related triazines. As might be expected, the reactivity of this assay with atrazine is substantial (41.6%) since the 2-chloro and 4-ethylamino groups of the hapten are retained. Other triazines with a 2-chloro group, terbuthylazine (23.8%), propazine (21.3%), and cyanazine (5.8%) also exhibit significant reactivity. Compounds which have methylthio or methoxy substituents at the 2-position display significantly lower reactivities (<1%). Although this simazine immunoassay system provides excellent sensitivity (LDD of 0.01 ppb, IC_{50} of 0.57 ppb), the specificity of the system is broader compared to the analogous atrazine system deriviatized through the 6-ethylamino position.

Simazine Immunoassay (2-Position Enzyme Conjugation and 6-Position Immunogen Substitution). The final hapten design evaluated (Table VI) uses the same antibody described in the previous simazine immunoassay (Hapten IV) but utilizes an enzyme conjugate synthesized from a hapten coupled through the 2-position (Hapten V, Table I). This heterologous system is less sensitive than the simazine assay described previously (LDD of 0.07 ppb, IC_{50} of 15.3 ppb) and the reactivity to deisopropylatrazine has increased. The decreased sensitivity using this heterologous system is in contrast to similar heterologous systems previously described which provided increased assay sensitivity (*10, 14-15*). The reactivity profile using the heterologous simazine assay is significantly different from the homologous simazine assay (Table V). The reactivity of the hapten with different antibody populations in the polyclonal antiserum may contribute to this observation. Others have reported that with a monclonal antibody this chnage would not be expected (*16*).

Altering the enzyme conjugate hapten structure dramatically affects the reactivity to related triazines. The atrazine reactivity has been significantly decreased from approximately 42% to less than 1%. Triazines with a 2-chloro group, such as atrazine, cyanazine and propazine, have a lower reactivity (<1%) then the previous simazine assay, with the exception of terbuthylazine. It is not apparent why this assay configuration is more reactive to terbuthylazine, which has a tert-butyl group at the 6-position, than atrazine, which has a isopropyl group at the 6-position. These simazine immunoassays illustrate the dramatic variations in sensitivity and specificity that can be achieved by changing one critical assay component, in this case the enzyme conjugate hapten design.

Table V. Specificity of the Simazine Immunoassay (Hapten IV)

Compound	LDD (ppb)	IC_{50} (ppb)	% Reactivity
Simazine	0.01	0.57	100
Metabolites			
Deisopropylatrazine	0.13	6.98	8.2
Desethylatrazine	0.25	13.4	4.2
Hydroxyatrazine	>100	>100	<1.0
Didealkylatrazine	25.7	>100	<1.0
Related Triazines			
Atrazine	0.05	1.37	41.6
Terbuthylazine	0.01	2.39	23.8
Propazine	0.03	2.68	21.3
Cyanazine	0.08	9.9	5.8
Terbutryn	0.50	>100	<1.0
Ametryn	1.90	>100	<1.0
Prometon	5.55	>100	<1.0
Prometryn	10.6	>100	<1.0

Table VI. Specificity of the Simazine Immunoassay (Hapten IV for Immunogen Synthesis and Hapten V for Enzyme Conjugation)

Compound	LDD (ppb)	IC_{50} (ppb)	% Reactivity
Simazine	0.07	15.3	100
Metabolites			
Deisopropylatrazine	0.11	56.3	27.2
Desethylatrazine	0.55	>100	<1.0
Didealkylatrazine	14.5	>100	<1.0
Hydroxyatrazine	>100	>100	<1.0
Related Triazines			
Atrazine	0.42	>100	<1.0
Ametryn	0.56	49.2	31.1
Terbuthylazine	0.01	58.5	26.1
Terbutryn	0.08	85.8	17.8
Cyanazine	1.15	>100	<1.0
Propazine	1.50	>100	<1.0
Prometryn	4.33	>100	<1.0
Prometon	3.64	>100	<1.0
Atrazine	0.42	>100	<1.0

Conclusion

Several immunoassay methods for the s-triazines (atrazine, cyanazine and simazine) have been developed using a single format; i.e., a magnetic particle-based solid-phase. The reactivity to various metabolites and related triazines of two atrazine methods, two simazine methods and a cyanazine method have been demonstrated. The cyanazine method described is both very sensitive and specific. The two atrazine methods described have different reactivity to various triazines depending upon the position the hapten is derivatized. The two simazine methods described illustrate the dramatic effect of variations to the enzyme conjugate hapten on both sensitivity and specificity. By manipulating the immunogen and enzyme conjugate hapten design, the specificity and sensitivity to metabolites and related triazines of the immunoassay can be modified. To develop a triazine immunoassay to meet the requirements of sensitivity and specificity for a particular application, both the immunogen and enzyme conjugate hapten design must be considered.

References

1. Hall, J.C.; Van Deynze, T.D.; Struger, J. *J. Environ. Sci. Health* **1993**, *5*, 577-598.
2. Thurman, E.M.; Goolsby, D.A.; Meyer, M.T.; Mills, M.S.; Pomes, M.L.; Kolpin, D.W. *Environ. Sci. Technol.* **1992**, *26*, 2440-2447.
3. Goolsby, D.A.; Coupe, R.C.; Markovchick, D.J. Water Resources Investigation Report 91-4163, U.S. Geological Survey, Denver, CO, 1991.
4. Cai, Z.; Sadagopa-Ramanujam, V.M.; Gilpin, D.E.; Gross, M.L. *Anal. Chem.* **1993**, *65*, 365-370.
5. Huber, S.J.; Hock, B. *Methods of Enzymatic Analysis, Drugs and Pesticides;* Vol. XII; Bergmeyer, H.U., Ed.: Weinheim VCH, 1986, p. 438.
6. Karu, A.E.; Harrison, R.O.; Schmidt, D.J.; Clarkson, C.E.; Grassman, J.; Goodrow, M.H.; Lucas, A.; Hammock, B.D.; Van Emon, J.M.; White, R.J. *Immunoassays for Trace Chemical Analysis: Monitoring Toxic Chemicals in Humans, Food and Environment;* ACS Symposium Series, Vol. 451; American Chemical Society: Washington, DC, 1991, p. 59.
7. Lawruk, T.S.; Lachman, C.E.; Jourdan, S.W.; Fleeker, J.R.; Herzog, D.P.; Rubio, F.R. *J. Agric. Food Chem.* **1993**, *41*, 747-752.
8. Van Emon, J.M.; Lopez-Avila, V. *Anal. Chem.* **1992**, *64*, 79-99.
9. Rubio, F.M.; Itak, J.A.; Scutellaro, A.M.; Selisker, S.Y.; Herzog, D.P. *Food Agric. Immunol.* **1991**, *3*, 113-125.
10. Goodrow, M.H.; Harrison, R.O.; Hammock, B.D. *J. Agric. Food Chem.* 1990, 38, 990-996.
11. Wüst, S.; Hock, B. *Anal. Letters*, **1992**, *25*, 1025-1037.
12. Langone, J.J.; Van Vanakis, H. *Res. Commun. Chem. Pathol. Pharmacol.* **1975**, *10*, 163-171.
13. Midgeley, A.R.; Niswender, G.D.; Rebar, R.W. *Acta Encrinol.* **1969**, *63*, 163-179.
14. Harrison, R.O.; Goodrow, M.H.; Hammock, B.D. *J. Agric. Food Chem.* **1991**, *39*, 122-128.
15. Schneider, P.; Hammock, B.D. *J. Agric. Food Chem.* **1992**, *40*, 525-530.
16. Muldoon, M.T.; Fries, G.F.; Nelson, J.O. *J. Agric. Food Chem.* **1993**, *41*, 322-328.

Chapter 5

Standardization of Immunoassays for Water and Soil Analysis

Bertold Hock[1,2], P.-D. Hansen[2], A. Krotzky[2], L. Meitzler[2],
E. Meulenberg[2], G. Müller[2], U. Obst[2], F. Spener[2], U. Strotmann[2],
L. Weil[2], and C. Wittmann[2]

[1]Technical University of München, Faculty of Agriculture
and Horticulture, Department of Botany at Weihenstephan,
D–85350 Freising, Germany
[2]Immunoassay Study Group, Deutsches Institut für Normung,
Normenausschuss Wasserwesen, Arbeitsausschuss I,4, Germany

Immunoassays have become a valuable tool in the field of environmental analysis, especially for screening a large number of samples within a short time. One of the approaches to guarantee the reliability and the comparability of the results is to prescribe a standardized method in detail. In Germany, the Immunoassay Study Group as a working group of the German Standards Institute DIN and the German Chemical Society is dealing with the standardization of immunoassays for pesticides and environmental pollutants. In 1994, a draft standard has been completed on selective immunoassays for the quantitative determination of plant treatment and pesticide agents in drinking, ground and surface water. In order to prove the feasibility of the standard procedure, round robin tests were simultaneously carried out for the herbicide atrazine and the environmental pollutant TNT including spiked and unspiked ground and surface water samples. The data show that the requirements of the guideline, the quantitative determination of a single pesticide, can be met, even in the presence of related compounds of environmental significance. An interlaboratory trial, carried out in 1995 with sixteen participants, provides the basis for the final DIN guideline for selective immunoassays.

The efficiency of environmental control depends on the quality of environmental analysis. Effective analysis is not only a matter of accurate and precise measurements, but also of the availability of screening methods which are highly sensitive, simple and cheap and which can be carried out in a short time for a large number of field samples. These demands can be fulfilled by immunochemical analyses *(1,2)*. This approach has been used with great success in the medical field for more than 35 years, but it has become available in environmental analysis only during the last years for a limited number of compounds such as approximately 80 pesticides and a few other substances of environmental concern *(e.g., 3-7)*.

Immunoassays as Screening Tools.

Several laboratories have provided evidence for the usefulness of immunoassays (IAs) as screening tools in environmental analysis (e.g., *2, 8-11*). The following example is taken from a laboratory *(12)* which has been involved during the recent years in a monitoring program to assay soil samples for the pesence of atrazine, one of the most important herbicides in Germany before its ban in April 1991.

In order to ensure the compliance with the ban on atrazine, more than 400 soil samples from corn fields in Southern Germany were collected and screened for atrazine residues using an enzyme immunoassay (EIA) for atrazine *(13)*. Although atrazine has been banned in Germany, thirty five soil samples had a concentration of atrazine greater than 100 µg/kg indicating recent application. The concentration of atrazine for all positive samples was verified by HPLC. Several of the samples for which the EIA response was negative were also verified by HPLC. Figure 1 shows a highly correlated, approximately 1-to-1 correspondence between atrazine concentrations determined from EIA and HPLC. These data demonstrate that immunoassay is not only a useful screening method, but also a quantitative method that may require only partial confirmation, for quality assurance purposes, by an independent method, such as HPLC.

The Role of Standardization.

In spite of the successful use of EIAs, it is not yet considered an established methodology in the field of environmental analysis, compared to classical methods such as GC, GC-MS or HPLC. This is especially true in Europe. There are two reasons for the slow acceptance of EIA. First is the the restricted number of analytes for which EIA tests exist. For example, EIAs have been developed for approximately 80 analytes; however, more than 300 active substances are applied and there may be several metabolites that would be useful to test for as well. More efficient methods of antibody production is the only way to alleviate this problem. The recombinant approach for antibody reproduction has been demonstrated to be an efficient method *(14,15)* and may be used to more rapidly increase the number of analytes available for analysis by EIA.

The second reason is the lack of standardization of EIA methodologies. Erroneous classification of EIA as a biological method has carried the stigma that EIA is imprecise and/or unreliable, thus, requiring complete verification by other more accepted methods. For researchers involved in long-term or long-range studies proper data interpretation is dependent upon the precision and accuracy of within and among the methods used by the monitoring laboratories analzing study samples. Thus, it is imperative that the monitoring laboratories can provide reliable and comparable results.

Schmidt *(16)* has considered two approaches that can be followed to ensure comparability of results among methods and laboratories: (1) The reliability and the comparability of the analyses is checked by regular interlaboratory testing. In this case, new approaches based upon progress in instrumental equipment and increase of knowledge are directly used for improving and therefore changing available methods. However, this policy is often unrealistic for laboratories in charge of water monitoring programs. These facilities usually lack the necessary capacity and staff required for the arduous process of proving equivalence of methods. (2) A standardized method is prescribed in detail. The analyst is encouraged to choose this method or a method that has been proven to be equivalent to the standard method. With this approach comparable results will be obtained by any that uses a method shown to be equivlent to the chosen standard. Standardization is especially useful in the case of long-term monitoring and the evaluation of long-term changes of water quality where several different scientists and laboratories may be involved over time. At present, the "standard" is in most cases the method of choice.

Standardization can have a powerful effect that should not be overlooked. For example, if progress is considered as as a steep slope leading to new achievements, standards may be implemented as steps in a staircase, which prevent sliding down to lower levels. This is illustrated in Figure 2. This figure shows that depending on the project different levels of precision, accuracy, and reproducibility may be required. Thus, a standard may be employed for each step to ensure that the level of precision and accuracy that will be required of any method employed for a particular step will be met. In this case standards in immunochemical techniques may be used to ensure that new technologies are associated with the proper level or step and that new methods are compared to the standard associated with that level. For instance, the requirements for immunosensors should not fall below the standards of immunoassays, especially in terms of sensitivity or robustness.

The DIN Approach.

In Germany the following procedure is used to employ standardization and to ensure that it is used by analytical laboratories *(cf. 16)*: The government, represented by the ministry of environment, charges the German Standards Institute DIN (Deutsches Institut für Normung e.V.) with the construction of these standards. DIN in close cooperation with the German Chemical Society (GDCh) has installed a committee for the construction of standards on water, waste water, and sludge (DIN NAW I,4; Normenausschuß Wasserwesen, Arbeitsausschuß I,4). Members of the comittee are selected from universities, chemical industry, and state authorities. This committee also serves as the German working group for all ISO (International Organization of Standardization) and European matters in water quality *(16,17)*. If this committee receives the request for standardization of a defined analytical method, it installs a special working group for the evaluation of standards. In 1989, the Immunoassay Study Group was established by the GDCh and DIN. This group is dealing with the standardization of immunoassays for pesticides and environmental pollutants. The tasks include the elaboration of a working document which, after the successful completion of round robin tests and interlaboratory trials, will lead to an official standard.

Categories of Immunoassays. The Immunoassay Study Group was intended to set up general guidelines and quality criteria for EIAs rather than to standardize individual EIAs for selected targets (a policy used by the AOAC and the EPA in the USA). However, the group had to decide which category of EIAs should be considered. Table I gives a survey on IA types with respect to their selectivities and whether they provide quantitative or qualitative data. For practical purpose of, for example, judging

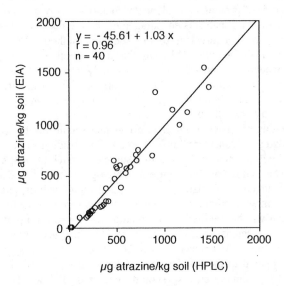

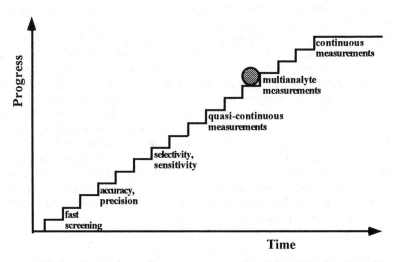

Figure 1. Comparison of atrazine concentrations of soil extracts determined by EIA and HPLC (from *12*). Extracts containing more than 300 μg atrazine/kg soil were diluted 1: 500 prior to the EIA, all other samples 1:150. Therefore matrix effects could be avoided.

Figure 2.   Progress in the technology of immunochemical methods. Standards are considered as steps preventing falling off at the appearance of new technologies. The time scale indicates an arbitrary measure.

water quality group-specific assays providing a sum parameter would be most beneficial. However, this goal is the most difficult one for immunochemical reasons. This is because a given antibody generally reacts with several related analytes. In addition, the affinity of the antibody toward each of the analytes is usually different (i.e. different levels of cross-reactivity). The assay would therefore provide analyte equivalents, which are of little analytical or toxicological relevance.

Furthermore, sum parameters can only be obtained if appropriate antibody mixtures or preferably, antibody arrays combined with multivariate analyses (e.g., *18)*, are employed. The latter alternative is usually beyond the scope of simple EIAs and better placed in the range of immunosensors, the former alternative usually fails because it is rare that a suite of antibodies that can complement each other in a suitable manner are available.

Therefore, standardization was restricted to selective assays. As in all EIA categories, quantitative assays have to be distinguished from qualitative assays, which provide threshold values (i.e. a yes/no answer on whether a predefined threshold is exceeded). In the latter case, the transgression of the threshold should be indicated, preferentially by a color change. On close inspection this alternative is difficult because threshold analyses require very steep calibration curves for the analyte. Since the threshold concentrations should be close to 0.1 µg/L to meet the demands of the European drinking water regulation, extremely sensitive antibodies are required. Sufficiently steep calibration curves are only obtained with monoclonal antibodies and, what appears to be more crucial, at high antibody concentrations. However, the latter demand excludes high assay sensitivity (for theoretical considerations cf. *19, 20)*. Therefore it was decided to restrict the general standard to selective EIAs for the quantitative determination of herbicides and pesticides in drinking, ground and surface water.

Draft Standard. In 1995, a draft standard was completed *(21)*. The guideline was constructed for the selective analysis of single compounds in concentrations ≥ 0.05 µg/L. The most crucial requirements of the guideline are as follows: The selectivity of the antibodies must guarantee that the analyzed sample concentrations are between 70 and 110% of the real concentration. Cross-reactivities have to be declared. The coefficient of variation for the parallels must not exceed 10% and the measuring range of the assay should be covered by at least 6 points, and a zero standard must be included.

Round Robin Tests. In order to prove the feasibility of the standard procedure, round robin tests were conducted in 1994 by the Immunoassay Study Group for the herbicide atrazine and for 2,4,6-trinitrotoluene (TNT) as an environmental pollutant. Figure 3 shows the results for atrazine. For this study, six samples from ground water and surface water (from the river Rhine) were provided by the water company Stadtwerke Mainz. Some of the samples were spiked with atrazine, simazine and terbuthylazine. Two different EIAs, provided by two companies (supplier 1 and 2), were used for the atrazine analyses. The water samples were also tested by the two companies using their respective EIA and also were tested by an independent

Table I. Grouping of Immunoassays

Assay Type	Selective Assays	Assays with Antibodies of Different Cross-reactivities	Group-specific Assays
quantitative assays	concentration of a single analyte (e.g. 0.1 ppb atrazine)	concentration of analyte equivalents (e.g. 0.1 ppb atrazine equivalents)	concentration of a group of analytes (e.g. 0.5 ppb s-triazines)
qualitative assays	threshold value for a single analyte (e.g. > 0.1 ppb atrazine)	threshold value for analyte equivalents (e.g. > 0.1 ppb atrazine equivalents)	threshold value for a group of analytes (e.g. > 0.5 ppb s-triazines)

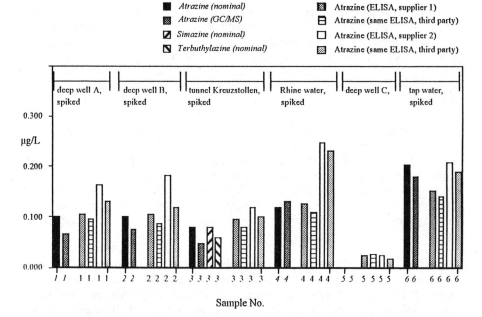

Figure 3.   Round robin test for atrazine, organized   in 1994 by the Immunoassay Study Group in DIN NAW I,4. The nominal concentrations (sample numbers in italics) indicate the spiking with atrazine, simazine and terbuthylazine.

analytical laboratory (third party). Each of the samples was also analyzed for atrazine by the water company using GC/MS. The spiked concentrations of atrazine, simazine and terbuthylazine are also shown in Fig. 3. These data demonstrate that the requirements of the guideline, the quantitative determination of a single pesticide, are best met by immunoassay No. 1. The guidelines for the standard were also met in the EIA study for TNT.

Interlaboratory Tests. After the round-robin tests an interlaboratory trial for atrazine and TNT was organized by the Immunoassay Study group according to the DIN regulations *(22,23)*. Sixteen laboratories responded to announcement released in several scientific papers. Each participant analyzed six atrazine samples with a test kit supplied by Riedel de Haën AG and six TNT samples with a test kit from r-Biopharm GmbH. Figure 4 shows the results for atrazine. Samples 1, 2, and 6 refer to the interlaboratory test and demonstrate an excellent correspondence between the nominal atrazine concentrations and the GC/MS and EIA data.

Three additional samples (No. 3-5) were were spiked with propazine, simazine, and terbuthylazine and analyzed to determine their influence on the performance of the EIA. Figure 4 shows that the concentration response for EIA is similar to the sum of GC/MS concentrations of atrazine and propazine. These results prove the strong cross-reactivity of the applied antibodies with the s-triazine propazine. However, since propazine is not currently used in Europe it is unlikely to pose a problem in using atrazine EIA as an test for atrazine concentrations.

Atrazine and propazine only differ in the 2'-substitution of their s-triazine ring, where atrazine exposes an ethyl group and propazine an isopropyl group. Since 2-aminohexane carboxylic acid-4-isopropylamine-6-chloro-1,3,5 triazine, where the original 2'-position is masked, was used as an immunoconjugate for the antibody production, it is not surprising that the resulting antibodies react with both atrazine and propazine. However, other substitutions for immunoconjugate synthesis will lead to analogous problems. Therefore it is very important to shift immunologically unavoidable cross-reactivities to compounds, which are not expected to cause interferences with the assays.

The results obtained for the TNT assay (not shown) indicate that the applied EIA provides information on TNT contaminations in a broad range of concentrations. Analysis over a broad range in concentrations is required for soil analysis. However, the TNT EIA is not applicable in its present form to monitor water quality according to the drinking water regulations.

Conclusions

Standardizing of analytical methods offers a means to preserve and enforce important analytical features such as accuracy, precision, sensitivity and robustness without unnecessarily impeding the implementation and use of new methods and new technologies. This paper presented an approach applied by DIN for the use of EIAs in the environmental field. Atrazine was taken as an example to provide evidence that EIAs can be applied for the quantitative and highly sensitive analysis of a single analyte in the presence of potentially cross-reactive compounds related to the structure of the target analyte.

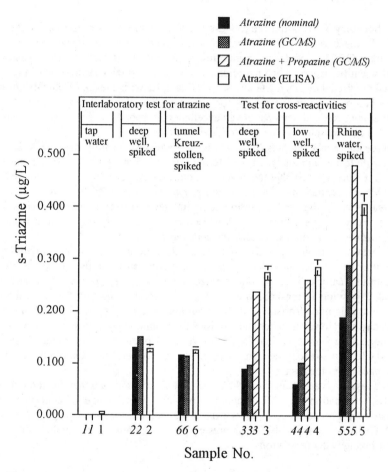

Figure 4. Interlaboratory test for atrazine (samples 1, 2, and 6), organized in 1995 by the Immunoassay Study Group in DIN NAW I,4), and test for cross-reactivities (samples 3-5). The nominal concentrations (sample numbers in italics) indicate the spiking with atrazine or atrazine + propazine.

At the same time, the limitations of the EIA technology cannot be overlooked. For example, the considerable time and effort required to produce new antibodies. This drawback is shared by all immunochemical methods. However, this limitation will eventually be overcome deriving recombinant antibodies from sufficiently large antibody libraries (14). However, the confinement of the analysis of single analytes is a drawback inherent to EIAs. Immunosensors, at least in theory, can better cope with the accomodation of multianalyte assays and sum parameters. It is expected that this technology will also enter the field of true continuous measurements, which EIAs cannot be adapted to do because they are bound to equilibrium measurements. Finally, there is no doubt that standardization will eventually reach the field of immunosensors and then will profit from the previous standardization work implemented for EIAs.

Acknowledgements. We would like to thank Dr. Sibylle Schmidt (Leverkusen, Germany) from the Deutsches Institut für Normung e.V. for providing valuable informations on DIN concepts and policies. The help of all participants in the round robin and interlaboratory tests for atrazine and TNT is gratefully acknowledged. We especially thank the companies, which supplied the samples and carried out the enzyme immunoassays for atrazine (Baker Chemikalien and Riedel de Haën AG) and TNT (Millipore GmbH and r-Biopharm GmbH), furthermore the water companies Stadtwerke Frankfurt and Stadtwerke Mainz, which supplied the samples and carried out HPLC and GC/MS analyses, respectively. Dr. Eline Meulenberg from ELTI Support, Nijmegen performed the independent immunoassay analyses for the round robin test. We thank Dr. Michael Meyer for reading the English manuscript.

References

1. Hock, B. together with the Immunoassay Study Group (Hansen, P.-D.; Kanne, R.; Krotzky, A.; Obst, U.; Oehmichen, U.; Schlett, C.; Schmid, R.; Weil, L.) *Analyt. Lett.* 24, **1991**, 529.
2. Dankwardt, A.; Wüst, S.; Elling, W.; Thurman, E. M.; Hock, B. *ESPR-Environm. Sci. & Pollut. Res.* **1994**, 1, 196.
3. Wittmann, C.; Hock, B. *Nachr. Chem. Techn. Lab.* **1991**, 39, M1-M40.
4. Sherry, J. P. *Crit. Rev. Analyt. Chem.* **1992**, 23, 217.
5. Van Emon, J. M.; Lopez-Avila, V. *Analyt. Chem.* **1992**, 64, 78A.
6. Meulenberg, E. P.; Mulder, W. H.; Stoks, P. G. *Environm. Sci. Technol.* **1995**, 29, 553.
7. Knopp, D. *Analyt. Chim. Acta* **1995**, 311, 383.
8. Hammock, B. D.; Gee, S. J.; Cheung, P. Y. K.; Miyamoto, T.; Goodrow, M. H.; Van Emon, J.; Seiber, J. N. *Pesticide Science and Biotechnology* (Greenhalgh, R.; Roberts, T.; eds.); Blackwell Scientific Publ.: Oxford, London, Edinburgh, 1987, pp 309-316.
9. Bushway, R. J.; Perkins, B.; Savage, S. A.; Lekousi, S. J.; Ferguson, B. S. *Bull. Environm. Contamination Toxicol.* **1988**, 40, 547.
10. Vanderlaan, M.; Watkins, B. E.; Stanker, L. *Environm. Sci. Technol.* **1988**, 22, 247.

11. Thurman, E. M.; Meyer, M.; Pomes, M.; Perry, C. A.; Schwab, P. A. *Analyt. Chem.* **1990,** 62, 2043.
12. Dankwardt, A.; Pullen, S.; Rauchalles, S.; Kramer, K.; Just, F.; Hock, B.; Hofmann, R.; Schewes, R.; Maidl, F. X. *Analyt. Lett.* **1995,** 28, 621.
13. Wüst, S.; Hock, B. *Analyt. Lett.* **1992,** 25, 1025.
14. Kramer, K.; Hock, B. *Lebensmittel- & Biotechnol.* **1995,** 12, 49.
15. Hock, B.; Dankwardt, A.; Kramer, K.; Marx, A. *Analyt. Chim. Acta* **1995,** 311, 393.
16. Schmidt, S. S *JWSRT - Aqua* **1989,** 36, 1150.
17. Schmidt, S. S. *JWSRT - Aqua* **1990,** 39, 6.
18. Karu, A. E.; Lin, T. H.; Breiman, L.; Muldoon, M. T.; Hsu, J. *Food Agricult. Immunol.* **1994,** 6, 371.
19. Ekins, R. P.; Newman, G. B.; O'Riordan, J. L. H. *Radioisotopes in Medicine: In vitro Studies* (Hayes. R. L.; Goswitz, F. A.; Murphy, B. E. P.; eds.); US Atomic Energy Commission: Oak Ridge, TN, 1968, pp 50-100.
20. Buechler, K. F.; Moi, S.; Noar, B.; McGrath, D.; Villela, J.; Clancy, M.; Shenhav, A.; Colleymore, A.; Valkirs, G.; Lee, T.; Bruni, J. F.; Walsh, M.; Hoffman, R.; Ahmuty, F.; Nowakowski, M.; Buechler, J.; Mitchell, M.; Boyd, D.; Stiso, N.; Anderson, R. *Clin. Chem.* **1992,** 38, 1678.
21. Deutsche Einheitsverfahren zur Wasser-, Abwasser- und Schlammuntersuchung. Suborganismische Testverfahren Gruppe 2), Teil 2: Rahmenbedingungen für selektive Immuntestverfahren (Immunoassays) zur Bestimmung von Pflanzenbehandlungs- und Schädlingsbekämpfungsmitteln (T 2). Vornorm DIN V, 38415-2. DIN Deutsches Institut für Normung e.V., Beuth Verlag GmbH, Berlin, pp 1-16.
22. Deutsche Einheitsverfahren zur Wasser-, Abwasser- und Schlammuntersuchung. Allgemeine Angaben (Gruppe A). Ringversuche, Planung und Organisation (A41). DIN 38404, Teil 41. Normenausschuß Wasserwesen (NAW) im DIN Deutsches Institut für Normung e.V., 13. Lieferung, Beuth Verlag GmbH, Berlin, 1984, pp 1-12.
23. Deutsche Einheitsverfahren zur Wasser-, Abwasser- und Schlammuntersuchung. Allgemeine Angaben (Gruppe A). Ringversuche, Auswertung (A42). DIN 38404, Teil 42. Normenausschuß Wasserwesen (NAW) im DIN Deutsches Institut für Normung e.V., Beuth Verlag GmbH, 13. Lieferung, Berlin, 1984, pp 1-24.

Chapter 6

In Situ Derivatization–Supercritical Fluid Extraction Method for the Determination of Chlorophenoxy Acid Herbicides in Soil Samples

Viorica Lopez-Avila[1], Janet Benedicto[1], and Werner F. Beckert[2]

[1]Midwest Research Institute, California Operations,
555–C Clyde Avenue, Mountain View, CA 94043
[2]National Exposure Research Laboratory, U.S. Environmental Protection Agency, 944 East Harmon Avenue, Las Vegas, NV 89119

This paper describes an in-situ derivatization/supercritical fluid extraction (SFE) procedure for 11 chlorophenoxy acid herbicides from a soil matrix. Soil samples (freshly spiked or spiked and weathered soils) are amended with pentafluorobenzyl bromide (PFBBr) and triethyl amine (TEA), prior to SFE, and are pressurized (static) with supercritical carbon dioxide at 400 atm/100°C for 60 min. During this time, the acids are converted to their corresponding PFB esters, which are very soluble in carbon dioxide, and thus, easily extracted from the soil matrix. During a 30-min dynamic extraction, the esters are collected in acetone or on a C_{18}-bonded silica trap and subsequently rinsed off the trap with acetone. The acetone extract is subjected to silica chromatography to remove the excess reagents and analyzed by gas chromatography with electron capture detection. Single-laboratory data and some results of a collaborative study are presented here.

Chlorophenoxy acid herbicides are of interest because of their widespread use in agriculture for weed control. Conventional methods for the analysis of chlorophenoxy acid herbicides involve extraction with organic solvents followed by derivatization with diazomethane and analysis by gas chromatography with electron capture detection. These procedures are time-consuming, use large volumes of organic solvents, and use diazomethane, which is explosive. Several approaches to the extraction of chlorophenoxy acid herbicides from soil samples with supercritical fluids were investigated by us and other researchers (*1-4*). In one approach, we investigated the feasibility of derivatizing the chlorophenoxy acid herbicides in situ with trimethylphenylammonium hydroxide (TMPA) or tetrabutyl ammonium hydroxide (TBA) and methyl iodide (MI). Our experiments, while failing to prove that the derivatization with TMPA was taking place during SFE, indicated that the presence of TMPA was necessary in order to recover these compounds by SFE. We then

0097–6156/96/0630–0063$15.00/0

found that the derivatization reaction with TMPA took place in the injection port of the gas chromatograph. Since not all chlorophenoxy acid herbicides could be recovered satisfactorily, we investigated TBA/MI as derivatization agents, mainly because they have been successfully used by Hopper (5) to derivatize 2,4-D, 2,4,5-T, and 2,4-DB extracted from food samples. The extractions with TBA/MI were performed at 400 atm and 80°C, 15 min static followed by 15 min dynamic. We used 0.5 mL of a 25-percent solution of TBA in methanol and 0.5 mL of neat MI for each 2-g soil sample. The extracted material was collected in 2 mL methanol and analyzed by gas chromatography/mass spectrometry. Sand, clay, and topsoil samples were spiked at 50 and 250 μg/g with seven chlorophenoxy acids and extracted as mentioned above (spiked samples were used because of lack of certified reference materials). Overall, the average recovery for the three matrices and the two spike levels was 95.5 percent. The individual recoveries ranged from 54.2 to 128 percent (4). Although our recoveries from freshly spiked soil samples were acceptable and the procedure was fairly simple to perform (when compared with those recommended by EPA in Methods 8150B and 8151), we decided to investigate another SFE procedure since not all Method 8151 compounds could be recovered quantitatively from freshly spiked soil samples.

In the second SFE approach, we investigated the extraction of the chlorophenoxy acids from soil samples with carbon dioxide modified with 10 percent methanol at 400 atm/80°C/30 min static followed by 30 min dynamic. Quantitative recoveries for most compounds were achieved when the soil sample (2 to 5 g) was amended with 1 mL glacial acetic acid prior to extraction. This second approach, which appeared quite promising as an alternative to the EPA Methods 8150B (which requires 60 mL acetone and 450 mL diethylether for a 50-g sample) and 8151 (which requires 900 mL methylene chloride for a 30-g sample), did not work with several commercial SFE systems. The acetic acid corroded the frits of the extraction vessels, and especially when multivessel systems were used, the acid was in contact with the vessel for prolonged periods of time causing severe corrosion.

Therefore, a third approach was investigated. In this third approach, the chlorophenoxy acid herbicides were converted to their corresponding pentafluorobenzyl (PFB) esters. The derivatization reaction took place at 400 atm and 100°C by pressurizing the soil sample, containing the chlorophenoxy acids, pentafluorobenzylbromide (PFBBr), and triethylamine (TEA), with supercritical carbon dioxide for 60 min, followed by a 30-min dynamic extraction. Both freshly spiked and weathered samples were extracted by this procedure.

In this paper we report on this latter in-situ derivatization/SFE procedure with PFBBr and TEA and report data from an interlaboratory study in which 12 laboratories (including our laboratory) extracted samples, prepared by us, following the SFE method described here. The SFE extracts were sent back to us for silica chromatography and analysis by gas chromatography with electron capture detection. The interlaboratory study was performed to evaluate the method with various operators using different commercially available SFE systems.

Experimental Section

Standards. Of the 11 chlorophenoxy acid herbicides used in the method development (Table I), compounds 1 through 6 and compound 11 were purchased from Chem Service, Inc. (West Chester, PA); compounds 7 and 8 from Crescent

Table I. Chlorophenoxy Acid Herbicides Investigated in This Study

Compound No.	Compound Name	CAS Registry No.	Abbreviation
1	2-(4-chloro-2-methylphenoxy)propanoic acid	7085-19-0	MCPP
2	3,6-dichloro-2-methoxybenzoic acid	1918-00-9	Dicamba
3	4-chloro-2-methylphenoxyacetic acid	94-74-6	MCPA
4	2-(2,4-dichlorophenoxy)propanoic acid	120-36-5	Dichlorprop
5	2,4-dichlorophenoxyacetic acid	94-75-7	2,4-D
6	2-(2,4,5-trichlorophenoxy)propionic acid	93-72-1	2,4,5-TP
7	2,4,5-trichlorophenoxyacetic acid	93-76-5	2,4,5-T
8	4-(4-chloro-2-methylphenoxy)butanoic acid	94-81-5	MCPB
9	4-(2,4-dichlorophenoxy)butanoic acid	94-82-6	2,4-DB
10	4-amino-3,5,6-trichloro-2-pyridine carboxylic acid	1918-02-01	Picloram
11	5-(2-chloro-4(trifluoromethyl)phenoxy)-2-nitrobenzoic acid	62476-59-9[a]	Acifluorfen
SU	3,4-dichlorophenoxyacetic acid	588-22-7	3,4-D
IS	4,4'-dibromooctafluorobiphenyl	10386-84-2	—

[a] The CAS Registry No. is that for the sodium salt of acifluorfen.

Chemical Co. (Hauppauge, NY), and compounds 9 and 10 from Ultra Scientific, Inc. (North Kingstown, RI). 3,4-D (purity 96 percent) was purchased from Aldrich Chemical Co. (Milwaukee, WI). All compounds were used as received without further purification (their purities were stated to be at least 99 percent). Stock solutions of the individual acids were prepared in acetone at 10 mg/mL and kept at 4 °C in the dark. A spiking solution of the chlorophenoxy acids was made by combining portions of the individual stock solutions and diluting them to 10 to 20 μg/mL with acetone.

4,4'-Dibromooctafluorobiphenyl, used as internal standard (IS), was purchased from Ultra Scientific as a solution in methanol at 250 μg/mL.

Reagents. PFBBr and TEA (purity >99 percent) were purchased from Sigma Chemical Co. (St. Louis, MO) and Chem Service, respectively. A fresh 25-percent PFBBr solution in acetone was prepared weekly by dissolving 2.5 g of PFBBr in acetone. A 10-mL volumetric flask was used for this purpose; 2.5 g PFBBr was weighed directly in the volumetric flask and acetone was added to bring volume to 10 mL. The volumetric flask was kept in a refrigerator at 4°C.

SFC/SFE grade carbon dioxide with a helium head pressure of 2,000 psi was obtained from Air Products and Chemicals, Inc. (Allentown, PA).

Soil Samples. One soil sample used in this study, obtained from the Sandoz Crop Protection Division (Gilroy, CA), was a clay loam (34 percent sand, 35 percent silt, 31 percent clay; pH 7.4; moisture content 10.6 percent; organic carbon content 1.8 percent). Three soil samples, identified as RT-801, RT-802, and RT-803, were sandy loam and clay loam samples. The physico-chemical properties of the soil samples are listed in Table II.

Table II. Physico-Chemical Properties of the
Soil Samples Prepared by RT Corporation

Parameter	Units	RT-801	RT-802	RT-803
Moisture	%	3.0	5.6	8.1
Chloride, soluble[a]	mg/kg	5.0	657	380
Fluoride, soluble[a]	mg/kg	0.5	1.0	4.5
Sulfate, soluble[a]	mg/kg	72	3,550	741
Ammonia as N, extract[b]	mg/kg	2.3	0.8	7.6
Nitrogen, total Kjeldahl	%	0.05	0.05	0.21
Nitrate as N, soluble[a]	mg/kg	6.4	28.2	0.7
Phosphorus, total	%	0.02	0.01	0.04
Phosphorus, extractable[c]	mg/kg	5.0	3.0	13.0
pH, saturated phase[d]	units	8.1	8.0	7.4
Cation exchange capacity	meq/100g	3.8	10.0	10.5
Carbonate, total	% as CaCO$_3$	2.95	12.7	1.64
Sulfur, total	%	<0.01	0.12	0.04
Neutralization potential	% as CaCO$_3$	5.4	15.8	4.5
Total organic carbon	mg/kg	7,600	4,000	24,800
Sand, 2.00 to 0.062 mm	%	66.0	29.0	63.0
Silt, 0.062 to 0.002 mm	%	26.0	50.0	19.0
Clay, <0.002 mm	%	8.0	21.0	18.0

[a] Saturated paste extraction.
[b] AB-DTPA extraction.
[c] Water extraction
[d] Potassium chloride extraction.

Soil Spiking Procedure. For spiking the soil matrices, two procedures were used. To prepare freshly spiked samples, portions of the soil matrices (2.5 g) were weighed into aluminum cups, and 350-μL portions of a concentrated stock solution containing the 11 chlorophenoxy acid herbicides in acetone (concentrations varied depending on the spike level) were added with a syringe. Adsorption losses were minimized by ensuring that the solution did not contact the aluminum cup. Mixing was performed by gently shaking the aluminum cup by hand. After the solvent had evaporated (approximately 5 min), the spiked sample was amended with PFBBr (0.2 mL of a 25-percent solution in acetone). Caution: This step must be performed in a fume hood. The amended samples were then loaded into the extraction vessels and sandwiched between two plugs of filter paper.

The soils identified as RT-801, RT-802, and RT-803 were spiked at 7,000, 700, and 70,000 μg/kg, respectively, as follows. In each case, a 30-kg portion of the soil, was spiked with a solution containing either 210 mg, or 21 mg, or 2,100 mg respectively, depending on the desired final concentration of each compound (after correction of the weight taking compound purity in consideration). For example, in the case of MCPB, the purity of the material was 95 percent, therefore, we adjusted the spiking level to 221 mg, 22.1 mg, or 2,210 mg in order to achieve concentrations of 7,000, 700, and 70,000 μg/kg, respectively. All compounds spiked were at purities above 95 percent, except for dicamba at 40 percent, MCPP at 38.3 percent, picloram at 21.1 percent and 2,4-DB at 19.5 percent (these four compounds were available only as formulations at these concentrations). The spiked samples were weathered over a period of 2 weeks. After weathering, the spiked soil was split into 50-g portions and stored in glass bottles with Teflon-lined caps (storage temperature 20-23°C) for approximately 11 months before being used in this study.

SFE Procedure. Most of our method development experiments were performed on an Isco (Lincoln, NE) SFX 2-10 extractor using two extraction vessels for simultaneous derivatization/extraction at 400 atm and 100°C for 30 to 60 min static and 30 min dynamic. The flow rate of carbon dioxide varied between 0.7 and 1.5 mL/min (as liquid). Some preliminary experiments were also performed on a Hewlett-Packard (Wilmington, DE) SFE system Model 7680T at 365 atm and a chamber temperature of 80 or 120 °C, for 30 min static and 30 min dynamic. The carbon dioxide flow rate was 3 mL/min (as liquid). The nozzle temperature was the same as the chamber temperature (80, 120 °C), and the trap temperature was 15 °C. The trap was packed with octadecyl-bonded silica (Hewlett-Packard). The extracted material was removed from the trap with two 1.5-mL acetone rinses. No compounds were detected in the second rinse. The trap temperature during the rinse step was 45 °C.

Of the 12 participants (there were initially 13 laboratories in the study but Lab. 4 did not submit any extracts) in the interlaboratory study, Lab. 1 used a Suprex SFE-50; Labs. 2, 5, and 6 used a HP 7680T; Lab. 7 used a HP 7680A; Labs. 3, 9, 10, and 11 used the Isco SFX 2-10; Labs. 8 and 13 used the Suprex Autoprep 44; and Lab. 12 used the Isco 3560. Details of their operating conditions are given in Table III.

Analysis. All extracts were analyzed in our laboratory with a Hewlett-Packard 5890 Series II gas chromatograph equipped with two electron capture detectors, autoinjector, electronic pressure controller, and a Hewlett-Packard DOS Chemstation. Quantification of the analytes was performed by internal standard calibration using 4,4'-dibromooctafluorobiphenyl as internal standard (this compound is added to the sample extract immediately prior to the gas chromatographic analysis). The details of the gas chromatographic conditions are given in Table IV.

Results and Discussion

Rejection of Outliers. Despite the effort put in by the participants in this collaborative study, we had to eliminate the results from all laboratories that used the HP SFE system, based on the poor reproducibility of the data and on difficulties with the system, as reported by the participants. The only acceptable data reported by the labs that used the HP SFE systems are those for the spiked Hydromatrix samples (Table V); Labs. 3, 9, 11, 12 used the Isco and the Suprex SFE systems and Labs. 2, 5, and 6 used the HP SFE systems. The results from the HP systems (when there were no overpressure problems and failure to rinse the trap) looked even better than those reported by Labs. 3, 9, 11, and 12. However, because the collaborators reported difficulties when extracting the samples and because the recoveries were low

Table III. SFE Operating Conditions Used in the Interlaboratory Study

Parameter	Isco SFX 2-10 (Lab. 3, 9, 10, 11)	Suprex SFE-50 (Lab. 1)	Isco-3560 (Lab. 12)
Fluid	Carbon dioxide (SFE/SFC grade)	Carbon dioxide (SFE/SFC grade)	Carbon dioxide (SFE/SFC grade)
Pressure (atm)	400	400	400
Temperature (°C)	100	100	100
Density (g/mL)	0.76	0.76	0.76
Carbon dioxide flow rate (as liquid; mL/min)	0.7-1.5	0.9-1.9	1.5
Direction of fluid flow	Downward	Upward	Downward
Static extraction time (min)	60	60	60
Dynamic extraction time (min)	30	25-40	30
Extraction vessel volume (mL)	2.5 or 10	4	10
Extraction vessel dimensions	7.5-mm x 5.5-cm length (2.5 mL) 15-mm ID x 5.5-cm length (10.0 mL)	10-mm ID x 5-cm length (4 mL)	15-mm ID x 5.5-cm length (10 mL)
Extraction vessel orientation	Vertical	Vertical	Vertical
Number of extraction vessels	2	1	24
Restrictor type	Stainless steel capillary (50-μm ID x 40.5-cm length)	Fused-silica capillary (50-μm ID x 40-cm length)	Stainless steel capillary (50-μm ID x 40.5-cm length)
Restrictor/nozzle temperature (°C)	100[a]	Room temperature	100
Collection solvent	Acetone	Acetone	Acetone
Volume of collection solvent (mL)	15[b]	15	8
Temperature of collection vial (°C)	Room temperature	Room temperature	-10

[a] Labs. 10, 11, and 12 used a coaxially-heated restrictor.
[b] The final volume of collection solvent after extraction is approximately 8 mL.

Table IV. GC/ECD Operating Conditions for the Analysis of the PFB Derivatives of the Chlorophenoxy Acid Herbicides

Parameter	Value
GC instrument	Hewlett-Packard GC Model 5890 Series II, equipped with two ECDs, autoinjector HP 7673, electronic pressure controller, and HPDOS Chemstation
Column 1	30-m length x 0.32-mm ID 0.25-μm film thickness DB-5 fused-silica open-tubular column
Column 2	30-m length x 0.32-mm ID 0.25-μm film thickness DB-1701 fused-silica open-tubular column
Column supplier	J&W Scientific (Folsom, CA)
Carrier gas flow rate	5.4 to 5.6 mL/min (helium)
Linear velocity	66 cm/sec
Makeup gas	47 mL/min (nitrogen)
Detector range	0
Detector attenuation	4
Temperature program	60°C (1-min hold) to 160°C at 10°C/min, 160°C to 250°C (5-min hold) at 4°C/min
Injector temperature	250°C
Detector temperature	320°C
Injection volume	1 μL
Solvent	Hexane-toluene (1:9)
Type of injector	Split/splitless; splitless mode; 60 sec
Detector type	ECD
Splitter type	Universal Y press-tight (Restek)

and the within-lab repeatability was poor, we had to eliminate the results from Labs. 2, 5, 6, and 7.

The results from Labs. 8 and 13 were also eliminated because of poor recoveries and poor repeatability overall; this may have been associated with the use of a fully automated system (i.e., Autoprep 44) on which the method has not been tested prior to the study.

Method Accuracy and Precision. Tables VI through VIII summarize the percent recoveries achieved by the participating laboratories from clay soil, RT-801 soil, and RT-803 soil samples. Most extractions were performed over a period of one month; the compounds did not appear to have degraded during this time based on the analytical values or "recoveries". Overall, the results look quite good. Of the 88

Table V. SFE Percent Recoveries for the Spiked Hydromatrix[a]

	Lab. 2	Lab. 3	Lab. 5	Lab. 6	Lab. 9	Lab. 11	Lab. 12
MCPP	104	92.1	111	106	113	87.8	79.8
Dicamba	118	90.8	122	102	111	80.7	91.0
MCPA	93.9	79.3	115	78.8	115	70.5	80.8
Dichlorprop	110	86.4	115	103	106	78.6	86.0
2,4-D	93.8	76.4	120	79.1	120	65.9	83.3
2,4,5-TP	106	88.0	109	110	100	88.4	82.0
2,4,5-T	92.5	69.6	109	85.3	97.4	71.9	78.6
MCPB	72.7	93.3	108	96.3	88.0	99.7	78.8
2,4-DB	84.2	97.6	112	102	107	94.3	84.4
Picloram	b	113	92.7	97.0	127	63.9	138
Acifluorfen	106	83.2	116	124	100	86.4	95.3
Range	72.7-118	69.6-113	92.7-122	78.8-124	88.0-127	63.9-99.7	78.6-138

[a] Hydromatrix is diatomaceous earth. Data from Labs. 1, 7, 8, 10, and 13 are not included. Labs. 1 and 7 reported that the restrictor got plugged during the extraction of the spiked Hydromatrix, and Lab. 10 reported that part of the extract was spilled. Labs. 8 and 13 were eliminated from the study because of poor performance overall. The spike level is 7,000 to 14,000 ng/g. The following compounds were spiked at 7,000 ng/g: MCPA, 2,4-D, 2,4,5-TP, 2,4,5-T, MCPB. The rest of the compounds were spiked at 14,000 ng/g. Single determinations.

b Not able to determine recovery because of matrix interference.

Table VI. SFE Percent Recoveries for the Freshly Spiked Clay Loam Soil[a]

Compound name	Lab. 1		Lab. 3		Lab. 9		Lab. 10		Lab. 11		Lab. 12	
	Rep. 1	Rep. 2	Rep. 1	Rep. 2	Average percent recovery	Percent RSD	Average percent recovery	Percent RSD	Average percent recovery	Percent RSD	Average percent recovery	Percent RSD
MCPP	89.0	102	84.4	81.6	106	10	75.0	7.4	101	20	117	4.3
Dicamba	74.7	86.7	73.4	77.4	87.9	21	84.1	11	86.8	17	120	3.1
MCPA	102	93.6	67.9	88.4	111	19	86.6	18	94.7	11	117	4.5
Dichlorprop	88.8	101	70.5	78.0	111	15	75.8	8.6	94.2	11	115	5.4
2,4-D	77.9	91.6	80.1	101	101	24	88.2	15	100	14	113	16
2,4,5-TP	83.4	95.0	85.0	88.2	103	9.4	86.4	16	101	11	108	3.9
2,4,5-T	95.1	112	79.6	96.4	120	7.1	99.4	6.9	112	29	102	8.5
MCPB	118	120	93.2	90.0	133	9.3	94.0	16	122	17	147	8.8
2,4-DB	89.1	99.9	80.2	74.2	120	0.8	74.0	10	99.4	11	109	9.6
Picloram	77.5	96.5	75.5	68.8	86.9	17	70.6	21	85.9	50	103	22
Acifluorfen	112	126	97.0	98.7	150	4.8	123	3.8	127	17	129	3.7

[a] The spike level for the clay soil is 1,000 to 2,000 ng/g. The number of replicate determinations was two for Labs. 1 and 3 and three for Labs. 9, 10, 11, and 12.

Table VII. SFE Percent Recoveries from Freshly Spiked RT-801 Soil Sample Using Various SFE Systems[a]

	Suprex SFE-50 (Lab. 1)	SFX 2-10 (Lab. 3)	SFX 2-10 (Lab. 9)	SFX 2-10 (Lab. 10)	SFX 2-10 (Lab. 11)	SFX 3560 (Lab. 12)	Average percent recovery	Percent RSD
MCPP	97.4	74.2	107	83.0	95.1	96.1	92.1	13
Dicamba	88.9	72.3	63.3	94.8	88.1	101	84.7	17
MCPA	89.1	72.0	74.9	70.4	86.8	96.8	81.7	13
Dichlorprop	97.2	73.4	101	80.7	96.3	101	91.6	13
2,4-D	85.6	84.9	62.5	67.1	108	86.0	82.4	20
2,4,5-TP	92.6	73.0	101	85.7	95.0	94.4	90.3	11
2,4,5-T	82.0	69.5	77.9	73.9	90.6	81.8	79.3	9.2
MCPB	75.7	78.9	104	67.3	95.7	85.2	84.5	16
2,4-DB	92.3	83.2	112	65.8	93.3	88.6	89.2	17
Picloram	99.6	126	88.2	113	27.1	117	95.2	38
Acifluorfen	87.1	91.9	155	101	116	92.4	107	24

[a] RT-801 soil was freshly spiked at 7,000 to 14,000 ng/g. Single determinations.

Table VIII. SFE Percent Recoveries from Freshly Spiked RT-803 Soil Sample Using Various SFE Systems[a]

	Suprex SFE-50 (Lab. 1)	SFX 2-10 (Lab. 3)	SFX 2-10 (Lab. 9)	SFX 2-10 (Lab. 10)	SFX 2-10 (Lab. 11)	Average percent recovery	Percent RSD
MCPP	94.5	87.8	95.0	104	94.7	95.2	6.1
Dicamba	39.9	33.3	31.7	92.1	40.8	47.6	53
MCPA	56.5	52.1	55.3	101	55.7	64.1	32
Dichlorprop	89.4	70.4	82.4	103	77.2	84.5	15
2,4-D	46.4	38.6	46.4	117	45.9	58.9	55
2,4,5-TP	91.2	70.7	88.4	99.4	83.2	86.6	12
2,4,5-T	57.8	43.5	60.2	104	61.4	65.4	35
MCPB	82.5	91.5	99.3	98.0	91.1	92.5	7.2
2,4-DB	102	98.1	105	102	92.0	99.8	5.0
Picloram	40.4	29.2	23.3	60.3	32.7	37.2	39
Acifluorfen	93.8	63.8	111	115	99.5	96.6	21

[a] RT-803 soil was freshly spiked at 70,000 to 140,000 ng/g. The data from Lab. 12 are not included because the surrogate recovery for this sample was only 6.5 percent. Single determinations.

recovery values that we report in Table VI for the eleven compounds, 83 values are within the 70 to 130 percent window, two values were slightly below 70 percent and three values were 133, 147 and 150 percent.

The recoveries from the RT-801 soil samples were almost similar to those from the clay soil samples with slightly more values outside the acceptance window (from the 66 values in Table VII, 59 values were between 70 and 130 percent, one value was 27.1 percent, five values ranged between 62.5 and 67.3 percent, and one value was 155 percent. Nonetheless, there was fairly good agreement among the six laboratories for which we report data (percent RSDs ranged from 9.2 to 24 percent, with one value at 38 percent because of the very low recovery reported by Lab. 11 for picloram). The results for the RT-803 soil are included in Table VIII. This soil is a sandy-loam type with a high organic content (Table II). Average recoveries across five laboratories were lower for dicamba, MCPA, 2,4-D, 2,4,5-T, and picloram, despite the fact that the levels at which these compounds were spiked were 10 times as high as those in Table VII. The higher organic carbon content of this soil as compared to that of the clay soil and the RT-801 soil samples may be responsible for the lower recoveries of some of the compounds.

The single-laboratory SFE method accuracy and precision data presented in Table IX were generated in our laboratory with weathered soil samples. The same samples were analyzed by Method 8150 and the results of those analyses are presented in Table X. While the SFE method appears to give acceptable recoveries for the freshly spiked soil samples, our study shows that for weathered samples compound recovery appears to be a function of the soil matrix. The organic carbon content and the clay content of the soil matrix appeared to influence compound recovery.

Table IX. Single-Laboratory SFE Method Accuracy and Precision[a]

| Compound name | RT-801 soil | | RT-802 soil | | RT-803 soil | |
	Average percent recovery	Percent RSD	Average percent recovery	Percent RSD	Average percent recovery	Percent RSD
MCPP	74.4	2.2	82.7	12	59.2	12
Dicamba	60.6	19	64.8	19	50.8	22
MCPA	155	8.7	199	7.2	57.7	15
Dichlorprop	146	2.1	160	5.0	83.0	13
2,4-D	140	12	165	19	42.2	7.3
2,4,5-TP	140	3.0	143	5.9	66.0	2.8
2,4,5-T	15.8	4.8	21.6	12	29.0	14
MCPB	17.5; 20.9[b]	--	26.7	4.4	49.8	17
2,4-DB	69.1	24	71.2	22	90.2	19
Picloram	40.6	28	48.8	35	56.1	24
Acifluorfen	103	4.1	100	13	100	13

[a] The number of determinations was three. The true spike values for the RT-801, RT-802, and RT-803 soil samples are 7,000; 700; and 70,000 μg/kg, respectively.
[b] Duplicate determinations.

Table X. Single-Laboratory Method 8150 Accuracy and Precision[a]

	RT-801 soil		RT-802 soil		RT-803 soil	
Compound name	Average percent recovery	Percent RSD	Average percent recovery	Percent RSD	Average percent recovery	Percent RSD
MCPP	b		b		84.6	14
Dicamba	91.4	3.7	101	6.5	123	5.4
MCPA	b		b		90.0	20
Dichlorprop	119	3.1	113	11	124	3.8
2,4-D	193	4.6	129	42	109	5.4
2,4,5-TP	164	2.6	92.9	18	111	4.0
2,4,5-T	21.6	6.3	8.6	45	73.1	4.6
MCPB	b		b		b	
2,4-DB	91.4	9.8	47.1	30	50.4	9.6
Picloram	82.9	5.4	71.4	8.9	160	3.4
Acifluorfen	30.0	2.6	21.4	12	101	3.1

[a] The number of determinations was four. The true spike values for the RT-801, RT-802, and RT-803 soil samples are 7,000; 700; and 70,000 μg/kg, respectively. These analyses were performed by ETC Santa Rosa Laboratory.
[b] Information not available.

Acknowledgment

The assistance of Isco, Inc., and Hewlett-Packard Co. for the loan of the SFE systems used in this study is gratefully acknowledged. RT Corporation (Laramie, WY) prepared the soil samples and ETC Santa Rosa (Santa Rosa, CA) analyzed them using EPA Method 8150. We especially thank the volunteer participants in the collaborative study for their willingness to undertake the effort. John Snyder of Lancaster Laboratories, Lancaster, PA; Bruce Benner of NITS, Gaithersburg, MD; Chuck Hecht of Chem Waste Management, Riverdale, IL; Steven Hawthorne, Soren Bowadt, and Yu Yang of the University of North Dakota, Grand Forks, ND; John Orolin of Inland Consultants, Skokie, IL; Richard Pfieffer and Tim Watts, USDA ARS, Ames, IA; Joe Levy of Suprex, Pittsburgh, PA; Laszlo Torma, Dept. of Agriculture, Montana State University, Bozeman, MT; Les Myers of Isco, Lincoln, NB; and Mark Bruce, Quanterra, North Canton, OH.

Notice

Literature Cited

(1) Hawthorne, S. B.; Miller, D. J.; Nivens, D. E.; White, D. C. *Anal. Chem.* **1992**, *64*, 405–412.

(2) Hills, J. W.; Hill, H. H., Jr.; Maeda, T. *Anal. Chem.* **1991**, *63*, 2152–2155.

(3) Rochette, E. A.; Harsh, J. B.; Hill, H. H., Jr. *Talanta* **1993**, *40*, 147–155.

(4) Lopez-Avila, V.; Dodhiwala, N.S.; Beckert, W. F. *J. Agric. Food Chem.*, **1993**, *41*, 2038-2044.

(5) Hopper, M. L.; *J. Agric. Food Chem.*, **1987**, *35*, 265-269.

Chapter 7

Application of In Vivo Fluorometry To Determine Soil Mobility and Soil Adsorptivity of Photosynthesis-Inhibiting Herbicides

Daisuke Yanase, Misako Chiba, Katsura Yagi, Mitsuyasu Kawata, and Yasushi Takagi

Naruto Research Center, Otsuka Chemical Company, Limited, 615 Hanamen, Satoura-cho, Naruto-city, 772 Japan

Using chlorophyll *a* fluorescence as an indicator of photosynthesis inhibition, two bioassay systems were devised to determine soil mobility and soil adsorptivity of newly synthesized photosynthesis-inhibiting imidazolidine derivatives, OK-8901 and OK-9201. One rather primitive method, fluorometry using cucumber cotyledon discs, showed accuracy comparable to GC or HPLC in assessing leachability of the two compounds in a soil column assay. Another more efficient method utilizes a microplate scanner to measure fluorescence intensity in suspended microalgal cells (*Chlorella vulgaris* Beijer), and was used in determination of the soil adsorption coefficient (K_{oc}). The results suggested that OK-9201 is highly soil-leachable, while OK-8901 showed intermediary soil adsorptivity between diuron and atrazine.

There has been growing environmental concern about leaching and consequent ground water contamination by soil-applied pesticides. Edaphic properties such as soil-adsorptivity and soil-mobility now bear significance comparable to pesticidal activity itself in rating candidate compounds. While quantification of pesticides in environmental assessments is usually done by instrumental analyses like gas-liquid chromatography (GC) or high performance liquid chromatography (HPLC), more maneuverable techniques are needed to sieve off unlikely compounds at an early stage of biological screening. This article describes tentative use of two entirely bioassay-based fluorometric techniques in assessing soil-leachability of two photosynthesis-inhibiting imidazolidine derivatives: OK-8901 (2'-fluoro-5-hydroxy-3-methyl-2-oxo-imidazolidine-1-carboxanilide) and OK-9201 (N-cyclohexyl-5-hydroxy-3-methyl-2-oxo-imidazolidine-1-carboxamide). Chemical structure, water solubility, and 1-octanol-water partition coefficient (K_{ow}) for the compounds tested are presented in Table I.

Fluorometric Techniques to Measure Aqueous Concentration of PS II Inhibitors

Chlorophyll *a* fluorescence, re-emittance of excess light energy from antenna chlorophyll *a* (*1, 2*), in photosynthesizing organisms is known to increase with the magnitude of inhibition of photosynthetic electron-transport (*3*). The yield of *in vivo* chlorophyll fluorescence then can serve as a sensitive bio-probe to detect PS (photosystem) II inhibitors in water samples. This section details two fluorometric techniques to measure aqueous concentration of photosynthesis-inhibiting herbicides.

0097–6156/96/0630–0077$15.00/0

Table I. Structure, log K_{ow}, and Water Solubility of Compounds Tested in the Present Study

Compound	OK-8901	OK-9201	diuron	atrazine
log K_{ow}	1.50	1.85	2.77	2.34
Water solubility (mg l⁻¹)	470	2200	42	30

Fluorometry with Cucumber Cotyledon Discs. Instruments specifically designated to measure intensity of *in vivo* chlorophyll fluorescence were not available for our earlier experiments. Fortunately, chlorophyll fluorescence in plant leaves, especially when it is enhanced by photosynthesis inhibition, is visible to the naked eye and can be recorded on ordinary photographs (*4-6*). In the present study, a primitive computer-aided system, originally devised to analyze distribution of PS II inhibitors in plant leaves (*7*), was used to analyze video-images of fluorescence in leaf discs.

Leaf discs measuring 10 mm in diameter were prepared from fully expanded cotyledons of 8-day-old cucumber seedlings (*Cucumis sativus* L. cv. Shiroibo) grown at 25°C under a 16-hr photoperiod of ca. 100 µE m⁻². Ten such leaf discs were bathed in 20 ml of serially diluted sample solutions overnight in the dark. They were then posted on black paper and illuminated by an excitation beam (380-620 nm, ca. 4 W m⁻²) from a 150 W slide projector equipped with a Corning 4-96 filter. Three minutes after the light onset, fluorescence images were recorded through a red cut off filter (Corning 2-64, >650 nm) by a CCD-video camera (SONY, SSC-M370), and captured by a computer (Apple Computer, Macintosh Classic) using an analog-digital converter (InterWare, Video Catcher VC-1) to be stored in the built-in hard disc as binary (black and white) bitmap files. Halftones, i.e. intermediately fluorescing areas, were expressed by dithering (an error diffusion algorithm). Fluorescence intensity at the terminal state (F_T) in cotyledon discs was approximated by the number of white pixels. As an example, the general scheme to obtain the dose-fluorescence curve for a PS II inhibitor, monuron (3-(4-chlorophenyl)-1,1-dimethylurea), is summarized in Figure 1.

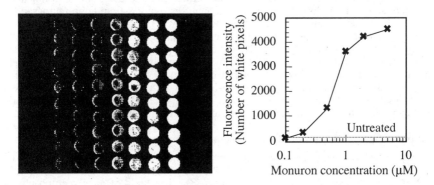

Figure 1. Dither images of chlorophyll fluorescence in cucumber cotyledon discs treated with monuron (left): from left to right, 0, 0.1, 0.2, 0.5, 1, 2, and 5 µM. In the chart (right), number of white pixels in each column of 10 cotyledon discs is plotted against herbicide concentration.

Aqueous concentration of PS II inhibitors was estimated by the dilution required to enhance fluorescence by 50%. The fluorescence increase was calibrated as

$$\text{calibrated fluorescence increase (\%)} = 100 \times \frac{\text{observed } F_T \text{ - } F_T \text{ in untreated discs}}{\text{saturated } F_T \text{ - } F_T \text{ in untreated discs}} \quad (1)$$
$$(F_T : \text{fluorescence intensity at the terminal state})$$

where the saturated fluorescence was represented by the fluorescence intensity observed in cucumber cotyledon discs treated with diuron at 10 μM. The fluorescence-enhancing ED_{50} for the tested solution was determined by the probit analysis and used to calculate the concentration:

$$\text{concentration in question} = C_{\text{ref}} \times \frac{ED_{50} \text{ for the reference solution}}{ED_{50} \text{ for the solution tested}} \quad (2)$$

where C_{ref} is the concentration of the compound to be tested in the reference solution.

In Figure 2, concentration of OK-8901 and OK-9201 in soil column leachates, measured by cucumber fluorometry, was plotted against the reference data obtained by HPLC (Shimadzu LC-6A with Zorbax BP ODS) or GC (Shimadzu GC-7AG equipped with FTD-8 detector). For both compounds, fluorometrically measured concentration agreed well with the reference data within a range from 0.1 to 100 ppm (r = 0.98).

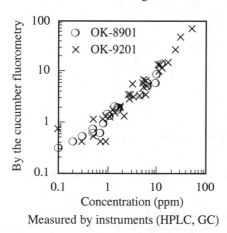

Figure 2. Validity of the fluorometric technique to quantify PS II inhibitors. Aqueous concentration of the two imidazolidine derivatives determined by cucumber fluorometry is plotted against the corresponding data obtained by HPLC (OK-8901) or GC (OK-9201).

The Microplate Fluorometry. In soil-adsorptivity measurement, the concentration at adsorption-desorption equilibrium was determined by another fluorometric system as described in earlier reports (8, 9). In this system, the increase of chlorophyll fluorescence due to photosynthesis inhibition in suspended *Chlorella vulgaris* Beijer cells, instead of cucumber cotyledon discs, was measured by a fluorescence microplate scanner (Corona, MTP-32 equipped with a fluorescence measuring unit MTP-F2, excitation: 440 nm, detection: 690 nm). The measuring principle and calculations were the same as in cucumber fluorometry.

Figure 3 explains the procedure to determine concentration of diuron in solutions equilibrated after 24 hrs of contact with soil (organic carbon content: 4.67%). In this experiment, ED_{50} for the reference 10 µM solution was 4.2 (µl in 300 µl), whereas two equilibrated solutions recorded ED_{50} of 44.4 and 41.0. Consequently, the concentrations of these two solutions were calculated by equation 2 as 0.95 µM and 1.02 µM, respectively. As shown in Figure 4, dilution of diuron solutions (10-100 µM) required for 50% increase of calibrated fluorescence intensity showed a close correlation with the concentration (r = 0.99).

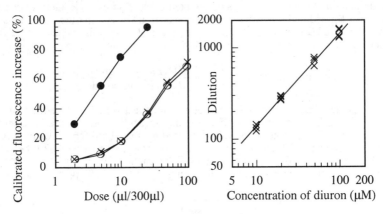

Figure 3 (Left). The dose-fluorescence curves for diuron solutions. The concentrations of two solutions (×, ○) were determined as 0.95 µM and 1.02 µM by comparing their apparent ED_{50} values with that of the reference solution at 10 µM (●).

Figure 4 (Right). Correlation between diuron concentration in aqueous solutions and dilution required for 50% increase of calibrated fluorescence intensity. Results of four independent trials were shown together.

Measurement of Soil-Leaching Parameters for the Two Imidazolidine Derivatives

With quantitative accuracy being established, the two fluorometric techniques were applied to assess compounds' soil-mobility and soil-adsorptivity, two major factors dictating the fraction of soil-applied herbicides to leach beyond the zone of weed roots.

Soil Mobility of the Two Imidazolidine Derivatives. Columns of sandy loam (pH 6.6, organic carbon content 0.46%) were prepared in glass cylinders (30 mm x 300 mm) and saturated to the maximum water holding capacity (378 ml kg^{-1}). Acetonitrile solutions of OK-8901 and OK-9201 were applied to the top surface of the soil columns at a rate equivalent to 20 kg ha^{-1}. The columns were then discharged with tap water and leachates were collected in a sampling volume of 50 ml up to the 9th fraction. Two such columns were used for each compound and the herbicide concentration in each fraction was determined by cucumber fluorometry as described earlier.
   The percolation profiles for OK-8901 and OK-9201 are shown in Figure 5. OK-9201 was rapidly discharged from the soil columns, with its eluviation peak around the 3rd to 6th fraction, while OK-8901 proved less soil-mobile as its concentration in the leachates was still increasing at the 9th fraction. Despite promising herbicidal performance, OK-9201 was judged unsuitable for practical use since its soil mobility

proved unacceptably great. Besides, low degradability of OK-9201 was predicted by a preliminary experiment in which this compound showed no sign of degradation in aqueous solutions at pH 4-9 through 39 days of incubation. Thus, OK-9201 was discarded as a herbicide candidate, and further research concentrated on OK-8901.

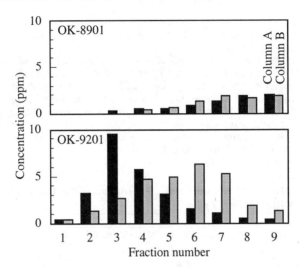

Figure 5. The percolation profiles for OK-8901 and OK-9201 in a soil-leaching simulation. Both compounds were applied to two (A and B) columns of sandy loam and leachates are collected as nine 50 ml fractions.

Soil-Adsorptivity of OK-8901 Compared with Diuron and Atrazine. The soil adsorption coefficient based on organic carbon content (K_{oc}) was determined with reference to the OECD-Guideline (*10*). Four soil samples with varying organic carbon content were obtained from various parts of Japan, viz. Hokkaido, Ibaragi, Wakayama, and Okayama, and served as the trial soils (Table II). An aliquot (10 ml) of the solutions of OK-8901, diuron, and atrazine prepared at 10, 20, 50, and 100 μM in aqueous 10 mM $CaCl_2$ was mixed with 2 g of the above-mentioned soils and shaken at 60 revolutions per minute for 24 hrs. In this procedure, usually 8 hrs of shaking time is sufficient to attain the adsorption-desorption equilibrium (*9*). After being centrifuged at 1000 G for 10 minutes, the supernatant was carefully collected and stored at 5°C until its concentration measurement by microplate fluorometry.

Table II. Properties of Four Soils Used for K_{oc} Determination

			The soil composition				
Soil origin	Soil texture	Organic carbon (%)	Clay (%)	Silt (%)	Sand (%)	pH	Exchange capacity (me/100g)
Ushiku, Ibaragi	LiC	6.8	34.1	35.4	31.5	6.2	34.0
Kamikawa, Hokkaido	LiC	4.7	25.6	30.4	44.0	5.8	22.0
Kishigawa, Wakayama	LiC	1.8	28.9	29.4	41.7	6.0	11.0
Sanyo, Okayama	SCL	0.7	22.0	17.5	60.5	6.7	8.7

The Freundlich adsorption constant, K, was extrapolated according to the Freundlich isotherm:

$$C_a = K \, C_w^{1/n} \quad \text{or} \quad \log C_a = 1/n \log C_w + \log K \tag{3}$$

where C_a is concentration in the soil and C_w is concentration in the solution, both at the equilibrium state (Figure 6). The K_{oc} was calculated as a function of the organic carbon content of the trial soils.

For all the three compounds tested, K showed a linear correlation with the organic carbon content of the trial soils (Table III and Figure 7). The log K_{oc} for OK-8901, atrazine, and diuron was determined as 2.82, 2.28, and 3.09, respectively. Briggs (*11*) and Kanazawa (*12*) suggested that there is a general correlation between K_{oc} and K_{ow} among many pesticides, while a reverse relationship was found between OK-8901 and atrazine. The present results suggest that OK-8901 is approximately 3 times more soil-adsorptive than atrazine. Therefore the threat of this compound to contaminate ground water seems comparatively small.

Table III. Determination of K_{oc} for Diuron, Atrazine and OK-8901 with 4 Different Soil Types

Soil origin:	Hokkaido	Ibaragi	Wakayama	Okayama		
	Organic carbon content (%)					
Compound	4.67	6.81	1.75	0.69	log K_{oc}	log K_{ow}
OK-8901	21.6	43.1	5.8	1.8	2.82	1.50
Atrazine	9.0	15.9	6.5	3.0	2.28	2.34
Diuron	46.1	79.5	11.8	3.6	3.09	2.77

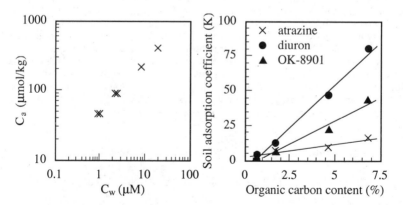

Figure 6 (Left). The Freundlich isotherm for diuron and soil from Kamikawa, Hokkaido. The K and $1/n$ were calculated as 46.1 and 0.72, respectively (r = 0.997).

Figure 7 (Right). Relationships between soil adsorption coefficient (K) for OK-8901, atrazine, and diuron and the organic carbon content of the trial soils.

Conclusions

In the present study, we tried to demonstrate that well-designed bioassays can partially substitute instrumental analyses in quantification of pesticides. Since we were working with photosynthesis-inhibiting herbicides, the yield of chlorophyll *a* fluorescence in plant materials was chosen as the probe. This idea is not new and there are already several reports on fluorometric techniques to detect PS II inhibitors in water samples (*13-15*). In these preceding studies, fluorometric measurement of photosynthesis inhibition was usually done by fluorescence induction kinetics (*13*), pulse-amplitude-modulated fluorometry (PAM) (*14*), or by flow-citometory (*15*), and therefore, elaborate instrumental systems were required. However, the measurement can be done much more simply by comparing fluorescence intensity at the terminal state (*3*) so long as uniform materials such as leaf discs or suspended microalgal cells are used.

As reported by Arsalane et al. (*13*), fluorometric techniques could not achieve the sensitivity required to detect herbicides at EC's maximum admissible concentration (0.1 μg l^{-1} for an individual pesticide) so long as the concentration steps are not included. Besides, working with crude solutions might not exclude artifactual fluorescence-quenching due to molecules released from some particular soil fractions. However, there are still some advantages for fluorometric bioassays. Of the two techniques described here, cucumber fluorometry is very simple and can be performed with minimum instrumentation even in a general household, while the microplate method is so productive as to measure 6 to 8 samples in one round without time-consuming extraction procedures. Fluorometric bioassays are especially advantageous in detecting compounds that lack aromatic rings, since such compounds are often hardly detectable by UV absorbance. This is the case with OK-9201 which, like some marketed herbicides with saturated rings such as hexazinone (3-cyclohexyl-6-dimethylamino-1-methyl-1,3,5-triazine-2,4($1H,3H$)-dione, water solubility: 33000 mg l^{-1}) and cyclurone (3-cyclo-octyl-1,1-dimethylurea, water solubility: 1100 mg l^{-1}), showed unusually high water solubility (2200 mg l^{-1}). As anticipated, OK-9201 proved alarmingly soil-leachable and was discarded accordingly, while OK-8901 survived showing adsorptivity greater than that of atrazine.

It is now obvious that some pesticides of current use are causing environmental contamination, and pesticide manufacturers have an urgent mission to develop more environmentally acceptable substitutes. As a strategy to minimize off-target dissipation of future pesticides, a screening for soil-leachability should be integrated into the whole research scheme; fluorometric techniques like ours are expected to serve as a quick screening to select safer candidates.

Acknowledgments

We are indebted to the Japan Association for Advancement of Phyto-Regulators for the soil material. Our gratitude is also extended to Mrs. H. Awata and Mr. E. Shibata for their patient and skillful technical assistance.

Literature Cited

1. Lichtenthaler, H. K. In *Applications of chlorophyll fluorescence in photosynthesis research, stress physiology, hydrobiology and remote sensing;* Lichtenthaler, H. K., Ed.; Kluwer Academic Publishers: Dordrecht, the Netherlands, 1988; pp 129-142.
2. Krause, G. H.; Weis, E. *Annual Rev. Plant Physiol. Plant Mol. Biol.* **1991**, *42*, 313.
3. Richard, E. P.; Gross, J. R.; Arntzen, C. J.; Slife, F. W. *Weed Sci.* **1983**, *31*, 361.
4. Garnier, J. Compt. *Rend. Acad. Sci.* **1967**, *265*, 874.
5. Miles, C. D.; Daniel, D. J. *Plant Sci. L.* **1973**, *1*, 237.

6. Yanase, D.; Andoh, A. *Pestic. Biochem. Physiol.* **1992,** *44,* 60.
7. Yanase, D.; Chiba, M.; Andoh, A.; Gotoh, T. *Pestic. Sci.* **1995,** *43,* 279.
8. Yanase, D.; Andoh, A.; Chiba, M.; Yoshida, S. *Z. Naturforsch.* **1993,** *48c,* 397.
9. Chiba, M.; Yanase, D.; Andoh, A.; Gotoh, T. *Pestic. Sci.* **1995,** *43,* 287.
10. Von Oepen, B.; Kordel, W.; Klein, W. *Chemosphere* **1991,** *22,* 285.
11. Briggs, G. G. *Phil. Trans. R. Soc. Lond. B* **1990,** *329,* 375.
12. Kanazawa, J. *Environ. Toxicol. Chem.* **1989,** *1,* 477.
13. Arsalane, W.; Paresys, G; Duval, J. C.; Wilhelm, C.; Conrad, R.; Buchel, C. *Europ. J. Phycol.* **1993,** *28,* 247.
14. Conrad, R.; Buchel, C.; Wilhelm, C.; Arsalane, W.; Berkaloff, C.; Duval, J. C. *J. Appl. Phycol.* **1993,** *5,* 505.
15. Weston, L. H.; Robinson, P. K. *Biotech. Techniques* **1991,** *5,* 327.

FATE AND TRANSPORT

Chapter 8

Interactions Between Atrazine and Smectite Surfaces

David A. Laird

National Soil Tilth Laboratory, Agricultural Research Service, U.S. Department of Agriculture, 2150 Pammel Drive, Ames, IA 50011

Smectites contribute much of the inorganic surface area of soils and therefore have a large potential for influencing the fate of atrazine in soil environments. The sorption capacity of smectites for atrazine varies widely depending on the surface charge density of the smectite, nature of the adsorbed cation, and pH of the equilibrating solution. Under neutral conditions, molecular atrazine is initially sorbed on smectites by a combination of water bridging between electronegative moieties on the atrazine molecule and adsorbed metal cations and hydrophobic bonding between the alkyl-side chains on the atrazine molecule and hydrophobic microsites on the smectite surface. Surface acidity, arising principally from enhanced hydrolysis of solvation water for adsorbed metal cations, catalyzes protonation and hydrolysis of atrazine sorbed on smectite surfaces. The hydrolysis product is protonated hydroxy/keto-atrazine, which may exist in any of 14 different tautomeric forms. Tautomerism and resonance allow protonated hydroxy/keto-atrazine to adapt to heterogeneous microsites on smectite surfaces which results in strong bonding.

Clay minerals are the dominant adsorbents of nonionic organic compounds in subsoils and sediments that contain less than 1% organic C. Clays may also contribute significantly to the sorption of nonionic organic compounds in surface soils (1-5). Dunigan and McIntosh (6), for example, studied atrazine sorption on various organic and inorganic fractions separated from the Walla Walla soil (16% sand, 68% silt, 16% clay and 2.5% organic matter). They found that the clay together with the clay associated organic matter sorbed 77.5 $\mu g/g$ while the clay with the organic matter removed sorbed 40.0 $\mu g/g$. In a similar study Laird et al. (7) quantified sorption of atrazine by various clay and organic matter fractions separated from a Webster soil. They used a mass balance approach to account for interactions between clay and organic matter and determined that the clay contributed about 32% of the atrazine sorption. Thus, a thorough understanding of the fate of nonionic organic compounds in soil environments must include an understanding of reactions between the compounds and the surfaces of soil minerals.

The heterogeneous nature of soil materials largely precludes identification of specific reaction mechanisms in studies involving whole soils. Smectites, however, contribute most of the inorganic surface area in soils and therefore dominate interactions between nonionic organic compounds and the inorganic constituents of soils. By comparison with whole soils smectites are relatively simple systems. The nature of smectite surfaces can be predicted from an understanding of their crystal and surface chemistry, and therefore smectites offer an opportunity to investigate interactions between nonionic organic compounds and inorganic surfaces at a mechanistic level.

Interactions between clay mineral surfaces and weak base herbicides have been investigated for many years (*8-10*). Until recently the mechanism for sorption of weak base herbicides such as atrazine on clay surfaces has been attributed to ion exchange of the protonated species. In 1992, however, Laird et al. (*11*) demonstrated that atrazine sorption is inversely correlated with the layer charge of smectites, and based on this evidence concluded that under neutral pH conditions atrazine is dominantly sorbed on smectites as a molecular species. In this chapter the nature of smectite surfaces is discussed and literature pertaining to reactions between smectite surfaces and atrazine is reviewed. A mechanistic model for the interactions between atrazine and smectite surfaces is proposed.

Nature and Properties of Smectites

Structure of Smectites. Smectites are 2:1 phyllosilicates. Elementary smectite particles (individual smectite layers) have an anionic framework consisting of four planes of close-packed oxygens and hydroxyls. Metal cations (Si, Al, Fe, and Mg) are present in the interstices of the anionic framework. The anions form coordination polyhedron about the metal cations (*12*) and are arranged with two Si-tetrahedral sheets sandwiching an Al/Mg-octahedral sheet (Figure 1).

Pyrophyllite ($Si_4Al_2O_{10}(OH)_2$) is the uncharged analog of dioctahedral smectites. In the pyrophyllite structure, positive charge due to structural Si and Al cations is exactly balanced by the negative charge of the anionic framework. With smectites, isomorphous substitution of lesser charged cations (e.g., Al^{3+} for Si^{4+} or Mg^{2+} for Al^{3+}) in the pyrophyllite structure leads to development of a deficit of positive charge at the site of substitution (ie., net negative layer charge). The negative layer charge of smectites ranges from 0.086 to 0.207 C/m^2 (0.89 to 2.1 $\mu mol_c/m^2$). Electrical neutrality is maintained by the presence of hydrated exchangeable cations adsorbed on surfaces of elementary smectite particles and excesses of cations in diffuse double layers adjacent to the external surfaces of smectite particles.

Smectite particles are poorly organized assemblages containing from one to several hundred elementary particles. The c-crystallographic axes of the elementary particles within such an assemblage are parallel, however the orientation of the a

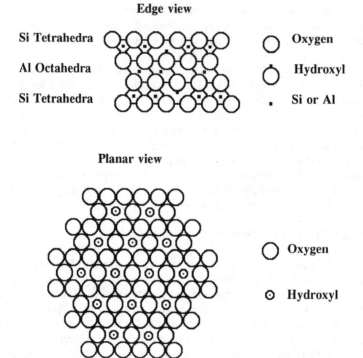

FIGURE 1. Two dimensional edge and planar views of the structure of 2:1 phyllosilicates.

and b axes varies randomly from one elementary particle to the next. Therefore, smectite particles poses only two dimensional order and are referred to as quasicrystals. Cation bridging provides a strong electrostatic force of attraction that holds elementary particles together within a quasicrystal.

Smectite quasicrystals have three types of surfaces; internal planar surfaces, external planar surfaces, and broken edges. In most soils and sediments, external planar surfaces represent only a small proportion of the total surface area of the smectite quasicrystals. However, external planar surfaces are important because diffuse double layers form in the solution adjacent to the external surfaces and interactions between double layers controls colloidal behavior. Broken edges also represent only a small proportion of the total surface area of smectite quasicrystals, but they are important because broken edges carry a variable charge. Under acidic conditions the variable charge may be positive which facilitates flocculation through edge-face bonding and provides sites for binding of anionic compounds. The vast majority of the surface area of smectite quasicrystals is due to internal planar surfaces. Internal surfaces provide most of the potential sites for sorption of organic compounds. Accessibility of internal surfaces, however, is sometimes limited by chemical or steric effects. For example, anionic organic compounds are excluded from the interlayers by electrostatic repulsion (*13*). The total surface area of a fully expanding smectite is about 750 m^2/g.

Hydration of Smectite Surfaces. The primary interaction between water and smectite surfaces is through the adsorbed metal cations. Water molecules form strong ion-dipole interactions with the adsorbed metal cations. The effect is similar to the hydration of metal cations in the bulk solution. Small multivalent cations (e.g., Al^{3+}, Mg^{2+}, and Ca^{2+}) retain both inner and outer hydration shells, whereas large monovalent cations (e.g., K^+ and Rb^+) retain only inner hydration shells. Confinement of cations and their associated waters of hydration in the interlayers of smectite quasicrystals reduces both the mobility (*14-16*) and the relative permittivity (*17-19*) of the hydration water relative to that of water in the bulk solution.

On a macroscopic scale smectite surfaces are strongly hydrophilic, however, the relative hydrophobic/hydrophilic character of smectite surfaces varies dramatically on a molecular scale. Heterogeneity results because negative layer charge is localized in basal oxygens proximal to sites of isomorphous substitution (*18, 20-23*). With tetrahedrally charged clays the negative surface charge is primarily carried by the three basal oxygens of the aluminate tetrahedra. With octahedrally charged clays the negative charge may be associated with as many as ten basal oxygens (*20*). Orbital interactions and electronic polarization spread the negative charge to second neighbor basal oxygens, but not much further (*23*). Due to the localization of charge, the basal oxygens associated with aluminate tetrahedra are nucleophilic and readily form hydrogen bonds with water molecules (*18, 23*). By contrast, the basal oxygens distal from the charge sites interact only weakly with water molecules. Thus, on a molecular scale the relative hydrophilic/hydrophobic character of smectite surfaces varies areally with the distribution and character of the surface charge sites. In low charge density smectites the surface charge sites are

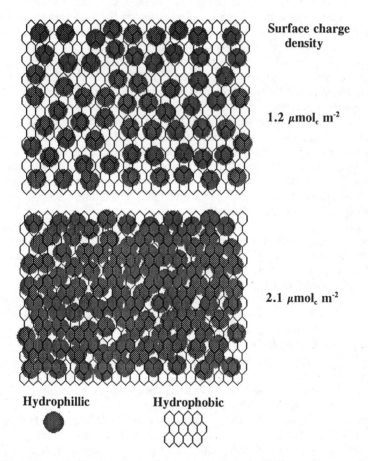

Surface charge density

1.2 μmol$_c$ m^{-2}

2.1 μmol$_c$ m^{-2}

Hydrophillic   **Hydrophobic**

FIGURE 2. Distribution of hydrophobic and hydrophilic regions on basal surfaces of smectites with surface charge densities of 1.2 and 2.1 μmol$_c$ m^{-2}. The hexagonal network shows the positions of basal oxygen atoms. The hydrophilic regions are 1.0 nm in diameter.

widely separated and a substantial portion of the smectite surface is hydrophobic. As the surface charge density increases there is less distance between charge sites and the hydrophilic regions began to coalesce. Assuming the radius of the hydrophilic region around a charge site is 0.5 nm then the hydrophilic area per charge site is 0.79 nm². A smectite with a surface charge density of 1.2 μmol_c m^{-2} has 1.38 nm² of surface area per charge site, hence 43% of the surface is hydrophobic. By comparison, a smectite with a surface charge density of 2.1 μmol_c m^{-2} has 0.79 nm² of surface area per charge site and therefore will have little or no hydrophobic surface area (Figure 2).

Surface Acidity.   Smectite surfaces exhibit both Lewis and Brönsted acidity. Lewis acidity depends on the electron-accepting properties of metal ions adsorbed on exchange sites or exposed on the lateral edges of elementary smectite particles. The Lewis acidity of clays dispersed in organic solvents may be substantial, however, the presence of water diminishes Lewis acidity, because hydroxyls or water molecules fill all of the coordination sites on the exposed metal cations. Weakly polar organic compounds are not competitive with water molecules for complexation with Lewis acid sites. Only strong organic ligands which can displace water (or hydroxyl) can effectively use the Lewis acid site. Reactions indicative of Lewis acidity in aqueous systems have been shown for some phenols and carboxylic acids adsorbed on metal oxide surfaces (24).

A Brönsted acid is a proton donor. Three types of Brönsted acid sites occur on smectite surfaces. First, hydronium ions retained by ion exchange on negative surface charge sites are a potential source of protons. Second, water molecules and even hydroxyls coordinated with the Lewis acid sites on exposed structural metal ions may donate protons. And third, water molecules associated with the hydration shells of metal cations adsorbed on ion exchange sites may hydrolyze and release protons,

$$Clay\text{-}[M(H_2O)_x]^{+n} = Clay\text{-}[MOH(H_2O)_{x\text{-}1}]^{+n\text{-}1} + H^+.$$

Enhanced hydrolysis of the solvation water of metal cations is also observed in bulk solutions. Clay surfaces increase this effect for three reasons. First, the concentration of metal cations is typically greater in the vicinity of clay surfaces than in the bulk solution. Second, water molecules adsorbed on clay surfaces are more polarized than water molecules in the bulk solution (25). And third, sorption of a compound on a clay surface restricts the mobility of the compound and thereby increases the residence time of the compound in the vicinity of the acidic functional groups.

The Brönsted surface acidity of smectites increases with decreasing water content of the clay and increasing electronegativity of the adsorbed metal cations (26-27). Electronegative cations withdraw electrons from oxygens of solvating water molecules facilitating proton release. Due to this effect the surface acidity of homoionic smecites increases for clays saturated with $K^+ < Na^+ < Ca^{2+} < Mg^{2+} < Al^{3+} < Fe^{3+}$. The effect of water content on the Brönsted surface acidity of smectites is less well understood, but is likely due to increasing polarization of

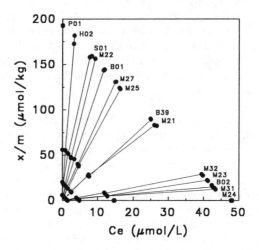

FIGURE 3. Isotherms for adsorption of atrazine on Ca-smectites (Reproduced with permission from ref. 11). Some properties of the smectite samples are listed in Table I.

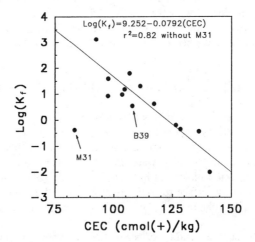

FIGURE 4. Relationship between cation exchange capacity of smectites and $Log(K_f)$ values for sorption of atrazine on the smectites (Reproduced with permission from ref. 11). Sample B39 was extracted from a typical agricultural soil and sample M31 was an interstratified smectite-illite.

surface water molecules with decreasing water content. Increasing surface acidity with decreasing residual water content has been demonstrated by quantifying protonation of NH_3 (*28*) and pyridine (*29*) adsorbed on 'dry' smectites. Although surface acidity is diminished in aqueous systems, Ca-smectites suspended in distilled water have been shown to promote significant protonation of weak bases as much as two pH units above the pKa of the base (*30*). Surface acidity is anticipated to increase with the surface charge density of smectites because both the concentration of metal cations in the interlayers and the polarization of surface water increase with surface charge density.

Reactions between Atrazine and Smectite Surfaces

Sorption of Atrazine. Tremendous variability in the affinity of reference and soil smectites for atrazine was observed by Laird et al. (*11*) (Figure 3). In this study, 14 different smectite samples (< 2 μm fractions) were treated for removal of organic matter, Ca-saturated, dialyzed free of excess salt, and freeze-dried before the sorption determinations. Duplicate sorption determinations were made using a 24 hr batch equilibration technique in 0.01 M $CaCl_2$ at four atrazine levels for each sample. Properties of the smectite samples are listed in Table I. The affinity of the various smectites for atrazine was evaluated by fitting the sorption data to Freundlich isotherms:

$$x/m = K_f \times Ce^{1/n}$$

where x/m is the amount of atrazine sorbed on the clay (μmol/kg), Ce is the amount of atrazine in solution (μmol/L), and both K_f and $1/n$ are empirical constants.

The pretreatments used by Laird et al. (*11*) largely eliminated confounding effects due to organic matter content, pH, and nature of the adsorbed cations, therefore the observed variability in affinity of atrazine for smectites was attributed to the nature of the smectite surfaces. The Freundlich sorption constants (K_f values) were significantly inversely correlated with the surface charge density and the surface area of the smectites but were not correlated with solution pH (pH range = 4.75 to 6.45). The cation exchange capacity (the product of surface charge density and surface area) accounted for 82% of the variability in the $\log(K_f)$ values (Figure 4). On the other hand, the Freundlich exponent ($1/n$) was significantly correlated with solution pH, which suggests that more than one mechanism contributes to atrazine sorption on smectites. Based on these results Laird et al. (*11*) concluded that for the pH range of the study that atrazine is dominantly sorbed on Ca-smectites as a neutral species. In a later study, Barriuso et al. (*31*) quantified desorption of atrazine for 12 of the 14 samples originally studied by Laird et al. (*11*). In general very little sorption/desorption hysteresis was observed, however the small amount of hysteresis that was observed increased with surface charge density of the smectites. These results indicate that the dominant mechanism for sorption of atrazine on smectites is reversible but also suggest a second nonreversible mechanism that is particularly important for sorption on high charge smectites.

Table I. Properties of samples used for analysis of atrazine sorption on smectites by Laird et al. (11)

Lab No.	Name/Location	Mineralogy	CEC[a]	SCD[b]	SA[c]	pH
H02	Hector, CA	Hectorite	102	1.38	743	5.80
B01	Wy. Bentonite	Montmorillonite	99	1.34	744	5.85
B02	Camp Berteau	Montmorillonite	122	1.63	749	5.45
M21	Polkville, MS	Montmorillonite	111	1.50	742	5.00
M22	Amory, MS	Montmorillonite	93	1.44	644	4.75
M23	Chambers, AZ	Montmorillonite	130	1.75	739	5.60
M24	Otay, CA	Montmorillonite	134	1.85	724	5.55
M25	Upton, WY	Montmorillonite	95	1.31	730	5.85
M27	Belle Fourche, SD	Montmorillonite	98	1.33	739	5.95
M31	Cameron, AZ	Smectite/illite	79	2.08	382	4.95
M32	IMV Bentonite	Montmorillonite	120	1.70	706	6.45
S01	IMV Saponite	Saponite	106	1.80	588	5.85
P01	Panther Creek, MS	Beidellite	88	1.42	619	4.80
B39	Webster soil clay	Montmorillonite	103	1.57	654	5.50

[a]Cation exchange capacity, cmol kg^{-1}.
[b]Surface charge density, μmol_c m^{-2}.
[c]Surface area m^2 g^{-1}.

Hydrolysis of Atrazine. The conclusion of Laird et al. (*11*) that atrazine is dominantly sorbed as a neutral species on smectite surfaces is in apparent contradiction with numerous earlier researchers who concluded that atrazine is retained on smectite surfaces as either protonated atrazine or protonated hydroxy-atrazine. Evidence for protonation and hydrolysis of atrazine on smectite surfaces is based on vibrational spectroscopy studies and on the relationship between sorption and both pH and the basicity of related s-triazines (*32-39*).

Russell et al. (*36*) used IR and NMR spectroscopy to evaluate abiotic degradation of various s-triazines sorbed on montmorillonites and clearly showed that surface acidity catalyzes hydrolysis of chlorotriazines. The first step is protonation of a ring N followed by cleavage of the C-Cl bond and formation of the protonated hydroxy species. The appearance of a strong carbonyl stretching band at 1740 cm^{-1} suggested that both enol and keto tautomers of protonated hydroxy triazines were formed. The 1740 cm^{-1} band was absent when the clays were prepared in chloroform but increased when water was added to the system. The 1740 cm^{-1} band was strong when the clays were prepared in alcohol. The results suggest that the keto form may be common under hydrated conditions.

Sorption of s-triazine on smectite increases with decreasing pH until the solution pH is equal to the pKa of atrazine (*32-33, 38*). Weber (*32, 38*) attributed decreased sorption of s-triazines for pH levels below their pKa's to displacement of the cationic forms by H_3O^+, however confounding effects due to clay hydrolysis under the very acidic conditions of the study were not considered. Armstrong et al. (*40*) reported both an increase in sorption and an increase in the rate of atrazine hydrolysis in soils with decreasing pH. Furthermore, they reported nearly a ten fold higher rate of atrazine hydrolysis in aqueous systems containing sterilized soil than in aqueous systems without soil at the same pH. The half-life for hydrolysis of atrazine in the aqueous system containing sterilized soil was 22 days (*40*). White (*39*) found that the nature of the substituents in the 2-position greatly influences both the ease of protonation and the ease of hydrolysis of s-triazines sorbed on Ca-montmorillonite.

In a recent study, Gilchrist et al. (*41*) used an on-line microfiltration-HPLC technique to study sorption equilibrium and sorption kinetics of atrazine on clay minerals, including Na- and Ca-montmorillonite. Most of the added atrazine was rapidly and reversibly sorbed on the montmorillonite. This "labile" fraction achieved equilibrium with the solution in a matter of minutes. A small portion of the added atrazine, however, could not be recovered under the conditions of the experiment. This residual fraction was termed "reversible but kinetically slow" on the assumption that it could eventually be recovered. For the Na-montmorillonite, the "reversible but kinetically slow" fraction increased from about 10% of the total atrazine on day one to about 30% of the total atrazine on day 15, thereafter the "reversible but kinetically slow" fraction remained constant at about 30% of the total added atrazine. Gilchrist et al. (*41*) speculated that the "reversible but kinetically slow" sorption was due to "diffusion of atrazine into mineral grain boundaries, lattice imperfections, or other structural features" and dismissed the possibility of hydrolysis because they were unable to detect hydroxy-atrazine in the solution or readily desorbable fractions.

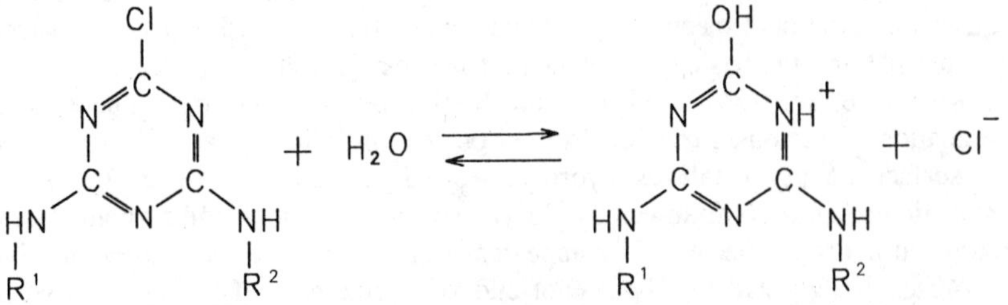

FIGURE 5.   Hydrolysis of atrazine to protonated hydroxy atrazine.

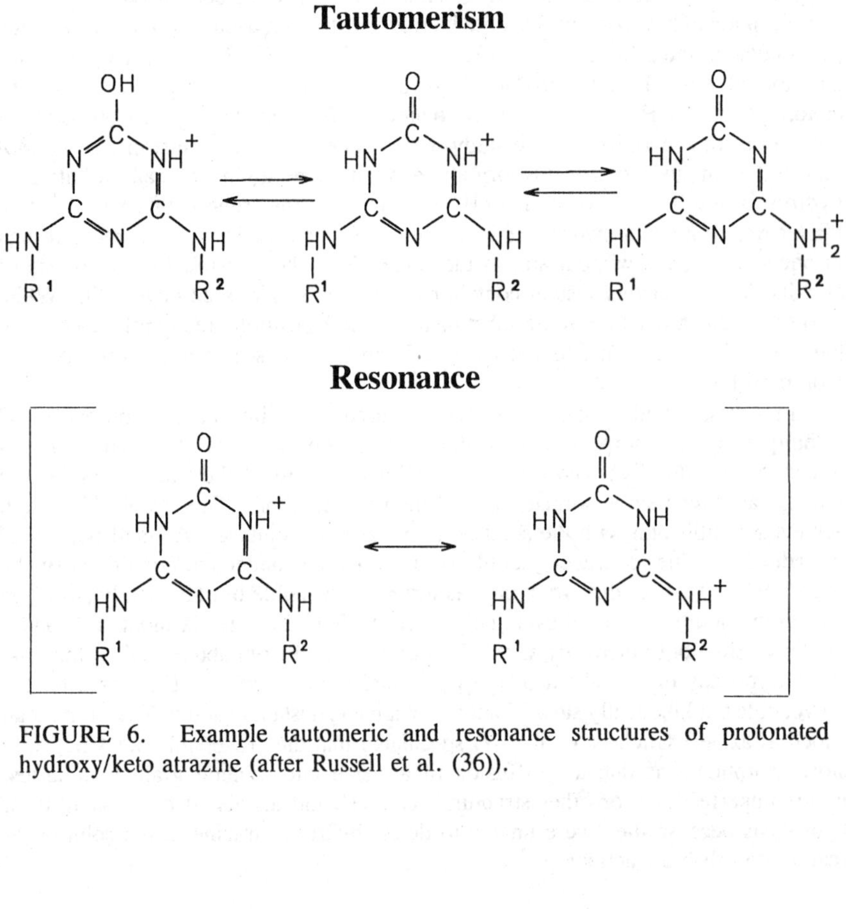

Tautomerism

Resonance

FIGURE 6.   Example tautomeric and resonance structures of protonated hydroxy/keto atrazine (after Russell et al. (36)).

Model for Reaction of Atrazine with Smectites. The various studies on sorption of atrazine by smectites are reconciled for neutral pH conditions if atrazine is retained on smectite surfaces by two different mechanisms. The first sorption mechanism is the rapid and reversible sorption of molecular atrazine. Sorption of molecular atrazine is consistent with the short equilibration time and reversible sorption of the "labile fraction" observed by Gilchrist et al. (*41*) and with the general lack of sorption/desorption hysteresis observed by Barriuso et al. (*31*). Neutral atrazine molecules are probably retained on smectite surfaces by both water-bridging and hydrophobic interactions. The lone-pair of electrons on the atrazine ring N atoms probably form H-bonds with the water molecules solvating the interlayer cations. At the same time, the hydrophobic alkyl-side chains of the atrazine molecule should readily compete with water molecules for retention on the hydrophobic microsites located between the charge sites on smectite surfaces (Figure 2). Because atrazine is capable of simultaneously interacting with both hydrophobic microsites on smectite surfaces and water molecules in the interlayer, sorption of atrazine should lower the surface free energy of the smectite-water interface. Increasing sorption of s-triazines on montmorillonite with increasing length of the alkyl-side chains (*32, 39*) is consistent with a contribution of hydrophobic bonding to sorption (Note, both Weber (*32*) and White (*39*) attributed this phenomenon to increasing basicity of the s-triazines not hydrophobic interactions). Hydrophobic bonding is also consistent with the observed decrease in sorption with increasing surface charge density of smectites (*11* and Figure 4), because the amount of hydrophobic surface on smectites decreases with increasing surface charge density (Figure 2).

The initial rapid and reversible sorption of neutral atrazine on smectite surfaces is apparently followed by a relatively slow process in which surface acidity catalyzes both protonation and hydrolysis of the sorbed atrazine. Atrazine is a weak base with a pKa of 1.7. Under acidic conditions the exposed lone pair of electrons on the ring N atoms may form a covalent bond with a proton. Surface acidity catalyzes such protonation by providing effectively a "lower pH environment" for sorbed atrazine than is experienced by atrazine in the bulk solution, and protonation facilitates hydrolytic dehalogenation of atrazine. The carbon atom in the number 2 position on the triazine ring is surrounded by electronegative N and Cl atoms. Protonation of one of the ring N atoms further withdraws electrons from the C atom facilitating nucleophilic attack (Figure 5). The net products of this reaction are Cl⁻ which is ejected from the vicinity of the clay surface by electrostatic forces and protonated-hydroxy-atrazine. The extra proton on protonated hydroxy-atrazine may be associated with any of the five N atoms giving rise to five different tautomeric structures, furthermore, the hydroxyl proton may shift to one of the N atoms forming protonated keto-atrazine (*36*). Nine different tautomeric structures of protonated keto-atrazine are possible, three that have protons on two of the ring N atoms, and six that have a proton on one ring N and two protons on one of the amine N atoms. The three tautomers that have protons on two of the ring N atoms each have four possible resonance structures (Figure 6).

The pKa of hydroxy/keto-atrazine is about 5.2 (*42*), hence in neutral to slightly acid systems most of the hydroxy/keto-atrazine sorbed on smectite surfaces will be protonated. Protonated hydroxy-atrazine has the potential to interact with smectite surfaces in three different ways. First, the protonated ring N may form an ionic bond with negative surface charge sites. Second, the hydroxyl protons may form hydrogen bonds with basal oxygens proximal to the surface charge sites. And third, the alkyl-side chains may form van der Waals bonds with the uncharged siloxane surfaces distal from the surface charge sites. In addition to the ionic and van der Waals bonds describe above, protonated keto-atrazine will likely form hydrogen bonds between the carbonyl O and H of interlayer water molecules and between the H associated with the amine and ring N atoms and basal O atoms proximal to surface charge sites. The heterogeneity of smectite surfaces on a molecular scale provides an opportunity for all of these bonding mechanisms to simultaneously contribute to the retention of protonated hydroxy/keto-atrazine on smectite surfaces. Furthermore, tautomerism and resonance allow protonated hydroxy/keto-atrazine molecules to adapt to unique micro environments and thereby achieve the strongest possible bonding to the clay surface. The net result is that protonated hydroxy/keto-atrazine is very strongly retained on soil and clay surfaces. Clay and Koskinen (43), for example, reported negligible desorption of added hydroxy-atrazine from two silt loam soils.

The "reversible but kinetically slow" fraction observed by Gilchrist et al. (41) was probably protonated hydroxy/keto-atrazine. Strong bonding between the protonated hydroxy/keto-atrazine and smectite surfaces likely kept solution levels of hydroxy-atrazine below the detection limit in their study. The small amount of hysteresis in desorption of atrazine from smectites observed by Barriuso et al. (31) was also likely caused by the hydrolysis of atrazine to protonated hydroxy/keto-atrazine. As discussed previously, surface acidity is expected to increase with the surface charge density of smectites and an increase in surface acidity should cause an increase in the rate of surface catalyzed hydrolysis of atrazine. Thus, the observed increase in atrazine sorption/desorption hysteresis with increasing surface charge density of the smectites is consistent with surface catalyzed hydrolysis.

Conclusions

Two mechanisms contribute to the sorption of atrazine on smectite surfaces under neutral pH conditions. The first mechanism is the rapidly and reversibly sorption of molecular atrazine by a combination of water bridging and hydrophobic interactions. The ring N atoms of atrazine molecules interact with water molecules solvating metal cations adsorbed on smectite surfaces. At the same time the alkyl-side chains of atrazine molecules out compete water molecules for retention on hydrophobic microsites on smectite surfaces. The second mechanism is the slow hydrolysis of sorbed molecular atrazine to protonated hydroxy/keto atrazine. The surface acidity of smectites, arising principally from enhanced hydrolysis of the solvation water of adsorbed metal cations, catalyzes both the protonation and hydrolysis reactions. Protonation of a ring N further withdraws electrons from the already electron deficient C in the number 2 position on the triazine ring, facilitating nucleophilic attack. The product of this reaction, protonated hydroxy/keto-atrazine, is vary strongly retained on smectite surfaces.

Literature Cited

(1)   Hassett, J. J.; Banwart, W. L. In *Reactions and Movement of Organic Chemicals in Soils;* Sawhney, B. L. et al., Eds.; SSSA Spec. Publ. 22; Soil Science Society of America: Madison, WI, 1989; pp. 31-44.

(2)   Harper, S. S. *Weed Sci.* **1988,** *36,* 84-89.

(3)   Sanchez-Martin, M. J.; Sanchez-Camazano, M. *Weed Sci.* **1991,** *39,* 417-422.

(4)   Konopka, A.; Turco, R. *Appl. Environ. Microbiol.* **1991,** *57,* 2260-2268.

(5)   Mingelgrin, U.; Gerstl, Z. *J. Env. Qual.* **1983,** *12,* 1-11.

(6)   Dunigan, E. P.; McIntosh, T. H. *Weed Sci.* **1971,** *19,* 279-282.

(7)   Laird, D. A.; Yen, P. Y.; Koskinen, W. C.; Steinheimer, T. R.; Dowdy, R. H. *Environ. Sci. Techno.* **1994,** *28,* 1054-1061.

(8)   Bailey, G. W.; White, J. L. *Residue Rev.* **1970,** *32,* 29-92.

(9)   Weber, J. B. *Residue Rev.* **1970,** *32,* 93-130.

(10)   Green, R. E. In *Pesticides in Soil and Water;* Guenzi, W. D. et al., Eds.; Soil Science Society of America: Madison, WI, 1974; pp 3-37.

(11)   Laird, D.A.; Barriuso, E.; Dowdy, R.H.; Koskinen, W.C. *Soil Sci. Soc. Am. J.* **1992,** *56,* 62-67.

(12)   Pauling, L. *The Nature of the Chemica Bond;* 3rd ed; Cornell University Press: Ithaca, N. Y., 1960.

(13)   Zhang, Z. Z.; Low, P. F.; Cushman, J. H.; Roth, C. B. *Soil Soc. Am. J.* **1990,** *54,* 59-66.

(14)   McBride, M. B., Pinnavaia, T. J.; Mortland, M. M. *J. Phys. Chem.* **1975,** *79,* 2430-2435.

(15)   Sposito, G.; Prost, R. *Chem Rev.* **1982,** *82,* 553-575.

(16)   Delville, A.; Grandjean, J.; Laszlo, P. *J. Phys. Chem.* **1991,** *95,* 1383-1392.

(17)   Conway, B. E. *Ionic Hydration in Chemistry and Biophysics;* Elsevier: Amsterdam; 1981.

(18)   Sposito, G. *The Surface Chemistry of Soils;* Oxford University Press, Inc.: New York; 1984.

(19)   Newman, A. D. C. In *Chemistry of Clays and Clay Minerals;* Newman, A. C. D., Ed.; Wiley-Interscience: New York, 1987; pp 237-274.

(20)   Farmer, V. C. In *The Chemistry of Soil Constituents;* D. J. Greenland; Hays, M. H. B. Eds.; Wiley: New York, 1978; pp 405-448.

(21)   McBride, M. B. In *Minerals in Soil Environments;* Dixon, J. B.; Weed, S. B. Eds.; 2nd ed.; Soil Science Society of America: Madison, WI, 1989; pp 35-87.

(22)   Bleam, W. F.; Hoffmann, R. *Phys. Chem. Minerals.* **1988,** *15,* 398-408.

(23)   Bleam, W. F. *Clays Clay Miner.* **1990,** *38,* 527-536.

(24)   McBride, M. B. *Environmental Chemistry of Soils,* Oxford University Press, Inc.: New York, 1994; pp 381-382.

(25)   Johnston, C. T.; Sposito, G.; Erickson, C. *Clays Clay Miner.* **1992,** *40,* 722-730.

(26)   Mortland, M. M. *9th Int. Congr. Soil Sci. Trans.* **1968,** *1,* 691-699.

(27) Mortland, M. M.; Raman, K. V. *Clays Clay Miner.* **1968**, *16*, 393-398.

(28) Mortland, M. M.; Fripiat, J. J.; Chaussidon, J.; Uytterhoeven, J. J. *Phys Chem.* **1963**, *67*, 248-258.

(29) Farmer, V. C.; Mortland, M. M. *J. Chem. Soc. Sec. A.* **1966**, 344-351.

(30) Feldkamp, J. R.; White, J. L. In *Proceedings of the International Clay Conference 1978;* Mortland, M. M.; Farmer, V. C., Eds.; Elsevier Scientific Publishing Co.: Amsterdam, 1979; pp 187-196.

(31) Barriuso, E.; Laird, D. A.; Koskinen, W. C.; Dowdy, R. H. *Soil Sci. Soc. Am J.* **1994**, *58*, 1632-1638.

(32) Weber, J. B. *Soil Sci. Soc. Am. Proc.* **1970**, *34*, 401-404.

(33) Yamane, V. K.; Green, R. E. *Soil Sci. Soc. Am. Proc.* **1972**, *36*, 58-64.

(34) Cruz, M.; White, J. L.; Russell, J. D. *Isr. J. Chem.* **1968**, *6*, 315-323.

(35) Bailey, G. W.; White, J. L.; Rothberg, T. *Soil Scil. Soc. Am. Proc.* **1968**, *32*, 222-234.

(36) Russell, J. D.; Cruz, M.; White, J. L.; Bailey, G. W.; Payne, W. R. Jr.; Pope, J. D. Jr.; Teasley, J. I. *Science,* **1968**, *160*, 1340-1342.

(37) Brown, C. B.; White, J. L. *Soil Sci. Soc. Am. Proc.* **1969**, *33*, 863-867.

(38) Weber, J. B. *Am. Miner.* **1966**, *51*, 1657-1670.

(39) White, J. L. In *Proceedings of the International Clay Conference 1975;* Bailey, S. W., Ed.; Applied Publishing Ltd.: Wilmette, IL, 1975; pp 391-398.

(40) Armstrong, D. E.; Chesters, G.; Harris, R. F. *Soil Sci. Soc. Am Proc.* **1967**, *31*, 61-66.

(41) Gilchrist, G. F. R.; Gamble, D. S.; Kodama, H.; Kahn, S. U. 1993. *J. Agric. Food Chem.* **1993**, *41*, 1748-1755.

(42) Weber, J. B. *Spectrochim. Acta.* **1967**, *23A*, 458-461.

(43) Clay, S. A.; Koskinen, W. C. *Weed Sci.* **1990**, *38*, 262-266.

Chapter 9

Estimation of the Potential for Atrazine Transport in a Silt Loam Soil

David A. V. Eckhardt[1] and R. J. Wagenet[2]

[1]U.S. Geological Survey, 903 Hanshaw Road, Ithaca, NY 14850
[2]Department of Soil, Crop, and Atmospheric Sciences, Cornell University, Ithaca, NY 14853

The transport potential of the herbicide atrazine (2-chloro-4-ethyl-6-isopropyl-s-triazine) through a 1-meter-thick root zone of corn (*Zea mays* L.) in a silty-loam soil in Kansas was estimated for a 22-year period (1972-93) using the one-dimensional water-flow and solute-transport model LEACHM. Results demonstrate that, for this soil, atrazine transport is directly related to the amount and timing of rain that follows spring applications of atrazine. Two other critical transport factors were important in wet years — [1] variability in atrazine application rate, and [2] atrazine degradation rates below the root zone. Results demonstrate that the coincidence of heavy rain soon after atrazine application can cause herbicide to move below the rooting zone into depths at which biodegradation rates are assumed to be low but are often unknown. Atrazine that reaches below the rooting zone and persists in the underlying soil can subsequently be transported into ground water as soil water drains, typically after the growing season. A frequency analysis of atrazine concentrations in subsurface drainage, combined with field data, demonstrates the relative importance of critical transport factors and confirms a need for definitive estimates of atrazine-degradation rates below the root zone. The analysis indicates that periodic leaching of atrazine can be expected for this soil when rainfall that exceeds 20 cm/mo coincides with atrazine presence in soil.

In 1990, about 29 million kilograms of atrazine were applied to soils in the United States for control of broad-leaf weeds in corn (*1*). These agricultural herbicides commonly are sprayed in liquid form just before spring planting when soils are relatively moist, and evapotranspiration rates are low. The fate of atrazine is largely controlled by the chemical, biological, and hydraulic properties of soil and by weather conditions immediately after application. When soil erosion is minimal and herbicides are mixed into shallow soil layers, most of the herbicide mass degrades biologically and hydrolyzes chemically within the root zone as the crop matures. There is evidence, however, that herbicides are often leached below the root zone, where travel times can be com-

0097–6156/96/0630–0101$15.00/0
© 1996 American Chemical Society

parable to those of conservative solutes (2). Thus, any rapid downward movement of water and dissolved herbicide residue that decreases the residence time within upper soil zone, where biodegradation is most active, will increase the potential for transport into ground water.

Much of the difficulty in studies of water and herbicide movement is due to the heterogeneous and anisotropic properties of soils (3-6). Hydraulic properties of soils that affect water movement, such as hydraulic conductivity and moisture-retention characteristics, can be determined by field experiments of infiltration and drainage and through laboratory analyses of representative soil cores. Chemical-transport properties can be estimated through measurements of water, conservative-tracer, and herbicide and metabolite contents in the soil during crop growth. The statistical distributions of the soil properties and the measured chemical and water contents through time, combined with reliable estimates of the rates and frequency of water and herbicide applications, provide a basis for evaluating the transport and fate of atrazine through solute transport models (6-9).

The problem addressed in this study was to consider the inherent variability in water and chemical applications and the uncertainty in measurements of soil and chemical properties in estimates of leaching potential for atrazine. In approaching this problem, a mechanistic model that represents Darcian unsaturated flow and advective-dispersive solute transport can be used with advantage over other modeling approaches if the uncertainty in transport variables is considered over a range in water and chemical applications (8, 10).

Objectives. In June 1990, the U.S. Geological Survey and Cornell University began a field and modeling study of subsurface transport and dissipation of atrazine in a silt-loam soil. There were two objectives of this work. The first was quantification of factors that affect water flow and atrazine transport at the study site. A second objective was to demonstrate the relative effects on atrazine transport of uncertainty in the critical transport factors for a 22-year period (1972-93) for which precipitation records were available for the study site.

This paper presents a comparison of atrazine concentrations simulated through the model to those measured in soil cores at the study site during a corn-growth season. Then, model predictions of atrazine transport are presented for the 22-year period. Lastly, results of a frequency analysis of atrazine concentrations in soil-water drainage are presented for the ranges of uncertainty in critical transport variables.

Methods

The study consisted of two phases — [1] field determination of hydraulic and chemical-transport properties of the soil, and [2] a series of model simulations to accomplish the objectives. The study combined field and laboratory methods and the numerical model LEACHM (11), which describes hydraulic and solute transport processes in the silt-loam plot in mechanistic terms. Figure 1 summarizes the activities in the approach.

Field Study. The 0.75-ha field site was on level soil at the Kansas River Valley Experiment Field operated by the Kansas State University in the southwestern region of the midwestern cornbelt near Topeka (Figure 2). The soil is a coarse-silty, mixed,

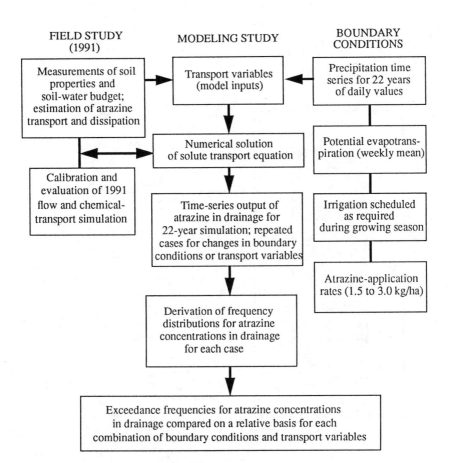

Figure 1. Method used to derive exceedance frequencies of atrazine concentrations in drainage for differing cases of boundary conditions and transport variables.

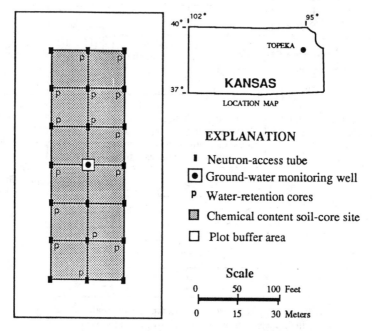

Figure 2. Locations of measurement sites at the field plot.

mesic Fluventic Hapludoll with a 30-cm silt-loam A horizon (0.8 percent organic carbon) over a silt-loam B horizon (0.2 percent organic carbon). The water table is in an alluvial-sand aquifer about 6 m the below land surface. Available weather data was recorded onsite for the 22-yr period. Mean annual rainfall for this period was 860 mm.

In a typical year, the soil is disked in March, and herbicides and anhydrous ammonia fertilizer are incorporated into shallow soil zones during April. Corn or soybeans are planted in late April. Irrigation is usually applied throughout July and August, until crop maturity. Crops are harvested in September, and the soil is chiseled and disked in preparation for the next growing season.

In this study, the soil-sample analysis was limited to the top 120 cm of soil because the most complex physical and biological processes occur in this zone. The site consists of two adjacent rows of six 15-m by 15-m plots, and the corners of each plot contained access tubes for monitoring soil-water content through neutron-attenuation (Figure 2). Unsaturated hydraulic conductivity was measured around the 20 access tubes by gravity-drainage experiments (*12*) at 15-cm intervals from land surface to a depth of 120 cm. Moisture retention and bulk density were determined by laboratory analyses of Tempe-cell cores (*13*) collected at similar depth intervals near the access tubes. Organic carbon contents and particle-size distributions were determined on samples collected at 15-cm intervals in soil cores from 12 random locations within the plots.

In April 1991, a bromide tracer (70 kg/ha) and atrazine (1.5 kg/ha active ingredient) were sprayed on the plots (*14*). The chemicals were immediately disked into the top 10 cm of soil to minimize loss in runoff, and the plot was planted with corn. Soil cores were collected from six random locations and composited into one sample for each depth interval for each of the 12 plots. Cores (15 cm long) were collected just before chemical application, immediately after application, and 30, 60, and 120 days thereafter at depth intervals of 0 to 15 cm, 15 to 30 cm, 45 to 60 cm, and 75 to 90 cm. The samples were split three ways; each split sample was analyzed for soil-water content, bromide concentration, and concentration of atrazine and two metabolites. Bromide was extracted through a washing procedure with a 0.005M $CaSO_4$ solution and analyzed by ion-chromatography. Atrazine, desethylatrazine, and desisopropylatrazine in soil samples were extracted by methanol and water washes, followed by C-18 solid-phase extraction and analysis by gas chromatography with mass spectrometry (*15, 16*). The analytical reporting limit for herbicide concentration in soil was 0.50 µg/kg.

During the 1991 growing season (May through August), volumetric water content was measured weekly at 15-cm intervals to a depth of 120 cm in the access tubes (Figure 2) to provide information on the water budget within the soil profile during crop growth. Additional water-content measurements were made during selected irrigation events to observe the relatively rapid infiltration and redistribution of applied water.

Modeling Study. The movement and mass balance of atrazine, desethylatrazine, bromide, and water were simulated by the LEACHM model (*11*) that solves the equations by finite-difference methods for one-dimensional transient-state unsaturated flow and advective-dispersive solute transport (*17*). The model included equilibrium sorption, first-order degradation, and root growth and uptake of water and chemical by plants. Growth of corn roots and the transpiration of water from the simulated profile during the growing season were estimated by the LEACHM root-growth subroutine, based on

Table I. Model formulation for baseline simulations based on 1991 field study

MODEL PROFILE
LEACHM (*11*), 1-m-thick soil profile with twenty 5-cm layers
Corn roots penetrate 0.5 m [root-growth model of Davidson *et al.* (*18*)]

INITIAL CONDITIONS
Chemical-free profile
Water content uniform at -20 kPa total hydraulic head

DATES
Model start date:	January 1, 1972
Model end date:	December 31, 1993
Chemical application:	April 24 (annually)
Corn emergence:	May 5 (annually)
Harvest:	August 17 (annually)

BOUNDARY CONDITIONS
Bottom boundary:	Free drainage under unit gradient
Precipitation:	Daily values recorded at field site
Evapotranspiration:	Weekly mean of hourly potential evapotranspiration at field site
Irrigation:	45-mm increments during crop growth
Atrazine:	1.5 kg/ha, active ingredient, applied annually; mixed into top 10 cm of soil
Bromide:	70 kg/ha (1991 only)

SOIL PROPERTIES
Bulk density:	1.40 g/cm^3
Hydraulic conductivity:	25 mm/d (at 65-percent saturation)
Dispersivity:	100 mm
Organic carbon (percentage of dry soil weight):	
0- to 30-cm depth:	0.8 percent
30- to 100-cm depth:	0.2 percent

ATRAZINE PROPERTIES
Equilibrium partitioning coefficient, K_{oc}:	170 L/kg
Half life:	
0- to 30-cm depth:	13 to 17 d
30- to 50-cm depth:	17 to 120 d
50- to 100-cm depth:	120 d

the model of Davidson *et al.* (*18*), in which water and chemical extraction from the soil were functions of root depth and density over time.

The modeling study had two phases: [1] model calibration followed by an evaluation of the accuracy of simulations for the movement and mass balance of water and atrazine, based on 1991 field measurements, and [2] extension of model predictions beyond the calibration period to derive frequency distributions of atrazine concentrations in soil-water drainage. Water and chemical movement and mass balances were simulated for a soil profile 1 m thick.

Model Calibration and Evaluation. Simulation of unsaturated water flow was based on hydraulic conductivity values estimated with field-based drainage experiments and water-retention analysis of soil cores (*12*). Model calibration for water flow consisted of [1] increasing the potential evapotranspiration rates estimated through the Penman equation (*19*) for the site, and [2] increasing the density and rate of growth of plant roots to reflect the uptake of water and bromide that was observed during crop growth in 1991. Irrigation was applied (45 mm/d) at the site when soil-moisture potential fell below -200 kPa during crop growth, and the model was modified to simulate this practice. The timing and applied volume of irrigation in the simulations were evaluated through direct comparisons of simulated and actual irrigation times and of simulated and measured water contents in soil before, during, and, after irrigation.

An evaluation of the simulation of atrazine and bromide fluxes was based upon chemical concentrations that were measured in soil cores during 1991 crop growth (*14*). Model calibration for solute transport consisted of adjusting the first-order degradation rate for atrazine to provide the best match between simulation results and measured values for atrazine distribution and mass balance in the soil plots over the 120-day growth period. Model evaluation then consisted of comparing simulated water contents and concentrations of atrazine, metabolites, and bromide with those measured in the field study.

Derivation of Frequency Distributions for 22-yr Precipitation Record. The interaction of random precipitation variability with the timing of atrazine application creates a dynamic boundary condition that is best evaluated by compiling results of model simulations into cumulative frequency distributions for a period that is representative of climatic trends. A daily precipitation record for 1972-93 and weekly means of adjusted Penman potential evapotranspiration (*19*) for 1986-93 at the site were used in a baseline simulation of the transport and fate of annual applications (1.5 kg/ha) of atrazine. The values of model parameters from the calibrated 1991 simulation (Table I) were used to define the baseline simulation. The degradation rates for atrazine in the 0- to 30-cm depth represent calibrated values that were based upon the dissipation rates that were measured in 1991. Degradation rates below 30 cm were estimated from values given in Wauchope *et al.* (*20*).

After the baseline simulation was conducted for the expected values of model parameters (Table I), the potential changes in atrazine concentrations in drainage were estimated for atrazine-application rates that were varied within the range of common agricultural usage. Also evaluated were the effects of uncertainty in two critical transport parameters -- the unsaturated hydraulic conductivity of the soil, and the degradation rates for atrazine. The effects of spatial variability in the hydraulic conductivity of

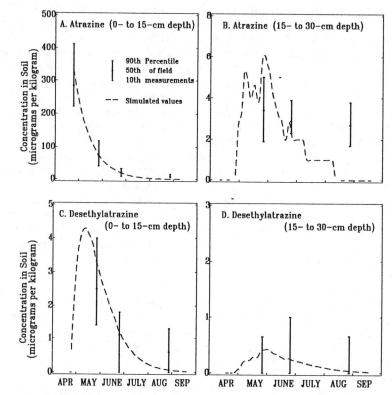

Figure 3. Simulated daily mean herbicide concentrations and concentrations measured in soil cores during 1991 crop growth.

soil and the uncertainty of degradation rates below the root zone were represented through discrete samples from probability-density functions. This approach represents a range of potential variability at the field site for these two transport variables and the differing application rates through a series of model simulations of the 22-yr period. Each simulation provided a cumulative frequency distribution of weekly mean atrazine concentrations in drainage, which were compared through graphs to rank the effects of the differing transport variables and application rates.

Results And Discussion

Field Study. Atrazine, its metabolite desethylatrazine (DEA), and bromide within the soil plots did not move below the top 60 cm during the 1991 growing season, mainly because rainfall was below average. Atrazine mass was rapidly decreased through biodegradation and hydrolysis within the upper 30 cm of soil during the first 60 days after application, but low concentrations of atrazine and DEA residues remained at 120 days after application (Figure 3). The persistence of residues in low concentrations at the site, which was observed in shallow soil for up to three years after application, indicated that atrazine and DEA were partitioned within solid-phase organic carbon or sequestered in soil micropores, thereby being unavailable for immediate bacterial degradation or transport. Increases in DEA concentration above the ambient persistence level were detected in soil cores between 30 and 60 days after atrazine application. DEA was present in nearly all samples of shallow soil (0 to 30 cm), where the atrazine was concentrated. The ratio of DEA to atrazine increased steadily for 60 days after application as DEA formed and atrazine was lost through de-alkylation, as noted also by Adams and Thurman (*21*) for this soil. Another de-alkylation product, desisopropyl-atrazine, was detected in a few samples at concentrations near the analytical reporting limit of 0.05 µg/kg.

Samples of corn grain and plant tissue from four 0.06-ha subplots were analyzed for Br⁻ after harvest. Mean plant densities and mean grain and tissue weights were then used to estimate Br⁻ uptake on both a plant and plot-area basis. Results indicate that more than half of the Br⁻ was taken up by the corn crop and was concentrated mainly in the stalks and leaves, rather than in the grain (*5*). Additional Br⁻ was probably in roots, but these were not analyzed. The results for the bromide concentrations in soil cores provided useful information in estimation of a dispersion coefficient (100 mm) for this soil and assessment of the accuracy of simulations for downward water flow during 1991 crop growth.

Model Calibration and Evaluation. The simulations of transient water contents and of atrazine, DEA, and bromide concentrations in the soil were nearly consistently within the range of the 10th to the 90th percentile of measured field values (Figure 3). The half life for atrazine within the upper 30 cm of soil was estimated through model calibrations to range between 13 and 17 days, which is a notably faster dissipation rate than that expected from the data in Wauchope *et al.* (*20*). Atrazine apparently degrades much more rapidly at the Kansas site than elsewhere in the Midwest, mainly because the Kansas site is warmer and farther south. Note that a first-order degradation process inadequately describes the observed persistence of atrazine and DEA at low concentrations in soil, but this was inconsequential because the bound residues are considered to be immobile and unavailable for transport (*22,23*).

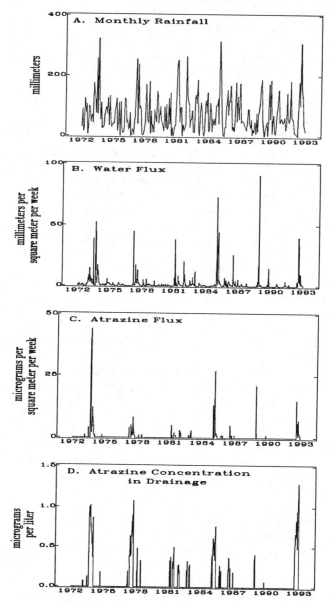

Figure 4. Monthly rainfall at the study area and baseline simulationsof subsurface drainage flux, atrazine mass flux, and atrazine concentrations in drainage, 1972-94.

The field-study results shown in Figure 3 and in Eckhardt *et al.* (*14*) support the application of the Darcian unsaturated-flow and advective-dispersive transport equations in this soil, mainly because the soil is uniformly textured and lacks significant layering or macroporosity. Accordingly, after calibration adjustments were made, the numerical model adequately simulated the water and solute contents that were measured in 1991. This growing season, however, was hotter and dryer than average for the Kansas River valley. Therefore, model simulations were extended to predict the transport and fate of atrazine in other years in which the potential was greater for transport and leaching.

Predictions of Atrazine Concentrations in Drainage. Predicted weekly values for total water flux, dissolved-atrazine flux, and dissolved atrazine concentrations in subsurface drainage past the 1-m depth were compared to evaluate the effects of the timing and rates of rain-water infiltration on the potential for downward transport of atrazine for 1972 through 1993. The monthly rainfall for this period and the baseline simulation (Table I) is shown in Figure 4 for water flux, atrazine flux, and atrazine concentrations in drainage. The simulation results indicate that heavy rains (sometimes exceeding 100 mm/d) soon after spring application of atrazine can produce rapid infiltration and downward leaching of atrazine in this soil. When atrazine moves deeper than the active rooting zone of the growing crop, its degradation rate sharply decreases (*24, 25*). Accordingly, atrazine residues that escape the root zone are susceptible to transport into ground water when excess soil water drains, typically after the growing season when evapotranspiration rates are low.

The baseline simulation shows an increased flux of drainage water after heavy rain that fell soon after atrazine application in 1973, 1977, 1981, 1985, 1989, and 1993, but little or no drainage in the other 16 years. In the absence of significant spring rain, most of the atrazine remained in the root zone and degraded. Simulations for the wettest years (1973, 1977, 1985, and 1993) indicate that heavy rainfall soon after atrazine application increased the flux of atrazine past the rooting zone, where it persisted until it was removed from the profile by drainage. Heavy rains in 1993 continued from the spring atrazine application through the summer and produced the 22-year maximum estimated concentration of 1.3 µg/L in drainage; this result supports evidence (*26*) that 1993 flooding caused increases in the frequency of herbicide detection in shallow ground water. Conversely, heavy rainfall in 1989 did not occur until several months after atrazine application and resulted in estimated concentrations of less than 0.5 µg/L, even though the drainage rates were the largest for the simulation period. The lower atrazine concentrations in the 1989 simulation than in the wetter-year simulations resulted from degradation during a relatively long residence time of residues in the root zone, followed by dilution by downward flowing water.

These results indicate that the potential for atrazine leaching in this soil is largely controlled by the timing of atrazine application in relation to periods of rain exceeding 20 cm/mo, when downward flux of water and atrazine can result. This supports the concept (*2,27*) that herbicide fate is strongly related to the downward water flow, particularly in soils with low organic-carbon content. Furthermore, it demonstrates that wet weather soon after the spring application of atrazine may have a greater effect on atrazine transport than the hydraulic properties of the soil.

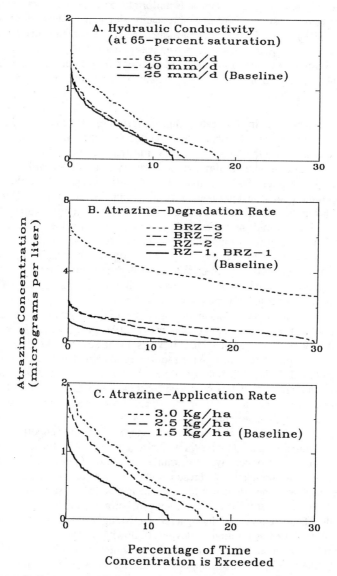

Figure 5. Concentrations of atrazine in drainage that exceed specified levels for differing values of unsaturated hydraulic conductivity, atrazine degradation rates, and atrazine application rates.

Effects of Uncertainty in Model Parameters. Uncertainty in model-input values were evaluated through time-series output for multiple 22-yr simulations. The outputs were used to derive separate cumulative-density distributions that describe the frequency that atrazine concentrations would exceed a given value for the selected levels of model inputs. Three sources of uncertainty were evaluated in these simulations: [1] spatial variability in unsaturated hydraulic conductivity, [2] uncertainty in atrazine-degradation rate, and [3] inconsistency in atrazine-application rates.

The first uncertainty factor — the soil's hydraulic conductivity — ranges over 2 orders of magnitude, from 2.3 to 104 mm/d, at 65 percent of saturation (*12*). Results of the frequency analysis (Figure 5a) suggest that plots with highly permeable soil allow high atrazine concentrations in drainage water, but these concentrations occur infrequently and represent only a small area within the study plots. The concentration-exceedance frequencies (Figure 5a) indicate that [1] atrazine concentrations in drainage water can exceed 1 µg/L in areas where the hydraulic conductivity exceeds 65 mm/d (upper 10 percent of the hydraulic-conductivity measurements, and [2] this concentration is exceeded less than 3 percent of the time.

The second uncertainty factor — atrazine-degradation rate — was tested over a range of degradation rates for the root zone and for depths below it. These rates decrease with depth (*24, 25*), mainly because organic carbon content decreases as root density decreases with depth. Degradation rates for depths below the rooting zone are often unknown because atrazine reaches these depths only infrequently. Atrazine degradation in soil below active rooting zones in some Midwestern soils may be negligible (*25*), which would result in conservative transport of any atrazine residue that escapes the root zone.

Baseline rates for atrazine degradation in this study are shown in Figure 6 and Table I. The baseline first-order degradation rates in the rooting zone (RZ-1) were based on dissipation rates that were measured within the upper 30 cm of soil and estimated values (*20*) for depths between 30 and 50 cm (bottom of the root zone). The baseline first-order degradation rate for depths below the rooting zone (BRZ-1), was assigned a constant value of 0.00575 d^{-1} (120-d half life).

For comparison with the above baseline estimates, decreased degradation rates representing longer residence times in the rooting zone were used (RZ-2 in Figure 6). Two estimates for below the rooting zone were also evaluated — 0.00192 d^{-1} (360-d half-life, BRZ-2) and 0.0 d^{-1} (infinite half life, BRZ-3).

The results in Figure 5b indicate that the estimated range in atrazine-degradation rate has a greater effect on predicted atrazine concentrations than does spatial variability in hydraulic conductivity (compare Figures 5a and 5b). For the decreased degradation rate in the rooting zone (RZ-2), a concentration of 1 µg/L in drainage water is exceeded 3 percent of the time, and for a half life of 360 days (BRZ-2) below the rooting zone, 1 µg/L is exceeded 15 percent of the time. When negligible degradation (conservative transport) was assumed for depths below the rooting zone (BRZ-3), a concentration of 3 µg/L is exceeded more than 30 percent of the time. This observation confirms the need for definitive estimates of atrazine-degradation rates below the root zone. It also confirms the importance of farming and irrigation practices that allow atrazine to remain within the rooting zone.

The third uncertainty factor -- the rate of atrazine application -- also has a significant effect on leaching potential (Figure 5c). For annual application rates of 1.5 kg/ha

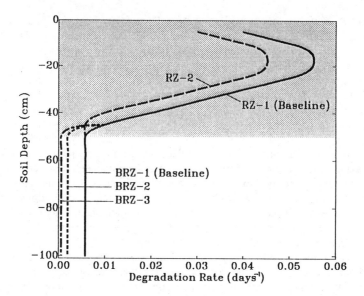

Figure 6. The expected range in atrazine degradation rates within the root zone (RZ, shaded), and estimates of its variability below the root zone (BRZ). [Note: Degradation rate $(d^{-1}) = 0.693/(half life).$]

(baseline), the estimated frequency of concentrations that exceed 1 µg/L is negligible. When the application rate is doubled to 3.0 kg/ha, the frequency of concentrations that exceed 1 µg/L increases to 7 percent. Thus, proper management of atrazine-application rate and timing is a critical factor in reducing the potential for atrazine contamination of ground water, especially in field areas that have the highest hydraulic conductivity.

The frequency analysis demonstrates that the leaching potential at the Kansas site is low, mainly because the degradation rate for atrazine in the root zone is rapid and because the hydraulic conductivity is low relative to other Midwestern soils. Atrazine can escape the root zone at this site, however, and the simulation results show that periodic leaching is most probable when heavy spring rain coincides with atrazine presence in the soil. Atrazine concentrations in leachate would be diminished by degradation, dispersion, and dilution as water moves downward through underlying deposits. Accordingly, the concentrations shown in Figure 5 are at least an order of magnitude greater than those observed in ground water in the Kansas River valley.

Summary and Conclusions

This study evaluated several components that affect estimates of the leaching potential for atrazine in soil:
(1) The spatial variability of hydraulic and chemical properties of the field soil, determined through water-redistribution experiments and soil-core analyses;

(2) The spatial and temporal distribution of atrazine, its metabolites, a bromide tracer, and water, measured at specified soil depths during corn growth;

(3) Mechanistic representations of unsaturated flow and solute transport, evaluated through comparisons of model simulations with field measurements.

The field study and model provided a basis for ranking the relative importance of transport factors, including boundary conditions and the physical and chemical properties of the soil and atrazine, in estimating the potential for atrazine transport through a silt-loam soil to ground water. The relative effects on atrazine transport of uncertainty in the transport factors clearly show that the boundary conditions — the time and volume of rainfall in relation to the time and rate of atrazine application — affect atrazine concentrations in subsurface drainage water. The modeling results confirm a need for accurate estimates of atrazine-degradation rates below the root zone, especially in regions where soil properties and climate can create a high leaching potential. The analysis further indicates that periodic leaching of atrazine can be expected for this soil when rainfall exceeds 20 cm/mo during the growing season.

The close correlation between periodic heavy rain at the time of atrazine application and the concentration of atrazine in drainage water demonstrates the benefit of incorporating a long-term record of precipitation in any evaluation of herbicide-transport potential. The approach described here provided useful comparisons of the relative effects of uncertainties in [1] estimates of transport parameters representing the unsaturated hydraulic conductivity of soil and the atrazine-degradation rates, and [2] boundary conditions for water and atrazine application rates on leaching potential. The modeling approach could have further use in regional comparisons of herbicide-transport potential in areas in which the soil type, agricultural practices, and precipitation patterns differ from those studied in the Kansas River Valley.

Acknowledgments

The authors thank P.L. Barnes and J.L. Hutson for their field assistance and modeling knowledge. We also thank S.E. Ragone, G.E. Mallard, and E. M. Thurman for support and encouragement. The report benefited greatly through comments by S.C. Komor, D.A. Haith, and two anonymous reviewers.

Literature Cited

1. Gianessi, L.P.; Puffer, Cynthia. *Herbicide use in the United States*; Resources for the Future, Quality of the Environment Division: Washington, DC, 1991; 128 pp.

2. Flury, M. *J. Environ. Qual.* **1996**, *vol. 25*, pp. 25-45.

3. Biggar, J.W.; Nielsen, D.R. *Wat. Res. Res.* **1976**, *vol. 12*, pp. 78-84.

4. Rao, P.S.C.; Wagenet, R.J. *Weed Sci.* **1985**, *vol. 33 (Suppl. 2)*, pp. 18-24.

5. Russo, D. *Soil Sci. Soc. Am. J.* **1981**, *vol. 45*, pp. 675-681.

6. Jury, W.A.; Gruber, J. *Wat. Res. Res.* **1989**, *vol. 25*, pp. 2465-2474.

7. Wagenet, R.J.; Rao, P.S.C. In *Pesticides in the soil environment: Processes, impacts, and modeling;* Cheng, H.H., Ed.; Soil Science Society of America: Madison, WI, 1990; pp. 351-399.

8. Addiscott, T.M.; Wagenet, R.J. J. Soil Sci. **1985**, *vol. 36*, pp. 411-424.

9. Carsel, R.F.; Parrish, R.S.; Jones, R.L.; Hansen, J.L.; Lamb, J.L. *J. Contam. Hydrol.* **1988**, *vol. 2*, pp. 111-124.

10. Woodbury, A.; Render, F.; Ulrych, T. *Ground Water,* **1995**, *vol. 33,* pp. 532-538.

11. Hutson, J.L.; Wagenet, R.J. *LEACHM -- Leaching estimation and chemistry model*; Cornell University: Ithaca, NY, 1992; SCAS Vol. 92-3.

12. Eckhardt, D.A.; Barnes, P.L. *U.S. Geological Survey Water Resources Investigations Report 91-4034*, **1991**, pp. 210-213.

13. Klute, A. In *Methods of soil analysis, 2nd ed.*; Klute, A., Ed.; American Society of Agronomy: Madison, WI, **1986**; pp. 635-662.

14. Eckhardt, D.A.; Wagenet, R.J.; Thurman, E.M.; Barnes, P.L. *U.S. Geological Survey Water Resources Investigations Report 94-4014*, **1994**.

15. Mills, M.S.; Thurman, E.M. *Anal. Chem.* **1992**, *vol. 64*, pp. 1985-1990.

16. Meyer, M.T.; Mills, M.S.; Thurman, E.M. *J. Chromatog.* **1993**, *vol. 629*, pp. 55-59.

17. Nielsen, D.R.; van Genuchten, M. Th.; Biggar, J.W. *Wat. Res. Res.* **1986**, *vol. 22*, pp. 89S-108S.

18. Davidson, J.M.; Graetz, P.S.; Rao, P.S.C.; and Selim, H.M. *U.S. Environmental Protection Agency Report 600/3-78-029*, **1978**.

19. van Bavel, C.H.M. *Water Res. Res.* **1966**, *vol. 2*, pp. 455-467.

20. Wauchope, R.D.; Buttler, T.M.; Hornsby, A.G.; Beckers, P.W.M. A.; Burt, J.P. *The SCS/ARS/CES pesticide properties database for environmental decision making*; Reviews of Environmental Toxicology; Springer-Verlag: New York, NY, 1992; Vol. 123, 164 pp.

21. Adams, C.D.; Thurman, E.M. *J. Environ. Qual.* **1991**, *vol. 20*, pp. 540-547.

22. Green, R.E.; Karickhoff, S.W. In *Pesticides in the soil environment: Processes, impacts, and modeling;* Cheng, H.H., Ed.; Soil Science Society of America: Madison, WI, 1990; pp. 79-101.

23. Novak, J.M.; Jayachandran, K.; Moorman, T.B.; Weber, J.B. In *Bioremediation: Science and Applications*; Skipper, H.D.; Turco, R.F., Eds.; Special Pub. No. 43; Soil Science Society of America: Madison, WI, 1995; pp. 13-31.

24. Lavy, T.L.; Roeth, F.W.; Fenster, C.R. *J. Environ. Qual.* **1973**, *vol. 2*, pp. 132-136.

25. Sinclair, J.L.; Lee, T.R. *U.S. Environmental Protection Agency Environmental Research Brief EPA/600/S-92/001*, **1992**.

26. Kolpin, D.W.; Thurman, E.M. *U.S. Geological Survey Circular 1120-G*, **1995**, 20 pp.

27. Hutson, J.L.; Wagenet, R.J.; Troiano, J.J. In *Proceedings of the International Conference and Workshop on the Validation of Flow and Transport Models for the Unsaturated Zone*; Wierenga, P.J.; Dominique Bachelet, Eds.; New Mexico State University: University Park, NM, 1988; pp. 158-166.

Chapter 10

The Effect of Ammonia on Atrazine Sorption and Transport

S. A. Clay[1], D. E. Clay[1], Z. Liu[1], and S. S. Harper[2]

[1]Department of Plant Science, South Dakota State University, Brookings, SD 57007
[2]Land and Water Sciences, Tennessee Valley Authority, Muscle Shoals, AL 35660

Ammonia application increased soil solution pH and decreased atrazine sorption to soil. K_f values for atrazine were 5.2 in nonamended soil (pH = 5.7) and 2.9 and 2.2 for ammonia-amended soils at pH 7.8 and pH 8.9, respectively. Atrazine sorption was affected differentially by base treatments of KOH and NaOH compared to NH$_4$OH. These results suggest that factors other than pH modification influenced atrazine sorption characteristics. Incubation of atrazine in NH$_4$OH extracts from soil indicated that hydroxyatrazine (HA) could be formed, however, in the presence of soil, HA was not detected. Partition coefficients for HA at pH values of 5.7, 7.8, and 8.9 were 130, 0.7, and 0.1, respectively. Ammonia application increased the potential of atrazine to leach through soil.

The prediction of herbicide and degradate mobility through soil is important in assessing the vulnerability of aquifers to chemical contamination. Herbicide mobility is inversely related to herbicide sorption values due to the soil retardation of the mass flux of the herbicide (*1, 2, 3*). Therefore, as sorption to soil decreases, the potential for herbicide movement increases.

Herbicide sorption to soil is often measured by laboratory batch equilibration techniques. The partition coefficient, K_D, quantifies sorption and is the quotient of the concentration in the solid soil phase divided by the concentration in the liquid.

Fertilizer applications may influence the sorption characteristics of ionizable herbicides because of increased or decreased soil pH or competition for exchange sites (*4, 5*). Sorption of some ionizable herbicides are not strongly influenced by soil pH and

0097–6156/96/0630–0117$15.00/0

therefore, not affected by fertilizer applications. For example, dicamba, a benzoic acid herbicide, had low K_D values (<0.2) at soil pH values from 4.1 to 6.2 and sorption was not affected by fertilizer treatment (4). However, the K_D of atrazine, a weakly basic herbicide, increased from 2.9 to 5.3 when a Waukegan soil was treated with ammonium sulfate. Atrazine sorption to a Plano soil also increased from 2.9 to 3.6 when soil was treated with elemental S (4). The increase of atrazine sorption to both soils was due to the decrease in soil pH from about 6 to about 4.5 caused by the application of the acidic fertilizers, ammonium sulfate or elemental S. Direct competition between herbicides and fertilizers for exchange sites also may occur (5). For example, sorption of 2,4-D, an anionic herbicide, decreased when phosphate fertilizer was applied to soil prior to 2,4-D application (5).

Fertilizers that temporarily increase soil pH include ammonia-based fertilizers such as anhydrous ammonia, aqua ammonia, and urea. These fertilizers quickly raise soil pH to 9 or greater (6, 7, 8). High soil pH may enhance chemical hydrolysis of atrazine to hydroxyatrazine (9), and/or change atrazine sorption and movement characteristics. Atrazine is sorbed to a lesser extent in high pH than low pH soils (10, 11) and has greater desorption from high than low pH soils (12).

Ammonia-based fertilizers are often used as N sources for corn and applied in the spring or early summer at about the same time as atrazine. Interactions between fertilizer and herbicides are obvious when applied to the same area with surface applications. However, interactions can occur even if these agrichemicals are physically separated at application. For example, when anhydrous ammonia was injected with fertilizer shanks 15 to 20 cm below the soil surface, subsurface ammonia bands were created and the unpacked fertilizer slot acted as a macropore that funneled water and herbicides to the fertilizer band (13). Atrazine movement through soil increased in areas directly below the ammonia band.

The objectives of the laboratory work reported here were to study atrazine adsorption/desorption, hydrolysis, and movement through soils under high pH conditions. Therefore, three bases, NH_4OH, $NaOH$, and KOH, were used as model systems to compare and contrast the effects of basic treatments on atrazine sorption to soil.

Materials and Methods

Atrazine Sorption. Atrazine sorption was determined on a Brandt silty clay loam. The Brandt soil had a particle size analysis of 16.5% sand, 56.2% silt, and 27.3% clay. Organic matter content was 2.4% and soil pH (determined in a 1:1 soil/0.01 $\underline{M}$ $CaCl_2$ w/v slurry) was 5.7.

The sorption experiments were conducted using batch equilibration techniques. Ten g soil and 5 mL of 0.01 $\underline{M}$ $CaCl_2$ or 0.01 $\underline{M}$ $CaCl_2$ containing NH_4OH solution at 1400 or 2800 ug of N g^{-1} soil was mixed at time 0 in 25-mL glass centrifuge tubes. Five ml of atrazine solution (spiked with ^{14}C-atrazine) containing 0, 2.1, 6.6, or 51 umol of atrazine L^{-1} in 0.01 $\underline{M}$ $CaCl_2$ were added immediately or 8 days after NH_4OH application.

To determine if only pH controlled atrazine sorption, 0.1 $\underline{M}$ NaOH or KOH in combination with 0.01 $\underline{M}$ CaCl$_2$ were added to soil to achieve similar solution pH values as the NH$_4$OH treatments. After shaking at 4 excursions s^{-1} for a 1-day equilibration period at 25 C, the slurry was centrifuged, and an aliquot was removed for herbicide, degradate, dissolved organic carbon (DOC) and pH analyses. The amount of ^{14}C remaining in supernatant was determined by liquid scintillation counting (LSC) techniques. About 100 uL of supernatant was spotted on thin layer chromatography (TLC) plates and developed with 1-butanol/acetic acid/water (11:5:4 v/v/v). Autoradiography was used to determine the distribution of radioactivity on the TLC plate and quantified by scraping bands of radioactivity and using LSC techniques. DOC content was determined by diluting 1 mL of supernatant with 20 mL of distilled, deionized water and using a Dohrmann DC-180 Carbon Analyzer.

Atrazine Desorption. Atrazine desorption from soil was determined from the soil slurry that was spiked initially with 51 umol atrazine L^{-1} by replacing 4 mL of the 1-day batch equilibration supernatant with 4 mL of the original pH adjusting solution that did not contain atrazine. The soil slurry was shaken for one day, centrifuged, 4 mL of supernatant removed, and once again replaced with 4 mL of original pH adjusting solution. The desorption equilibration was conducted over 5 consecutive days.

Atrazine Sorption/Desorption Data Analysis. The amount of atrazine sorbed was the difference between initial solution and final solution values. The linear form of the Freundlich equation

$$\log (C_s) = \log K_f + 1/n \log (C_e)$$

was used to describe atrazine sorption and desorption. In the Freundlich equation, C_s is micromoles of herbicide sorbed per kilogram of soil, C_e is micromoles of herbicide per liter of supernatant solution after equilibration, and K_f and $1/n$ are empirical constants.

Hydroxyatrazine Sorption. Hydroxyatrazine (HA) sorption to the Brandt soil was determined in a 1-day batch equilibrium study using ^{14}C-hydroxyatrazine. Ten mL of solution containing 9.1 umol HA L^{-1} in 0.01 $\underline{M}$ CaCl$_2$ alone or with NH$_4$OH solution at concentrations similar to those reported above was added to 10 g of Brandt soil. Radioactivity in solution was determined prior to and after batch equilibration by LSC techniques. The partition coefficient (K_D) between soil and solution was calculated.

Soil Extract Effects on Atrazine. The effect of soil solution extracted with NH$_4$OH or KOH on atrazine also was determined in the absence of soil. Soil was extracted with 0.01 $\underline{M}$ CaCl$_2$ alone, containing 5200 or 1300 ug NH$_4$OH-N g soil^{-1}, or using 0.01 $\underline{M}$ CaCl$_2$ in combination with KOH to obtain pH values similar to those obtained with NH$_4$OH. Atrazine (containing ^{14}C-atrazine) was added to the extract and incubated in the dark at 25 C for 2 h to 8 days. After incubation, an aliquot was spotted on TLC plates and developed and bands quantified as reported above.

Atrazine Movement Through Ammonia-Amended Soils. In a laboratory study, the influence of NH_4OH on atrazine movement through soil was measured in intact soil columns of the Brandt soil collected in August from noncompacted rows of corn fields that had been moldboard plowed in the spring. The spring moldboard plowing disrupts previously formed channels that may have been created by insects, earthworms, and roots. Ammonia and atrazine (containing ^{14}C-atrazine) or $CaCl_2$ and atrazine were applied to the surface of soil columns (15 cm diam. by 15 cm deep). One day after chemical application, the columns were leached with 5.4 L of water (3 pore volumes) over a 2 hr period. Leachate was collected in 50-ml aliquots and, after allowing the column to drain, soil was removed in 2-cm increments. [14]C in each depth was determined by LSC techniques.

Results and Discussion

Atrazine Adsorption. The supernatant pH after one day of equilibration with 0.01 $\underline{M}$ $CaCl_2$ was about 5.7 (Table I). The addition of 1400 and 2800 ug NH_4OH-N g^{-1} raised soil solution pH to about 7.8 (medium pH) and 8.9 (high pH), respectively.

Table I. Solution pH and DOC Content (mg L^{-1}) 1 or 8 Days After Base Addition

Treatment	Day 1 pH	Day 1 DOC	Day 8 pH	Day 8 DOC
$CaCl_2$	5.74 (0.01)[a]	76 (2)	5.69 (0.01)	30 (1)
Medium pH				
NH_4OH	7.81 (0.01)	144 (2)	7.70 (0.01)	207 (4)
KOH	7.97 (0.01)	144 (2)	7.02 (0.06)	156 (21)
NaOH	8.19 (0.04)	261 (6)	7.56 (0.01)	329 (17)
High pH				
NH_4OH	8.92 (0.03)	390 (6)	8.07 (0.37)	720 (5)
KOH	8.68 (0.05)	360 (42)	7.54 (0.07)	610 (3)
NaOH	8.93 (0.05)	2640 (244)	8.69 (0.04)	1770 (78)

[a] Numbers in parentheses are the standard deviation of the mean; n=6.

Supernatant DOC concentration in $CaCl_2$ treatments after one day of equilibration was about 76 mg C L^{-1} (Table I). The amount of DOC solubilized in the medium pH treatment after one day equilibration was similar with NH_4OH or KOH. When NaOH was added, DOC solubilization was twice that of the other bases. In the high pH treatment after one day of equilibration, NH_4OH and KOH again solubilized similar amounts of organic C. However, NaOH solubilized 7 times more than the other bases.

The supernatant pH values after Day 8 were slightly lower than the respective Day 1 values. The DOC content in the CaCl$_2$ supernatant decreased after 8 days. In the medium pH treatment, DOC concentrations increased from Day 1 to Day 8. In the high pH treatments, DOC concentrations increased from Day 1 to Day 8 when NH$_4$OH and KOH were used and decreased in the NaOH treatment. The decrease in solution pH after the 8 day incubation may have been due to soil buffering capacity, an increase in solution acidity due to increased DOC content, or both.

Generally, base addition at day 1 decreased atrazine sorption to soil (Table II). At the medium pH, atrazine sorption was similar between all base treatments. K$_f$ values decreased from 5.2 in the CaCl$_2$ treatment to about 3 at the medium pH. At the high pH, atrazine sorption using NH$_4$OH or KOH was similar, with K$_f$ values of about 2.3. However, the NaOH treatment had a K$_f$ sorption value that was similar to the CaCl$_2$ treatment. Atrazine K$_f$ sorption values decreased from Day 1 to Day 8 in the CaCl$_2$ control but were similar in base treated samples, with the exception of high pH NaOH treatment.

Table II. Atrazine sorption isotherm coefficients at Day 1 and Day 8

	Day 1		Day 8	
Treatment	K$_{f\,ads}$[a]	1/n$_{ads}$[b]	K$_{f\,ads}$	1/n$_{ads}$
CaCl$_2$	5.16$_{(4.89-5.44)}$	0.86	3.53$_{(3.12-3.98)}$	0.85
Medium pH				
NH$_4$OH	2.89$_{(2.43-3.44)}$	0.86	2.79$_{(2.56-3.04)}$	0.85
KOH	3.32$_{(3.06-3.61)}$	0.84	2.84$_{(2.56-3.16)}$	0.85
NaOH	3.10$_{(2.65-3.61)}$	0.84	2.54$_{(2.27-2.86)}$	0.85
High pH				
NH$_4$OH	2.15$_{(1.97-2.35)}$	0.88	2.57$_{(2.18-3.02)}$	0.86
KOH	2.56$_{(2.28-2.87)}$	0.90	2.56$_{(1.83-2.49)}$	0.89
NaOH	5.46$_{(4.36-5.84)}$	0.88	2.34$_{(1.86-2.95)}$	0.87

[a] Units of K$_f$ are umol$^{1-1/n}$ L$^{1/n}$ kg^{-1}. Numbers in parentheses are the 95% confidence interval (CI) for K$_f$; antilogs of log K$_f$ - CI log K$_f$ and log K$_f$ + CI log K$_f$.
[b] The standard deviation of the slope (1/n) of the best fit line for each treatment was < 0.01 and r^2 = 0.99 for all regression lines.

These data show that pH alone did not control atrazine sorption. If pH alone controlled sorption, then as solution pH decreased from Day 1 to Day 8, atrazine sorption would have been expected to increase. However, this was not observed. Therefore, changes in sorption must have been a function of several factors including solution pH, reactivity of solubilized organic pools, solution ionic strength, and equilibration time.

The amount of DOC solubilized from soil by the medium concentration of base did not influence atrazine sorption. These results are in agreement with those of Clay et al. (*14*) and Pennington et al. (*15*) who reported that atrazine sorption to soil was not affected by the presence of low levels of DOC in solution. However, at the high pH, soil treated with NaOH exhibited increased atrazine sorption, and may be explained by NaOH dispersing soil and exposing more soil interlayer surface areas and exchange sites for atrazine sorption.

Atrazine Desorption. Base addition increased atrazine desorption as indicated by lower K_f values and higher $1/n$ values compared to $CaCl_2$ treatment (Table III). The K_f and $1/n$ values for the high pH NaOH treatment were not determined because of soil dispersion.

The desorption isotherm of atrazine at the medium pH after one day equilibration with base was similar to the adsorption isotherm (ie the $1/n$ values for sorption and desorption were equal). This implies that very little, if any, of the sorbed atrazine was irreversibly bound to soil. However, at Day 8, hysteresis was observed in all treatments except when soil was raised to the high pH with KOH.

Table III. Atrazine desorption isotherm coefficients at Day 1 and Day 8

Treatment	Day 1		Day 8	
	$K_{f\,des}$[a]	$1/n_{des}$[b]	$K_{f\,des}$	$1/n_{des}$
$CaCl_2$	$11.86_{(10.74-11.40)}$	0.42	$11.12_{(10.60-11.67)}$	0.27
Medium pH				
NH_4OH	$3.69_{(3.29-4.31)}$	0.75	$4.62_{(4.43-4.81)}$	0.61
KOH	$2.10_{(1.97-2.24)}$	1.10	$4.39_{(3.96-4.85)}$	0.67
NaOH	$2.17_{(1.93-2.44)}$	1.04	$4.96_{(4.70-5.23)}$	0.56
High pH				
NH_4OH	$4.00_{(2.85-4.15)}$	0.61	$5.55_{(4.94-6.23)}$	0.49
KOH	$0.95_{(0.88-1.02)}$	1.37	$2.07_{(1.91-2.25)}$	0.92
NaOH	Nd	Nd	Nd	Nd

[a] Units of K_f are $umol^{1-1/n}\,L^{1/n}\,kg^{-1}$. Numbers in parentheses are the 95% confidence interval (CI) for K_f; antilogs of log K_f - CI log K_f and log K_f + CI log K_f.
[b] The standard deviation of the slope $(1/n)$ for the best fit line was < 0.08 and $r^2 > 0.81$ for all regression lines.

The adsorption and desorption data suggest two possibilities when investigating the influence of base amendments on atrazine movement in soil. The first possibility is that atrazine mobility will be decreased because there is a higher proportion of the applied atrazine irreversibly bound to soil. However, base treatments may increase atrazine mobility due to increased desorption from soil.

Solution Effects on Atrazine.   The alkalinity of the NH_4OH and KOH sorption solutions used in these studies may have catalyzed the hydrolysis of atrazine to HA (9). TLC analyses of sorption and desorption solutions extracted from soil did not contain any detectable levels of HA. However, HA has been shown to have strong affinity for soil (12, 16) with $K_{f\,ads}$ values of 26 (pH 6) to 60 (pH 4.5) and no desorption detected at either pH (11). Therefore, HA may not have been present in solution even if formed due to strong adsorption to soil. Another possibility is that HA may have been in sorption solutions but at concentrations below detection limits. Therefore, the effects of NH_4OH and KOH soil solution extracts on atrazine in the absence of soil were investigated. Only parent atrazine was detected by TLC techniques (R_f value = 0.8) after a 1-day incubation in soil solution extracted with $CaCl_2$. However, when atrazine was incubated for 8 days in solutions extracted with NH_4OH or KOH, about 10 to 20% of the ^{14}C was determined to be HA (R_f value = 0.53).

Hydroxyatrazine Adsorption. The results of the above study indicate that HA may form in alkaline solution extracts and that amounts may be quite high if incubation time is long enough. HA sorption to soil decreases as pH increases (*11*), however, sorption of HA to high pH soils is not well documented. Therefore, a 1-day batch equilibration study was conducted to determine the sorption of HA to ammonia-treated soils. The HA sorption coefficients for the Brandt soil at pH 5.7 was 130. The K_D values for HA sorption using NH_4OH solution were 0.7 at pH 7.9 and 0.1 at pH 9 which indicates that HA sorption was even less than atrazine sorption at these same pH values.

Evaluation of both the incubation and sorption data indicates that HA indeed may form and if formed may be leached due to it's weak sorption to soil at high pH. HA most likely was not detected in the basic sorption solutions because it was not formed during one-day batch equilibration with atrazine. HA was probably not detected in desorption solutions (even if formed over the 5-day desorption period) due to dilution with "clean" desorption solutions and the low amounts that would have been formed.

Atrazine Movement Through Ammonia-Amended Soil. A study was conducted to determine if decreased sorption and increased desorption of atrazine due to NH_4OH addition increased the leaching potential of atrazine. About 23% of the applied ^{14}C was leached from unfertilized soil columns compared to 33% leached from ammonia-treated columns. The leachate did not contain detectable levels of HA. The amount of ^{14}C in the surface 2 cm of soil from ammonia-treated columns was 11% less than the amount measured in the untreated columns.

Conclusions

The results of this study indicate that the application of ammonia fertilizer changed soil chemical characteristics, including increasing solution pH and DOC content, that influenced atrazine sorption, desorption, and movement. However, comparison with other basic treatments indicate that pH modifications do not influence atrazine sorption characteristics equally. The increase in soil pH generally decreased atrazine sorption and increased desorption isotherms. Hydroxyatrazine formed from atrazine in soil solutions extracted with NH_4OH. Although not detected in sorption solution extracts, sorption of hydroxyatrazine at pH 7.9 and 9 was very low. Ammonia application also increased the amount of atrazine leached through soil columns. These data suggest that ammonia-based fertilizers that increase soil pH have the potential to increase atrazine movement in the soil. Field research has shown that DOC and atrazine movement below NH_4OH bands was much greater than in areas where NH_4OH bands were not present (*13, 17*). Two mechanisms are working in concert to increase atrazine movement below fertilizer bands. These mechanisms are: i) the effect of the fertilizer slot on water flow; and ii) the effect of increased pH on atrazine sorption. The leaching study shows that soil changes induced by ammonia application are sufficient by themselves to increase atrazine movement through soil.

Acknowledgments

Partial support for this projects was provided by Tennessee Valley Authority and USDA-CSRS. Mention of products in this publication does not represent an endorsement of the authors or of the South Dakota State Experiment Station or criticism of similar ones not mentioned.

Literature Cited

1. Helling, C.S. *Soil Sci. Soc. Am. Proc.*, **1971**, *35*, 737-743.
2. Scheunert, I. *In Fate and Prediction of Environmental Chemicals in Soils, Plants, and Aquatic Systems.* Mansour, M., Ed.; Lewis Publishers: Boca Raton, FL, **1993**, pp. 1-22.
3. Weber, J. *In Agrochemical Environmental Fate: State of the Art.* Leng, M.L.; Leovey, E.M.K.; Zubkoff, P.L., Eds.; Lewis Publishers: Boca Raton, FL, **1995**, pp. 99-116.
4. Clay, S.A.; Koskinen, W.C.; Allmaras, R.R.; Dowdy, R.H. *J. Environ. Sci. Health* **1988**, *B23*, 559-573.
5. Madrid, L.; Morillo, E.; Diaz-Barrientos, E. *In Fate and Prediction of Environmental Chemicals in Soils, Plants, and Aquatic Systems.* Mansour, M., Ed.; Lewis Publishers: Boca Raton, FL, **1993**, pp. 51-59.
6. Kissel, D.E.; Cabrera, M.L.; Ferguson, R.B. *Soil Sci. Soc. Am. J.*, **1988**, *52*, 1753-1796.
7. Norman, R.J.; Kurtz, L.T.;, Stevenson, F.J. *Soil Sci. Soc. Am. J.*, **1987**, *51*, 809-812.
8. Liu, Z.; Clay, S.A.; Clay, D.E.; Harper, S.S. *J. Agric. Food Chem.*, **1995**, *43*, 815-819.
9. Esser, H.O; Dupuis, E.; Ebert, E.; Marco, G.J.; Vogel, C. *In Herbicides - Chemistry, Degradation, and Mode of Action.* Kearney, P.C.; Kaufmann, D.D., Eds., Marcel Dekker: New York, **1975**, Vol 1; pp 129-208.
10. McGlamery, M.D.; Slife, F.W. *Weeds*, **1966**, *14*, 237-239.
11. Goetz, A.J.; Walker, R.H.; Wehtje, G.; Hajek, B.F. *Weed Sci.*, **1988**, *37*, 428-433.
12. Clay, S.A.; Koskinen, W.C. *Weed Sci.*, **1990**, *38*, 262-266.
13. Clay, S.A.; Scholes, K.A.; Clay, D.E. *Weed Sci.* **1994**, *42*, 86-91.
14. Clay, S.A.; Allmaras, R.R.; Koskinen, W.C.; Wyse, D.L. *J. Environ. Qual.*, **1988**, *17*, 719-723.
15. Pennington, K.L.; Harper, S.S.; Koskinen, W.C. *Weed Sci.*, **1991**, *39*, 667-672.
16. Schiavon, M. *Ecotox. Environ. Safety*, **1988**, *15*, 46-54.
17. Clay, D.E.; Clay, S.A.; Liu, Z., Harper, S.S. *Biol. Fertil. Soils*, **1995**, *19*, 10-14.

Chapter 11

Fate of a Symmetric and an Asymmetric Triazine Herbicide in Silt Loam Soils

W. C. Koskinen[1], J. S. Conn[2], and B. A. Sorenson[3,4]

[1]Department of Soil, Water, and Climate, Agricultural Research Service,
U.S. Department of Agriculture, University of Minnesota,
1991 Upper Buford Circle, St. Paul, MN 55108
[2]Agricultural Research Service, U.S. Department of Agriculture,
University of Alaska, 309 O'Neil Building, Fairbanks, AK 99775
[3]Department of Agronomy and Plant Genetics, University of Minnesota,
St. Paul, MN 55108

The formation of ^{14}C-atrazine and ^{14}C-metribuzin degradation products and their distribution in the top 90 cm of silt loam soils under field conditions was determined over a 16-month period. The metribuzin experiment was conducted on Tanana (3.8% organic carbon (OC), pH 6.5, 14% clay) and Beales (6.4% OC, pH 4.7, 8% clay) silt loam soils while the atrazine experiment was conducted on a Port Byron silt loam soil (2.4% OC, pH 5.4 and 23% clay). Metribuzin degraded rapidly in both soils: 12% remaining at 1 MAT; 2% remaining by 16 MAT. The maximum amount of metabolites desamino-, diketo-, or desaminodiketo-metribuzin found was 4%; most of the ^{14}C remaining was in the form of unextractable bound residues. The majority of the metribuzin residue remained in the surface soil. Some metribuzin residues leached to 45 cm: 0.1 and 0.3 % of the applied ^{14}C remained in the 6.4 and 3.8% OC soils, respectively. In the atrazine experiment, 68% of the applied ^{14}C was present 16 months after treatment (MAT). Radioactivity moved to 30-40 cm by 1 MAT. Atrazine accounted for 20% of the ^{14}C applied 16 MAT, and was the predominant ^{14}C-compound in soil below 10 cm through 16 MAT. Hydroxyatrazine (HA) (13% of the ^{14}C present 1 MAT) was the major degradation product in the top 10 cm of soil. Predominant degradation products at depths greater than 10 cm were HA and deethylatrazine. Deisopropylatrazine and deethyldeisopropylatrazine were also detected 1 MAT.

[4]Current address: QMAS, P.O. Box 756, Walhalla, ND 58282

In field studies, some herbicides have been shown to degrade slower at high latitudes than at lower latitudes. For instance, Conn and Cameron (*1*) found that 61% of triallate [*S*-(2,3,3-trichloro-2-propenyl)bis(1-methylethyl)carbamothioate] applied was present at the end of the growing season in Alaska, while only 10 to 19% remained 5 mo after application in England (*2*). Conn and Knight (*3*) found that approximately twice as much trifluralin [2,6-dinitro-*N*,*N*-dipropyl-4-(trifluoromethyl)benzenamine] remained at the end of 1 yr in Alaska than in the same time in Saskatchewan (*4*) and Manitoba (*5*) Canada. Metribuzin persisted longer in Alaska (12% remaining after 115 d) (*1*) than at Watkinville, Georgia (*6*) where all metribuzin had degraded within 40 d after application.

Long herbicide persistence in the high latitudes of the subarctic is due to cold soils. Subarctic soils can be frozen for 6 to 7 months and summer soil temperatures at 5 cm and deeper are below 15°C most of the time (*7,8*). Soil temperature plays a major role in determining rates of microbial degradation. For instance, dissipation half-life of metribuzin [4-amino-6-(1,1-dimethyl)-3-(methylthio)-1,2,4-5(4H)-one], which is microbially degraded (*9*), was 16 days at 35°C and 377 days at 5°C (*10*). Degradation in subsurface soils is generally reported to be slower than in surface soils. Temperature and moisture have been shown to affect degradation in subsurface soils (*11*). Decreased microbial populations and activities in subsurface soils compared to surface soils result in a decreased rate of degradation (*9*). Cold climates in the North Central United States can also result in greater herbicide persistence than in milder climates. For instance, atrazine residues appear to be very resistent to degradation following field application. Following field application of ^{14}C-atrazine to a sandy loam (*12*), more than 75% of the radioactivity, of which 27 % was atrazine, was still present 16 mo later.

As herbicide persistence increases, the potential for herbicide leaching also increases. Thus, the potential for herbicide leaching in the subarctic and North Central United States soils is higher than soils in warmer climates at lower latitude locations with similar soil texture and organic carbon content and rainfall. The extent of leaching is also dependent on the nature of the residues and the physical and chemical properties of the soil. For instance, deeper atrazine movement in a silt loam than in a sand, even though OC content of the silt loam was nearly two times higher and had greater atrazine sorption than the sand, was attributed to wetter soil conditions caused by the higher water holding capacity, and slower infiltration rate allowing atrazine more time to desorb and move with the water through the soil (*13*).

No studies have been conducted under subarctic conditions of leaching of pesticides such as metribuzin, or its metabolites, commonly used for weed control in subarctic potato and spring barley production systems. Therefore, research was conducted to determine the extent of metribuzin and metribuzin metabolite leaching in two soils under subarctic conditions near Fairbanks and Delta Junction, Alaska. Studies were also initiated to determine the extent of leaching of major atrazine degradation products in the top 90 cm of a Minnesota silt loam soil over a 16-mo period.

Materials and Methods.

Study Sites. Metribuzin experiments were conducted at the University of Alaska Agricultural and Forestry Experiment Station farms located at Fairbanks and at Delta Junction. The climate at both sites is strongly continental and is characterized by temperature extremes from -50 to 35°C. The mean annual temperature is -2.2°C at Fairbanks and is -3.3°C at Delta Junction. Soils are frozen for 6 to 7 mo each year and when not frozen exhibit a sharp temperature gradient with depth. Temperatures at 15 cm seldom reach 15°C (*8*). The average annual precipitation at Fairbanks and Delta Junction is 298 and 390 mm with over 60% of the precipitation falling as rain. Potential evaporation has been calculated at 466 mm for Fairbanks and 450 mm for Delta Junction (*14*). The Fairbanks soil was Tanana silt loam (loamy, mixed, nonacid Pergelic Cryaquept) and the Delta Junction soil was a Beales silt loam (mixed Typic Cryopsament). Soil particle size, organic carbon, and pH are summarized in Table I.

Table I. Soil physical and chemical properties of Tanana, Beales and Port Byron silt loam soils

Soil	Soil Depth	pH	Organic Carbon	Clay	Silt
	(cm)		----------------------- % ------------------		
Tanana	0-15	6.5	3.8	14	71
	30-45	7.4	0.69	11	63
Beales	0-15	4.7	6.4	8.0	50
	30-45	5.1	0.74	8.0	25
Port Byron	0-10	5.4	2.4	23	68
	10-20	5.9	2.2	24	67
	30-60	6.0	1.3	26	68
	60-90	5.9	0.5	25	64

The atrazine experiment was conducted at the Olmstead County Water Quality Research Plots at the Steve Lawler Farm, Rochester, MN on a Port Byron silt loam (Hapludoll). In contrast to Alaska, southern Minnesota has a mean annual temperature of 6°C and the soils are only frozen for 4 months. The annual precipitation is 640 mm while the annual pan evaporation is 1000 mm (*15*).

Alaska Experiment Design. In 1990, 'Datal' barley (78 kg ha⁻¹) was planted into barley stubble using a no-till drill on May 9 at Delta Junction and on May 11 at Fairbanks. The plots were 4.6 x 9.1 m. Fertilizer was banded between the rows at a rate of 101 kg ha⁻¹ N, 56 kg ha⁻¹ P₂O₅, and 45 kg ha⁻¹ K₂O. Potassium bromide was applied at a rate of 98 kg ha[-1] to the plots at Fairbanks on May 17 and at Delta Junction on May 18, 1990.

Four 10.2 cm-diameter, 91 cm long ABS pipes were pushed into the soil of each plot using downward hydraulic pressure supplied from the blade of a crawler-type tractor. These pipes were installed on June 5 at Delta Junction and on June 6, 1990, at Fairbanks. An aliquot (1.0 mL) of ^{14}C-labeled metribuzin (2.26 kBq) (Mobay Corp., Agric. Chem. Div., Kansas City, MO) was applied to each pipe at Delta Junction on June 6 and at Fairbanks on June 8, 1990. The specific activity for metribuzin [5-^{14}C] was 770 MBq mmol[-1] and was >97% pure, as determined by thin-layer chromatography. Commercial metribuzin formulation (Sencor 75 DF) (Mobay Corp., Agric. Chem. Div., Kansas City, MO) was applied to barley plots on the same day that the lysimeters were installed, at the 3- to 5-leaf stage. Barley was planted at Delta on May 21-22 and at Fairbanks on May 23-24, 1991, using the same seeding and fertilizer rates and implements described for 1990. No additional potassium bromide or herbicide were applied. Weeds were controlled by hoeing.

Replicate soil columns were removed on July 11, August 15, September 19, 1990; and May 21 and September 17, 1991. The tubes were capped and kept at -20°C until analysis. The soil columns were thawed for 16 hr. A table saw was used to make two longitudinal cuts most of the way through the plastic and a utility knife was used to complete the cuts. The top half of the plastic column was removed. The soil in the column was sectioned into depth intervals: 0-7.5; 7.5-15; 15-30; 30-45 and 45-60 cm. Separate soil samples were taken from each column for determination of soil moisture and total radioactivity, and for extraction of metribuzin and metabolites. Samples were kept at -15°C until analysis. Triplicate soil cores (22 mm X 90 cm) were obtained for bromide analysis on the same dates as lysimeter removal using a "zero contamination" JMC probe (Clements Associates Inc., Newton, IA).

Minnesota Experiment Design. Sections of 0.3 m diameter poly-vinyl chloride (PVC) pipe (wall thickness of 1.2 cm) 0.9 m long were inserted into the soil with the front end loader of a 955L Caterpillar trackscavator in October 1987. Compression of the soil inside the columns following insertion was less than 5 cm. Additional columns 0.7 m deep were inserted into the soil at each location and equipped with a water collection device (16) to monitor movement of ^{14}C from the bottom of the column. Water sampling was conducted following each rainfall or irrigation event and prior to each sampling.

After insertion, an additional 9-cm section of pipe was attached to the top of each column to prevent water from flowing across the treated soil. A fence was constructed to control access to the plots. ^{14}C-atrazine (0.85 MBq) in 5 ml methanol was applied to the center 10 cm of the column, 2.5 cm below the soil surface on May 20, 1988 as previously described (12) to minimize volatilization losses. Following application, three corn (Zea mays L.) seeds were planted in the center of each column. Corn was planted in the rest of the plot area at 60,000 seeds ha[-1], and atrazine, alachlor, and dicamba were applied as a tank mix to the entire plot area at 2.2, 3.4 and 0.7 kg ha[-1], respectively. After emergence, corn was thinned to 1 plant per column. Irrigation was applied to the columns in 2.5 cm increments as needed. Soil columns were removed immediately following ^{14}C-atrazine application, and 1, 2, 4, and 6 mo after ^{14}C-atrazine treatment (MAT). Prior to removing each column, the corn plants were removed and bagged. The column was capped with a PVC pipe cap. The columns were removed as previously described (*12*), transported to a freezer, and stored at -15°C until processing.

During the second yr, columns were removed 12 and 16 MAT. Following the 12 MAT sampling, 3 corn seeds were planted into each of the remaining columns on May 22, 1989. After emergence, corn was thinned to 1 plant per column. Corn was planted in the plot area at 60,000 seeds ha[-1], the remaining columns were covered and a tank mix of atrazine, alachlor and dicamba was applied to the entire plot area, as done during the first year. At 16 MAT the plants and the columns were removed as previously described.

Chemical Extraction and Analyses. For metribuzin and metabolite extraction, all centrifugation steps and sample shaking was performed with a platform shaker at 325 RPM. Initially, duplicate 12-g soil samples were mixed with 12 mL of 0.01 N $CaCl_2$ in 50-mL glass centrifuge tubes and shaken for 48 hr. After centrifuging (759 x g for 30 min), the supernatant was weighed and transferred to a 20-mL scintillation vial. To determine total ^{14}C activity in this supernatant, one mL of supernatant was transferred to a 7-mL scintillation vial, 5 mL of scintillation counting solution was added, and ^{14}C activity determined.

Soil, after $CaCl_2$ extraction, was reextracted with 36 mL 4:1 methanol:water (v:v). After shaking for 6 h, the tubes were centrifuged and the supernatant transferred to a 50-mL conical tube. An additional 33 mL of 4:1 methanol:water (v:v) was added to the soil and was shaken for 16 h. Following centrifugation, the two supernatants were combined and the methanol evaporated using compressed air and a 50°C water bath. [14]C in the supernatant after evaporation was determined on a 1-mL aliquot as previously described.

[14]C in the soil remaining after extraction, determined by combustion as described later, was considered bound residue. Percent moisture of this soil was determined using another sample dried to constant weight at 60°C after allowing methanol to evaporate.

Aqueous supernatants from $CaCl_2$ and methanol:water extractions were extracted with chloroform. $CaCl_2$ or methanol:water supernatants were extracted three times using 5 mL of chloroform each time. Chloroform extracts were dried by passing through phase separatory paper into 50-mL conical tubes. To determine ^{14}C activity in the water, total weight of water was determined and then 1 g water was transferred to a 7-mL scintillation vial, 5 mL of cocktail added, and counted using LSC. Similarly, ^{14}C activity in chloroform was determined from total chloroform weight and counting 1 g solution with 15 mL of scintillation cocktail.

One drop of decanol was added to chloroform extracts, the chloroform was evaporated, 1 mL of methanol was added, and the resulting solution was passed through a 0.45 μm filter. The samples were then evaporated to near dryness, and redissolved in 0.5 mL of a 1:1 (v:v) methanol:water mixture. Standards of metribuzin, desaminometribuzin (DA), diketometribuzin (DK), and desaminodiketometribuzin (DADK) were used to determine retention times on a Hewlett Packard 1090 HPLC (Hewlett-Packard Co., Avondale, PA) equipped with a diode array detector and

Adsorboshere-C18-5 micron column (Alltech Assoc. Inc., Deerfield, IL) (250 x 4.6 mm i.d.). The solvent was methanol:0.05% acetic acid (43:57, v:v) with a flow rate of 1.0 mL min^{-1}. Wavelengths used were 245-275 nm and 285-315 nm. Injection volume was 100 µl. Retention times for DK, DADK, metribuzin, and DA were 10.4, 13.1, 21.4, and 23.2 min., respectively. A fraction collector was used to collect metribuzin and metabolites after passing through the HPLC column based on their retention times. ^{14}C activity of each fraction was determined by LSC.

    In the atrazine experiment, each column was sectioned into 10-cm depth increments (12). Soil samples were weighed and mixed for 15 min in a jar mill type mixer. Separate subsamples were taken to determine soil moisture, total radioactivity, and degradation products. Soil moisture was determined by oven drying a 20 g subsample at 110° C for 24 hr. ^{14}C residues were extracted from soil as previously described (12) by refluxing three times in methanol and water. ^{14}C-compounds were separated and quantified using thin layer chromatography (TLC) with 20 x 20 cm 0.25 mm silica TLC plates. Rf values of analytical standards following two elutions in 110:2:2 chloroform:methanol:formic acid (v:v:v) were 0.88, 0.66, 0.59, 0.23, and 0.03 for atrazine, deethylatrazine (DEA), deisopropylatrazine (DIA), deethyldeisopropylatrazine (DEDIA), and hydroxylated derivatives, respectively. To quantify the chlorinated products and total hydroxylated residues, plates were scanned for 10 min on a Berthold linear plate analyzer (Berthold Scientific Instruments Company, Pittsburg, PA). To isolate hydroxylated derivatives the plates were developed in a second solvent system (75:20:4:2, chloroform:methanol:water:formic acid v:v:v), Rf values of analytical standards were 0.98, 0.93, 0.90, 0.71, 0.62, 0.38, 0.32 for atrazine, DEA, DIA, DEDIA, hydroxyatrazine (HA), deethylhydroxyatrazine (DEHA), deisopropylhydroxyatrazine (DIHA), respectively. Statistical analysis of the data was conducted by calculating the mean and standard error of the mean from the three replicates.

    For total ^{14}C in soil, 3 subsamples were combusted from each depth and replicate using a Packard 306 Sample Oxidizer (Packard Instruments Co., Downers Grove, IL). Soil (0.4 g) was mixed with 0.1 g of microcystalline cellulose and placed in a paper cone and combusted for 2 min. CO_2 was trapped in CARBO-SORB and then mixed with PERMAFLUOR E+ (Packard Instruments Co., Downers Grove, IL). Radioactivity (^{14}C) was determined by liquid scintillation counting (LSC) using a Packard Tri-carb scintillation analyzer (Packard Instruments Co., Downers Grove, IL). The efficiency of oxidation was determined to be 0.95 $\pm$ 0.03.

Results and Discussion: Metribuzin.

^{14}C **Dissipation.** Overall dissipation of ^{14}C was slower in the Tanana silt loam soil than in the Beales soil over the first 105 d after application. At 105 d after application, 80% of applied ^{14}C remained in Tanana soil compared to 45% in Beales soil (Table II). The loss of ^{14}C probably is due to mineralization to $^{14}CO_2$ rather than volatilization or leaching. Moorman and Harper (9) reported up to 20% of ^{14}C-ring-labelled metribuzin mineralized to $^{14}CO_2$ in as silty clay loam soil. While volatilization was not measured in our experiment, metribuzin has not been reported to be susceptible to volatilization.

Table II. Dissipation of ^{14}C in Beales and Tanana silt loam soils

Soil	Depth	Days after Application				
		35	70	105	349	468
	cm	----------------------- % of ^{14}C applied[†] ---------------------				
Beales	0-7	68.1	31.3	41.2	27.9	44.3
	7-15	0.31	0.13	2.45	1.77	2.38
	15-22	0.20	0.01	0.38	0.01	1.05
	22-30	0.05	0.03	0.14	ND	ND
	30-45	0.05	0.06	0.10	ND	ND
	Total	68.7	31.5	44.3	29.7	47.7
Tanana	0-7	103.1	58.3	78.1	81.65	59.4
	7-15	2.0	5.67	ND	2.73	5.85
	15-22	0.22	0.38	1.22	0.15	0.41
	22-30	0.15	0.24	ND	0.03	0.07
	30-45	0.19	0.30	0.39	ND	ND
	45-60	0.10	0.03	0.24	ND	ND
	Total	106.0	64.0	79.9	84.6	65.8

[†] average C.V. = 31%.

Leaching, discussed in detail below, also did not appear to be a major process contributing to ^{14}C loss from the soils. Averaged over both soils, less than 1.0% of applied ^{14}C was detected below 22 cm from 35 to 468 days after application.

Leaching of ^{14}C. Throughout the 468 d of the experiment, the majority of the ^{14}C was found at the 0 to 7.5 cm depth. However, at 35 d after application, ^{14}C was found to depths of 60 and 45 cm in Tanana and Beales soils, respectively. Amounts of ^{14}C, too small to positively identify as metribuzin or specific metabolites, were also found throughout the soil profile during the course of the experiment. For instance, through 105 d after application, less than 2 percent of the applied ^{14}C was found below 15 cm in both soils. In contrast, by day 105, 45% of applied bromide was at the 22.5- to 45-cm depth (data not shown). At 349 and 468 d after application, the greatest amounts of ^{14}C detected deeper than in the top 7.5 cm of soil were found at depths 7.5 to 15 cm in both soils but did not exceed 6% of applied. Less than 1% of the applied ^{14}C was found below 22 cm.

These results agree with Burgard et al. (17), who found metribuzin remained in the top 15 cm of loamy sand soil under irrigated potato production in east-central Minnesota in spite of water movement to the bottom of the analyzed profile (90-105 cm). In other studies, metribuzin did not leach past 30 cm in non-irrigated sandy loam (18,19) or furrow-irrigated clay (10) and reached 38 cm in irrigated sandy loam soil (20). These studies determined movement of parent metribuzin only, while our study determined movement of total ^{14}C-metribuzin residues, i.e. parent and metabolites.

Metribuzin Degradation. Initial degradation of metribuzin was rapid in both soils. At 35 d after application, an average of 31% of ^{14}C remaining in the soil was metribuzin (Table III). Of ^{14}C remaining at 105 d after application, an average of 19% was metribuzin. There was no further metribuzin degradation over the winter months while the soil was frozen; the same amount of ^{14}C remaining in the soil was metribuzin at 349 d after application. Degradation resumed during the spring, and at the end of the experiment 9% of remaining ^{14}C was metribuzin. These data are similar to those of Conn and Cameron (9) who found 13% of applied metribuzin remaining

Table III. Product distribution of ^{14}C in Beales and Tanana silt loam soils as a function of time.

Soil	Days after application	Metri- buzin	DA	DK	DADK	Un- known	Unextr actable
	Days	---------------------------- % of ^{14}C remaining[†] -------------------					
Beales	35	35.0	8.0	1.9	1.4	7.8	46.0
	70	28.1	8.9	1.8	1.9	10.2	49.2
	105	20.4	6.3	1.4	1.8	10.2	49.2
	349	18.7	8.1	0.8	2.7	9.9	60.0
	468	11.5	3.7	0.9	2.4	9.6	72.0
Tanana	35	28.2	9.5	1.3	2.0	9.7	49.4
	70	19.3	5.1	1.3	3.1	13.7	57.5
	105	17.7	5.2	1.5	2.8	13.8	59.1
	349	24.1	2.5	1.6	2.0	11.3	58.5
	468	6.5	5.2	1.3	2.8	14.1	70.1

[†] average C.V. = 33%.

in the surface 15 cm of Volkmar and Beales silt loam soils 1 yr after application. Burgard et al. (*17*) found 17% of metribuzin applied to a loamy sand remaining in the top 45 cm of soil 1 yr after application. More of the remaining ^{14}C was metribuzin in the Beales than in the Tanana silt loam (Table III) on 4 of 5 sampling dates. The lower pH of the Beales soil (4.7) versus the Tanana soil (6.5) may have contributed to slower metribuzin degradation. Ladlie et al. (*21*) found that dissipation of metribuzin decreased with a decrease in soil pH from 6.7 to 4.6.

Metribuzin degraded to a number of metabolites, including DA, DK, and DADK. There were also unknown nonpolar and polar metabolites. In the surface soils, the greatest amount of ^{14}C was in the form of unextractable ^{14}C residues. Greater amounts of DA were found in both soils than DK or DADK. After 70 days in the field there was generally more DADK than DK in both soils. In contrast, Moorman and Harper (*9*) and Webster and Reimer (*22*) recovered more DK than DA in surface soils. However, the differences between our study and those of Moorman and Harper (*9*) and Webster and Reimer (*22*) are probably not significant.

Results and Discussion: Atrazine.

^{14}C **Dissipation.** Recovery of total ^{14}C from the soil 1 and 2 mo after ^{14}C-atrazine application was 96 and 81% of that applied, respectively (Table IV) (*23*). ^{14}C

Table IV. Dissipation of ^{14}C in Port Byron silt loam soil

Soil depth	Months after application						
	0	1	2	4	6	12	16
cm	------------------------------- % of applied[†] -------------------------						
0-10	100.0	90.7	69.9	53.1	56.2	60.2	50.1
10-20	ND	4.5	8.1	5.9	7.6	12.5	12.1
20-30	--[‡]	0.3	2.1	3.3	1.6	2.9	2.7
30-40	--	0.2	0.8	1.2	0.8	1.8	1.2
40-50	--	ND	0.1	0.1	0.1	1.2	0.9
50-60	--	ND[§]	ND	ND	0.1	0.9	0.5
60-70	--	ND	ND	ND	0.1	0.1	ND
70-80	--	ND	ND	ND	ND	0.1	ND
80-90	--	ND	ND	ND	ND	ND	ND

[†] average C.V. = 27%.
[‡] --. not analyzed.
[§] ND, not detected.

remaining in the soil decreased to 64% of that applied 4 MAT and did not decrease further during the remainder of the study; 68% remained 16 MAT. A small amount of ^{14}C is presumeably lost due to volatilization and plant uptake. Potential volatilization losses were minimized by applying the ^{14}C-atrazine below the soil surface. The maximum amount of applied ^{14}C recovered in the corn plant was 5.2% 2 MAT (*23*). The rest of the ^{14}C was assumed to be lost by mineralization to ^{14}CO$_2$ rather than by leaching out of the lysimeter, discussed below.

Leaching of ^{14}C. Between application and 1 MAT, ^{14}C leached to the 30- to 40-cm depth Table IV). Between the 1- and 2- MAT sampling, ^{14}C in the 10- to 20- cm depth increased from 4.5 to 8.1%, but was not detected deeper than 40 to 50 cm. No increase in ^{14}C in the 10- to 20- cm depth or leaching beyond 40 to 50 cm was observed at 4 MAT. ^{14}C was detected to the 60- to 70-cm sampling depth by 6 MAT (Table IV). ^{14}C collected in leachate between the 4- and 6-MAT indicated that some ^{14}C (<0.002% of applied) had moved through the soil with the water during this period. ^{14}C was detected in leachate (700 ml) collected from the monitoring column between 6 and 12 MAT. Atrazine was the only compound detected in the leachate. ^{14}C at the 10- to 20-cm depth increased to 12.5% of applied at 12 MAT, and ^{14}C was detected throughout the 80-cm deep soil column. Distribution of ^{14}C in the soil profile at 16 MAT was the same as at 12 MAT except ^{14}C was only detected to 50-60 cm.

Atrazine degradation. Atrazine degradation products in soil result from chemical and microbial processes. Chemical hydrolysis results in the formation of hydroxyatrazine (HA) (*24-26*). Microbial degradation results in *N*-dealkylation (*27-30*) to form deethylatrazine (DEA), deisopropylatrazine (DIA), and deethyldeisopropylatrazine (DEDIA) (*31*). Further degradation of dealkylated products results in formation of 4-amino-2-chloro-1,3,5-triazine (*32*). Hydroxylation of dealkylated products or dealkylation of hydroxylated products results in the formation of deethylhydroxyatrazine (DEHA), deisopropylhydroxyatrazine (DIHA), and deethyldeisopropylhydroxyatrazine (DEDIHA) (*28,29*). Complete degradation of atrazine to CO$_2$ has been observed through continued hydroxylation of the triazine ring to form ammeline, ammelide, and cyanuric acid prior to ring cleavage (*31*).

Compounds detected in soil during the course of the present study were atrazine, HA, DEA, DIA, DEDIA, an unidentified product (UP), possibly 4-amino-2-chloro-1, 3, 5-triazine (*32*), and some unidentified polar metabolites (Table V). Each compound is represented as a percentage of radioactivity recovered at each depth to evaluate relative proportions of each compound at each depth over time. Atrazine present in the top 10 cm of soil decreased from 88% of ^{14}C recovered immediately following application to 58 and 37% 1 and 2 MAT, respectively (Table V). The proportion of ^{14}C as atrazine decreased to an average 27% at 4 to 16 MAT. Decreased proportion of atrazine over time was accounted for primarily by increased levels of unextracted residues, HA, and DEA and to a lesser degree by DIA, DEDIA, and the UP. Unextractable residues in the top 10 cm of soil increased from 4% of total ^{14}C measured immediately following application to 33% at 2 MAT and remained constant through the remainder of the study. The proportion of HA in the top 10 cm increased from 5.6% immediately following application to a high of 26% at 12 MAT. The maximum proportion of microbially dealkylated metabolites DEA, DIA, and DEDIA in the top 10 cm were 7, 3, and 3%, respectively, throughout the 16 mo incubation period. The maximum proportion of ^{14}C present as UP in the top 10 cm of soil was 1.7% at 16 MAT.

At depths greater than 10 cm, the proportion of atrazine decreased with time indicating the potential for degradation of atrazine in the soil below the plow layer. For instance, the proportion of atrazine present in the 10- to 20-, 20- to 30-, and 30- to 40-cm depths was 56, 75 and 68%, respectively, 1 MAT (Table IV). The proportion of atrazine decreased to 18, 37, and 20% for the same depth increments, 16 MAT. The decreased proportion of atrazine at depths greater than 10 cm was associated with increased proportion of DEA, HA, and unextracted residues.

Leachate collected from monitoring columns in the silt loam in this study and a clay loam (*32*) and sandy loam (*12*) indicated that water moved through the soils and was capable of leaching ^{14}C-compounds. Greater leaching of ^{14}C in the silt loam and clay loam (*32*) than in the sandy loam (*12*) appears to be a function of the greater water holding capacity of these soils. Greater clay and silt contents in the silt loam and clay loam soils compared to the sandy loam maintained higher soil moisture longer, allowing more time for desorption of atrazine residues to occur and leach with the water. Rapid water movement in the sandy loam did not allow sufficient time for residues to desorb from soil and move with water. Previous results (*13*) also showing greater atrazine leaching from columns packed with silt loam than sand may be due to similar reasons. In the silt loam, initial leaching resulted in greater concentrations of residues in the 10- to 20-cm depth, but leaching was limited to the top 40 cm. In contrast, initial leaching in the clay loam (*32*) was rapid and characteristic of macroporous flow as observed in previous research (*33-35*).

Atrazine residues were degraded more rapidly in the silt loam than in the previously studied clay loam or sandy loam soil during the first 4 MAT. Radioactivity in the silt loam decreased to 64% of applied 4 MAT, compared to 75 and 80% in the sandy loam (*12*) and clay loam (*32*) soils, respectively. Total radioactivity in the silt loam and sandy loam soils did not decrease after the 4-MAT sampling and was 68 and 78%, respectively, 16 MAT. In the clay loam, total radioactivity decreased from 81 to 64% between the 12- and 16-MAT sampling. During this time, HA in the top 10 cm of the clay loam decreased from 20 to 11%, polar metabolites increased from 5 to 18% and DEA increased from 6 to 12%.

HA was the major degradation product in the upper 10 cm of soil at all three locations. Microbial degradation resulted in the formation of DEA, DIA, DEDIA, and UP. As residues leached below the top 10 cm of soil, the proportion of HA decreased while the proportion of DEA increased. This suggested that DEA was more mobile in soil or degradation of atrazine at depths greater than 10 cm favored formation of dealkylated products over hydroxylated products. Increased levels of DEA suggest that microbial breakdown was occurring and microbial *N*-dealkylation of hydroxylated residues to more polar compounds and ^{14}CO$_2$ may be

Table V. Distribution of ^{14}C following application of ^{14}C-atrazine to Port Byron silt loam

Time after application (months)	Soil depth (cm)	AT[†]	DEA	DIA	DEDIA	HA	Unknown	Unextracted
					% of recovered[‡]			
0	0-10	87.5	0.9	0.5	ND	ND	1.5	4.0
	10-20	ND[§]	ND	ND	ND	ND	ND	ND
1	0-10	58.5	3.5	1.5	ND	13.0	8.8	16.0
	10-20	55.5	11.0	3.6	ND	6.2	4.7	20.0
	20-30	74.9	6.7	2.2	ND	1.2	ND	15.0
	30-40	67.8	2.6	3.5	ND	3.0	ND	23.0
	40-90	ND	ND	ND	--	--	ND	ND
2	0-10	36.7	3.7	2.2	0.5	16.6	10.2	32.7
	10-20	53.5	11.3	4.7	0.5	7.1	10.3	14.0
	20-30	62.9	12.8	3.8	1.1	6.4	5.0	9.5
	30-40	58.5	14.5	4.2	ND	12.2	9.0	6.5
	40-50	79.6	8.2	ND	ND	ND	ND	ND
	50-90	ND	ND	ND	ND	ND	ND	ND
4	0-10	27.5	5.2	1.9	1.8	21.8	10.7	34.5
	10-20	48.0	17.0	1.8	1.0	6.2	13.5	13.5
	20-30	47.1	15.6	4.3	1.1	12.2	10.1	10.7
	30-40	46.9	14.5	3.8	nd	17.5	3.0	14.7
	40-50	89.3	10.7	ND	ND	ND	ND	ND
	50-90	ND	ND	ND	ND	ND	ND	ND
6	0-10	30.9	7.4	2.7	2.3	17.7	14.7	28.3
	10-20	33.2	14.6	3.3	1.8	9.5	17.3	23.7
	20-30	51.2	10.8	0.6	2.6	19.2	7.0	10.3

Time	Depth							
	30-40	57.1	11.3	2.8	ND	13.2	ND	15.7
	40-50	39.8	14.4	9.9	ND	35.8	ND	ND
	50-60	60.0	20.0	ND	ND	20.0	ND	ND
	60-70	80.0	ND	ND	ND	20.0	ND	ND
	70-80	ND	ND	ND	ND	--	ND	ND
12	0-10	25.2	5.3	3.0	2.8	26.2	8.9	35.0
	10-20	25.7	17.3	5.8	3.4	6.6	14.7	21.3
	20-30	29.0	14.5	2.7	4.4	16.0	3.3	38.7
	30-40	34.9	20.7	5.1	ND	8.3	ND	34.0
	40-50	43.3	6.9	2.1	ND	16.7	ND	33.5
	50-60	41.8	4.6	1.8	ND	19.0	ND	30.0
	60-70	44.2	9.0	1.7	ND	14.1	ND	ND
	70-80	44.2	9.3	1.7	ND	13.9	ND	ND
16	0-10	24.0	5.9	2.2	3.0	13.2	12.4	34.7
	10-20	18.0	16.8	2.4	3.5	5.7	16.9	17.7
	20-30	37.0	3.9	0.6	1.1	9.3	6.9	30.3
	30-40	20.3	12.6	1.8	ND	21.3	ND	30.3
	40-50	26.1	18.9	3.0	ND	8.3	ND	22.5
	50-60	25.8	12.0	ND	ND	18.2	ND	ND
	60-90	ND	ND	ND	ND	ND	ND	ND

† AT = atrazine, DEA = deethyatrazine, DIA = deisopropylatrazine, DEDIA = deethyldeisopropylatrazine, HA = hydroxyatrazine.
‡ average deviation = 25%.
§ ND = not detected.

occurring. These data are supported by Bowman (*13*) who also observed greater DEA production with increasing soil moisture. Other research also indicates lower adsorption and greater mobility of DEA than atrazine and other degradation products (*36-38*).

In summary, degradation of metribuzin appeared to proceed rapidly in subarctic soils and does not appear to more readily leach downward. It also appears that metabolites of metribuzin do not leach in appreciable quantity. Significant levels of atrazine residues remained 16 MAT. Detection of these residues in leachate collected from 70-cm deep monitoring columns indicates leaching did occur. Even though the amount of leaching is extremely small, it appears that contamination of water supplies may occur by leaching through the soil. The extremely low atrazine levels detected in ground water supplies would require leaching of only < 0.1 % of applied atrazine. It appears that contamination of ground water by atrazine and DEA can occur in silt loam soils. The question remains in both cases, however, as to whether the bound residues will be slowly available in the future. Additional research is needed on long-term bioavailability of bound residues.

Acknowledgments.

This research was supported in part by University of Minnesota Center for Agricultural Impacts on Water Quality, Minnesota Legislative Commission on Minnesota Resources, and USDA North Central Pesticide Impact Assessment Program.

Literature Cited.

1.   Conn, J. S.; Cameron, J. S. *Can. J. Soil Sci.* **1988**, *68*, 827-830.
2.   Fryer, J. D.; Kirkland, K. *Weed Res.* **1970**, *10*, 133-158.
3.   Conn, J. S.; Knight, C. W. *An evaluation of herbicides for broadleaf-weed control in rapeseed.* Res. Bull. 62. Agr. For. Expt. Sta., Univ. Alaska, Fairbanks., AK, 1984; 22 pp.
4.   Smith, A. E.; Hayden, B. J. *Can. J. Plant Sci.* **1976**, *56*, 769-771.
5.   Pritchard, M. K.; Stobbe, E. H. *Can J. Plant Sci.* **1980**, *60*, 5-11.
6.   Banks, P. A.; Robinson, E. L. *Weed Sci.* **1982**, *30*, 164-168.
7.   Siddoway, F. H.; Lewis, C. E.; Cullum, R. F. *Agroborealis* **1984**, *16*, 35-40.
8.   Knight, C. W.; C. E. Lewis. *Soil Tillage Res.* **1986**, *7*, 341-353.
9.   Moorman, T. B.; S. S. Harper. *J. Environ. Qual.* **1989**, *18*, 302-306.
10.   Hyzak, D. L.; R. L. Zimdahl. *Weed Sci.* **1974**, *22*, 75-79.
11.   Bouchard, D. C., T. L. Lavy; D. B. Marx. *Weed Sci.* **1982**, *30*, 629-632.
12.   B. A. Sorenson, D. L. Wyse, W. C. Koskinen, D. D. Buhler, W. E. Lueschen; M. D. Jorgenson. *Weed Sci.* **1993**, *41*, 239-245.
13.   B. T. Bowman. *Environ. Toxic. Chem.* **9**; 453-461 (1990).

14. Patrick, J. H.; P. E. Black. Potential evapotranspiration and climate in Alaska by Thornthwaite's classification. U. S. Dept. Agric., Forest Serv. Research Report PNW-71, 1968.
15. Baker, D. G.; Nelson, W. W.; Kuehnast, E. L. Climate of Minnesota, Part XII. Agric. Expt. Sta., Univ. Minnesota, St, Paul, MN, 1979, 23pp.
16. Barbee, G. C.; K. W. Brown, K. W. *Soil Sci.* **1986**, *141*, 149-154.
17. Burgard, D. J.; Dowdy, R. H.; Koskinen, W. C.; Cheng, H. H.. 1994. *Weed Sci.* **1994**, *42*, 446-452.
18. Nicholls, P. H.; Walker, A.; Baker, R. J. *Pest. Sci.* **1982**, *12*, 484-494.
19. Walker, A. *Weed Res.* **1978**, *18*, 305-311.
20. Jones, R.E.; Banks, P.A.; Radcliffe, D.E. *Weed Sci.* **1990**, *38*, 589-597.
21. Ladlie, J. S.; Meggitt, W. F.; Penner, D. *Weed Sci.* **1976**, *24*, 477-481.
22. Webster, G. R. B.; Reimer, G. J. *Weed Res.* **1976**, *16*, 191-196.
23. Sorenson, B. A.; Koskinen, W. C.; Buhler, D. D.; Wyse, D. L.; Lueschen, W. E.; Jorgenson, M. D. *Intern. J. Environ. Anal. Chem.* **1995**, in, press
24. Armstrong, D. E.; Chesters, G.; Harris, R. F.. *Soil Sci. Soc. Am. Proc.* **1967**, *31*, 61-66.
25. Armstrong, D. E.; Chesters, G. *Environ. Sci. Tech.* **1968**, *9*, 683-689 .
26. Skipper, H. D.; Gilmour, C. M.; Furtick, W. R. *Soil Sci. Soc. Am. Proc.* **1967**, *31*, 653-656.
27. Giardina, M. C.; Giardi, M. T.; Filacchioni, G. *Agric. Biol. Chem.* **1982**, *46*, 1439-1445.
28. Giardi, M. T.; Giardina, M. C.; Filacchioni, G. *Agric. Biol. Chem.* **1985**, *49*, 1551-1558.
29. Kaufman, D. D.; Blake, J. *Soil Biol. Biochem.* **1970**, *2*, 73-80.
30. Wolf, D. C.; and Martin, J. P. *J. Environ. Qual.* **1975**, *4*, 134-139.
31. Cook, A. M.; Hutter, R. *J. Agric. Food Chem.* **1981**, *29*, 1135-1143.
32. Giardina, M. C.; Giardi, M. T.; Filacchioni, G. *Agric. Biol. Chem.* **1980**, *44*, 2067-2072.
32. Sorenson, B. A.; Koskinen, W. C.; Buhler, D. D.; Wyse, D. L.; Lueschen, W. E.; Jorgenson, M. D. *Weed Sci.* **1994**, *42*, 618-624.
33. Hall, J. K.; Murray, M. R.; Hartwig, N. L. *J. Environ. Qual.* **1989**, *18*, 439-445.
34. Isensee, A. R.; Nash, R. G.; Helling, C. S. *J. Environ. Qual.* **1990**, *19*, 434-440.
35. Starr, J. L; Glotfelty, D. E. *J. Environ. Qual.* **1990**, *19*, 552-558.
36. Adams, C. D.; Thurman, E. M. *J. Environ. Qual.* **1991**, *20*, 540-547.
37. Muir, D. C. G.; Baker, B. E. *J. Agric. Food Chem.* **1976**, *24*, 122-125.
38. Muir, D. C. G.; Baker, B. E. *Weed Res.* **1978**, *18*, 111-120.

Chapter 12

Fate of Atrazine and Atrazine Degradates in Soils of Iowa

Ellen L. Kruger and Joel R. Coats

Pesticide Toxicology Laboratory, Department of Entomology,
112 Insectary, Iowa State University, Ames, IA 50011–3140

Several studies have been conducted to investigate the fate of atrazine (ATR, 2-chloro-4[ethylamino]-6[isopropylamino]-*s*-triazine) and major degradation products of ATR in soils of Iowa by using laboratory radiotracer studies, field lysimeters, and a field-scale approach. Complete metabolism studies of uniformly ring-labeled ^{14}C-chemicals revealed some major trends. Persistence of ATR, deethylatrazine (DEA, 2-chloro-4[amino]-6[isopropylamino]-*s*-triazine), and deisopropylatrazine (DIA, 2-chloro-4[ethylamino]-6[amino]-*s*-triazine) was greater in subsurface soils than in surface soils. In surface soil of Ames, DEA and didealkylatrazine (DDA, 2-chloro-4,6-[diamino]-*s*-triazine) were predominant degradates of ATR after 60 d, and hydroxyatrazine (HYA, 2-hydroxy-4[ethylamino]-6[isopropylamino]-*s*-triazine) was the predominant degradate of ATR after 180 d. The persistence of ATR, DEA, and DIA was significantly reduced under saturated soil moisture conditions than in soils held at a moisture near field capacity. Relative mobilities of ATR and degradates in five Iowa soils (surface and subsurface), determined by soil thin-layer chromatography, indicate that DEA is more mobile than ATR. The relative mobilities of DIA, DDA, and ATR were similar. Also, laboratory studies with undisturbed soil columns are supportive of greater mobility of DEA than ATR. In a field-scale study investigating the mobility of ATR and its degradates, it was indicated that ATR degradation products by themselves, or in combination with the parent compound can exceed the maximum contaminant level (MCL) of 3 µg/L currently set for ATR alone. In ATR-applied field plots, DEA and DIA were detected along with ATR in tile drain water samples, with concentrations of DEA exceeding DIA. In Extrazine®- (a herbicide mixture of 67% cyanazine [CYA; 2-chloro-4-ethylamino-6-(1-cyano-1 methylethylamino)-*s*-triazine] and 21% ATR) - applied field lysimeters, the concentrations of DIA exceeded those of DEA.

0097–6156/96/0630–0140$15.00/0

Detection of pesticide degradates in surface and groundwater has prompted increased interest in the fate of these compounds in soil. Controlled laboratory studies using radiotracers can provide important information about the behavior of pesticides and their breakdown products under various conditions. Numerous studies have been conducted using radiolabeled compounds. Results are greatly influenced by many factors including experimental design, temperature, moisture content, soil type, and analytical techniques. Direct comparisons to literature data are difficult to make due to these factors.

Laboratory studies using radiotracers provide a very sensitive means for determining degradation, mobility, and persistence. In our laboratory, several approaches have been used to study environmental fate of ATR and ATR degradates. Degradation pathways have been presented in the literature [1,2]. We have conducted soil microcosm studies to investigate the degradation of *s*-triazine herbicides as influenced by soil characteristics, soil moisture, and soil pesticide history. We have also been involved in mobility studies using large undisturbed soil columns, field box lysimeters, and field-scale tile drain systems. It is the intent of this chapter to summarize the studies conducted in this laboratory (and in collaboration with other laboratories) and to make comparisons of the fate of ATR and major degradation products of ATR in Iowa soils.

Fate of ^{14}C-ATR and Degradates in Soil Microcosm Studies

Soils from five locations in Iowa were used for studies conducted in this laboratory. Pertinent soil characteristics are shown in Table 1; the five locations represent a wide range in soils. All ^{14}C-compounds used in these studies were uniformly-ring labeled, and labeled chemicals were supplied by Ciba Crop Protection (Greensboro, NC).

Comparative Fates of ATR, DEA, DIA, and HYA in Ames Soils.
Two independent radiotracer studies conducted in our laboratory looked at the fate of ATR, DEA, DIA, HYA in some or all of the soils listed in Table 1 [2,3]. The comparative fates of these four compounds can only be made for the Ames surface (0 to 30 cm) and subsurface (65 to 90 cm) soils, as these were the soils common in both studies. Soils were treated with ^{14}C-ATR, DEA, DIA, or HYA. Soil moisture tensions were adjusted to near field capacity (-33 kPa soil moisture tension) and incubated for 60 d in the dark at 24 °C. During the incubations, [14]CO_2 was trapped in NaOH (1 N). Soils were extracted and analyzed using thin-layer chromatography, autoradiography, soil combustion, and liquid scintillation techniques [2,3].

Some major trends were seen in the persistence and degradation of these chemicals in the Ames soils. Persistence of ATR, DEA, and DIA was greater in subsurface soil than in surface soil (Figure 1) [2, 3]. HYA was equally persistent in surface and subsurface soils. Degradation studies were not carried out in the 90-to 120-cm depth for DEA and HYA.

Significantly greater amounts of DIA and ATR were mineralized in the surface (0 to 30 cm) soil than in subsurface soil (90 to 120 cm) from Ames (Figure 2), as

Table 1. Characteristics of Iowa soils with no previous pesticide history

Site	Depth (cm)	Texture	Sand %	Silt %	Clay %	O.M.[a] %
Ames	0-30	Sandy Loam	64	22	14	2.5
	65-90	Sandy Clay Loam	54	22	24	1.1
	90-120	Sandy Clay Loam	46	28	26	1.6
Treynor	0-30	Silty Clay Loam	14	58	28	2.8
	65-90	Silty Clay Loam	10	54	36	0.8
Chariton	0-30	Clay Loam	26	40	34	2.0
	65-90	Clay	2	34	64	0.4
Fruitland	0-30	Sand	90	4	6	1.1
	65-90	Sand	90	4	6	0.4
Nashua	0-30	Loam	42	34	24	3.6
	65-90	Sandy Clay Loam	46	26	28	0.4

[a] Organic matter content of soil

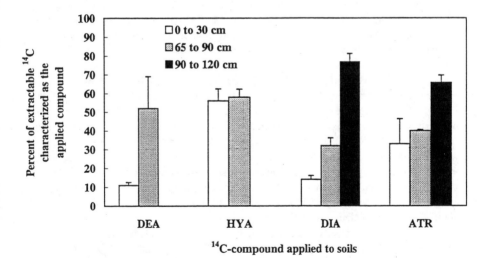

Figure 1. The percent of extractable ^{14}C remaining as the applied compounds in Ames soil treated with either ^{14}C-ATR, -DEA, -DIA, or -HYA after a 60-d incubation. Degradation studies for DEA and HYA were not carried out in the 90- to 120-cm depth. Bars (with standard deviations) represent the mean of two replicates.

determined by the amount of $^{14}CO_2$ in NaOH traps for the 60-d incubation period. Baluch et al. [3] did not find significant differences in mineralization of DEA and HYA in Ames soils among depths from 0 to 90 cm (Figure 2). DIA was the degradate most readily mineralized, with approximately 7.5% of the applied ^{14}C-DIA recovered as $^{14}CO_2$ in surface soils after 60 d. This is consistent with research by Skipper and Volk [4], which indicated that the ethyl side chain is more readily removed from ATR compared to the isopropyl side chain. Minimal mineralization of ATR occurred in soil from either depth (< 1% of applied ^{14}C).

The major degradates of ATR after a 60-d incubation in surface soil of Ames, held at a soil moisture near field capacity, were DEA (2.4% of the applied ^{14}C) and DDA (4%) [2]. After 120 d, however, the quantity of HYA exceeded that of DEA, DIA, and DDA in ATR-treated surface soil. The greater persistence of HYA compared with ATR, DEA, and DIA may explain why HYA is sometimes reported to be the major degradate of ATR.

Soil-bound residues, as determined by combusting subsamples of extracted soil by using a Packard Sample Oxidizer (Packard Instrument Co.), were formed to a greater extent in surface soils than in subsurface soils, with the exception of HYA-treated soils (Figure 3) [2, 3]. Sixty-one percent of the applied ^{14}C-DIA was bound after 60 d in surface soils, significantly greater than in soil from the 90 to 120-cm depth. Forty-five percent of the applied ATR and 46% of the applied DEA were bound in surface soil after 60 d.

Influence of Soil Moisture on Degradation of ATR, DEA, and DIA in Ames Soils. Soil moisture influenced the fate of ATR, DIA, and DEA. Soils were treated with either ^{14}C-ATR, -DEA, or -DIA, and soil moistures were adjusted to either saturated or a specific unsaturated soil moisture condition. Unsaturated soils were those which had a soil moisture tension of -33 kPa (near field capacity). Saturated soils had a layer of water over the surface of the soil, with a soil moisture content of 30% (gravimetric soil moisture) greater than the soil moisture tension at -33 kPa [2,5]. Soils were incubated in the dark at 24 °C for 60 d. $^{14}CO_2$ was trapped during the incubation period, and soils were extracted and analyzed at the end of the 60-d period as discussed previously.

These studies indicated more rapid degradation of ATR, DEA, and DIA under saturated soil moisture conditions than in unsaturated soils. After a 60-d incubation in Ames soils, 45% of the applied ^{14}C-DIA remained in saturated subsurface soil (from the 90- to 120-cm depth), significantly less than that remaining in subsurface soils (66%) held at near field capacity [2]. Additionally, ATR and DEA were significantly less persistent in saturated surface soil (0 to 30 cm) than in unsaturated surface soil [5]. A comparison of ATR and DEA fate in sterile and nonsterile soil indicated that this decreased persistence was due to biological degradation (Figure 4).

Influence of Pesticide History on the Degradation of DEA in Nashua Soils. The pesticide history of soil can influence the fate of chemicals subsequently applied to

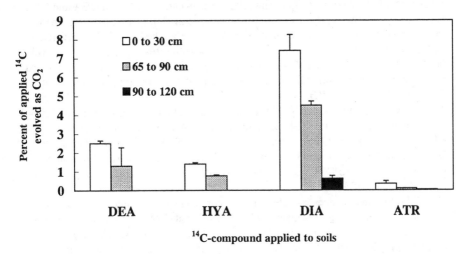

Figure 2. Evolution of $^{14}CO_2$ from Ames soils treated with either ^{14}C-ATR, -DEA, -DIA, or -HYA after a 60-d incubation. Degradation studies for DEA and HYA were not carried out in the 90- to 120-cm depth. Bars (with standard deviations) represent the means of two replicates.

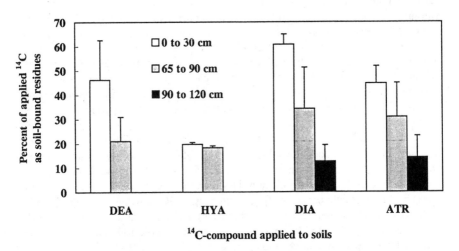

Figure 3. Soil-bound residues formed in soil treated with either ^{14}C-ATR, -DEA, -DIA, or -HYA after a 60-d incubation. Degradation studies for DEA and HYA were not carried out in the 90- to 120-cm depth. Bars (with standard deviations) represent the means of two replicates.

it. Enhanced degradation can result from microbial adaptation, which leads to more rapid degradation of the chemical [6]. Upon repeated exposure to a chemical, adapted microbes may degrade the chemical more rapidly, using it for a source of carbon and/ or nitrogen. Although this phenomenon has not been documented for ATR, a study in our laboratory indicated that this does occur with DEA. Soils were compared from adjacent plots which had two different ATR histories. Soils from the plot that had received 15 consecutive years ATR application was designated ATR-history soil, while soil from the plot that had not received ATR application for 15 years were called no-history soils. Soils from the surface (0 to 30 cm) and subsurface (65 to 90 cm) were treated with uniformly ring-labeled ^{14}C-DEA. Soil moistures were adjusted to and incubated in the dark at 24 °C for 60 d. $^{14}CO_2$ was trapped during the incubation period, and soils were extracted and analyzed as stated previously [2-3,5].

DEA was more rapidly degraded in ATR-history soil compared with no-history soil [7]. After a 60-d incubation, 25% of the applied DEA remained in ATR-history soils compared with 35% in no-history soils (ANOVA; p < 0.05). The amount of $^{14}CO_2$ evolved from complete mineralization of DEA was significantly greater in the ATR-history surface soil than in the no-history surface soil (ANOVA; p < 0.05), with 34% and 17% mineralization, respectively [7].

Mobilities of ATR and ATR Degradates

Mobilities of ATR and DEA in Large Undisturbed Soil Columns from Ames. The mobilities of ATR and DEA were investigated in our laboratory by using large undisturbed soil columns. Large intact soil columns (15 cm in diameter by 60 cm long) were removed from a field with no pesticide history. Soil columns were brought back to the laboratory and prepared for experimentation as previously described by Kruger et al. [8]. A modification to the top of the soil column allowed for monitoring of $^{14}CO_2$, ^{14}C-organic volatiles, and aerobicity of the headspace [9]. Soil columns were treated with either ^{14}C-ATR or ^{14}C-DEA (2 replications each) and then leached weekly with 3.8 cm of simulated rainfall [8,9]. Leachate was collected at the bottom of the soil columns, and the amount of radioactivity recovered in the leachate was determined using liquid scintillation techniques.

For ATR-treated columns, the quantity of radioactivity recovered in the leachate was not sufficient enough to be characterized as parent and degradation products; thus, only the total percentage of ^{14}C in the leachate could be reported. For DEA-applied soil columns, the level of radioactivity was increased so that the radioactivity recovered in the leachate could be characterized as DEA or degradation products. The leachate from DEA-applied soil columns was extracted by using solid-phase extraction techniques [9]. Liquid scintillation techniques were used to quantify radioactivity in the eluant. Thin-layer chromatography and autoradiography were used to characterize the components in the eluant (DEA or degradates).

In these studies, DEA was more mobile than ATR. Total ^{14}C-residues leached from soil columns for the 12-wk study are shown in Figure 5. From the DEA-applied

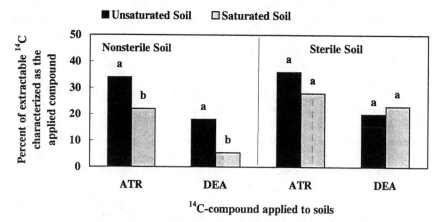

Figure 4. Persistence of ATR and DEA in Ames soils under unsaturated (-33 kPa soil moisture tension) and saturated soil moisture (30% above -33 kPa) conditions (60-d incubation). Pairwise comparisons (by ANOVA; $p < 0.05$) made among moisture conditions for each chemical. Means with the same letter are not statistically different. Bars (with standard deviations) represent the means of two replicates.

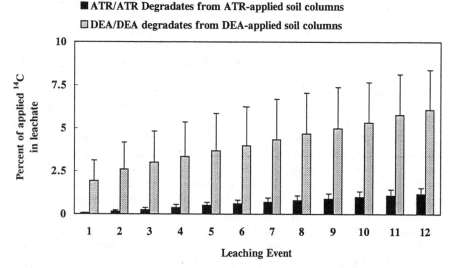

Figure 5. Percentage of applied ^{14}C in leachate of soil columns (representing total ^{14}C-residues) applied with either ^{14}C-ATR or -DEA. Soils were leached with 3.8 cm simulated rainfall weekly for 12 weeks. Bars (with standard deviations) represent the means of two soil columns.

soil columns, 6% of the applied ^{14}C was leached after 12 wk (cumulatively), while only 1% of the applied ^{14}C from ATR-treated columns had leached after that period of time. From DEA-applied columns, 3.6% of the total radioactivity recovered in the leachate was characterized as DEA. Less than 0.2% of the applied ^{14}C was characterized as DDA and deethylhydroxyatrazine (DEHYA, 2-hydroxy-4[amino]-6[isopropylamino]-*s*-triazine), whereas 3.8% was leached as unidentified polar degradates, as determined by the amount of radioactivity in the eluate after solid-phase extraction [9].

Relative Mobilities of ATR and ATR Degradates in Soil Thin-Layer Chromatography (STLC) Experiments with Five Iowa Soils. Relative mobilities of ATR, DEA, DIA, DDA, and HYA were also assessed in our laboratory by using STLC [10]. Surface and subsurface soils from the five locations in Iowa mentioned previously (Table 1) were used in this study. Soil slurries were made by using pulverized soils and ultrapure water. Slurries were poured onto glass plates (20 cm by 20 cm) which had two opposing edges wrapped with masking tape to a thickness of 1 mm. A glass rod was used to roll the slurries evenly across the plates at a depth of 1 mm by using the masking tape as a resting place for both ends of the glass rod. Plates were dried, and approximately 0.5 µCi of ^{14}C-ATR, -DEA, -DIA, -DDA, and -HYA were spotted individually 3 cm from the bottom of the STLC plates. Ascending chromatography was carried out by placing the plates in developing chambers containing 2 cm ultrapure water. After the water had travelled to within 3 cm of the top of the STLC plates, the plates were removed and allowed to dry. Autoradiography was used to visualize the movement of the ^{14}C-triazines. The movement of the solvent front and the chemical fronts were determined and used to calculate the R_f values (relative frontal movement which equals the distance travelled by the front of the radioactive spot divided by the distance travelled by the water).

The mobilities of each chemical tested increased in soil from the subsurface (65 to 90 cm) compared with the surface soil (0 to 30 cm). DEA exhibited the greatest mobility, compared with those of ATR, DIA, DDA, and HYA (Figure 6). HYA was nearly immobile in all soils (R_f values < 0.25). ATR, DIA, and DDA were equally mobile in most soils studied (ANOVA; $p < 0.05$).

Mobilities of ATR and ATR Degradates in a Field Plot Study. To further study the mobility of ATR and degradates under field conditions, tile drains were sampled from ATR-applied fields plots [11]. Water samples were collected from tile drains which were separate for each field plot. Solid-phase extraction techniques were used to isolate ATR, DEA & DIA. Samples were analyzed by liquid chromatography. The general trend of concentrations of ATR and degradates in tile drains was ATR > DEA > DIA. It was also noted that the degradation products, by themselves or in combination with ATR, could exceed the maximum contaminant level set for ATR (3 µg L^{-1}).

Mobilities of ATR, DEA, and DIA in Field Box Lysimeters Treated with Extrazine®. Additionally, field box lysimeters were used to study the movement of

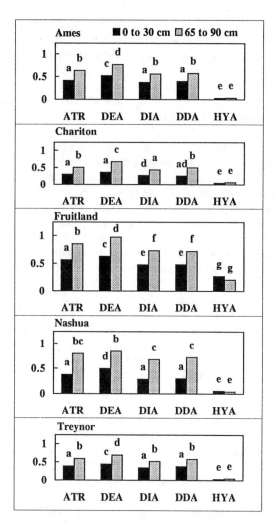

Figure 6. Relative mobilities determined by soil thin-layer chromatography in soils from five locations in Iowa. Mean R_f values (by location) with the same letter are not statistically different (ANOVA; $p < 0.05$).

constituents of Extrazine® II, a herbicide that is a combination of 67% CYA and 21% ATR [12]. Each lysimeter was 229 cm long, 91 cm wide, and 152 cm deep. Each lysimeter was equipped with a porous drainage tile at the 152-cm depth which was connected to a drainage-tile sump apparatus. After each rainfall event, tile drain water was removed, and a subsample was extracted by using solid-phase extraction techniques. Samples were analyzed by gas chromatography and mass spectrometry (GCMS).

From tile drain of lysimeters applied with Extrazine® at a rate of 3.4 kg ha^{-1}, concentrations of DIA exceeded those of DEA during the 1993 growing season (Kruger et al., unpublished). For the first 30 rain events of 1993, the mean concentrations of DEA and DIA in tile drain samples were 0.09 and 0.3 µg/L, respectively. For the last 13 rain events the mean concentrations of DEA and DIA were 0.11 and 1.2 µg/L, respectively. From ATR degradation, it has been reported that DEA concentrations exceed those of DIA [2, 4, 8, 13, 14]. Since CYA can also be degraded to DIA [14], it seems that the contribution of CYA to the DIA concentrations in the tile drain of these lysimeters was very significant.

Summary

In general, degradation is more rapid in surface soils and under saturated soil moisture conditions. Mobility of DEA is greater than that of the parent compound. DIA and DDA mobilities are similar to the mobility of ATR. Levels of DIA were less than DEA in the soil profile of ATR-applied soil columns. Also, DEA was detected in higher concentrations than DIA in the tile drains of ATR-applied field plots. DIA concentrations exceeded those of DEA in tile drains of field lysimeters applied with CYA and ATR. Thus, CYA may contribute to concentrations of DIA in ground water.

Risk assessment of ATR and degradates necessarily depends on their environmental fate, and their environmental fate is a function of several factors, including mobility, persistence, and degradation products formed as well as their degradability. For example, DEA is more mobile than ATR, but it is also more rapidly degraded. Any potential importance of DEA residues must, of course, be balanced by the probabilities and concentrations of its occurrence. Most of the transformation products of ATR are more degradable than the parent under most environmental conditions, but their relative mobilities vary considerably. It is clear that the fate and significance of herbicides and their degradates in the environment depend on multiple factors and require extensive investigation to understand the complete picture.

Conclusions

1. Persistence of ATR, DEA, and DIA was greater in subsurface soils than in surface soils.
2. Mineralization of ATR and DIA was greater in surface soil compared with subsurface soil.
3. DIA was most susceptible to mineralization, compared with ATR, DEA, and HYA.
4. ATR, DEA, and DIA were less persistent in saturated soils than in unsaturated soils.
5. DEA degradation was enhanced in an ATR-history soil compared with a no-history soil.
6. DEA was more mobile than ATR in studies using soil thin-layer chromatography and undisturbed soil columns.
7. ATR, DIA, and DDA were equally mobile in soil thin-layer chromatography experiments.

8. Concentrations of DEA exceeded those of DIA in tile drains from ATR-applied field plots.
9. Concentrations of DIA exceeded those of DEA in tile drains from field box lysimeters treated with Extrazine® (67% CYA and 21% ATR).

Acknowledgments

This research was supported by grants from the Leopold Center for Sustainable Agriculture, and USDA's Management System Evaluation Area Program (MSEA) and North Central Region Pesticide Impact Assessment Program. Ciba Crop Protection provided analytical standards and radiolabeled chemicals. The editorial comments of Todd Anderson and anonymous reviewers were appreciated. This is Journal Paper J-16556 of the Iowa Agriculture and Home Economics Experiment Station, Ames, IA, Project 3187.

Literature Cited

1. Erickson, L. E.; Lee, K. H. *Crit. Rev. Environ. Contam.* **1989**, 19, 1-14.
2. Kruger, E. L.; Somasundaram, L.; Kanwar, R. S.; Coats, J. R. *Environ. Toxicol. Chem.* **1993**, 12, 1959-1967.
3. Baluch, H. U.; Somasundaram, L; Kanwar, R. S.; Coats, J. R. *J. Environ. Sci. Hlth.* **1993**, B28, 127-149.
4. Skipper, H. D.; Volk, V. V. *Weed Sci.* **1972**, 20, 344-345.
5. Kruger, E. L.; Rice, P. J.; Anhalt, J. C.; Anderson, T. A.; Coats, J. R. *J. Environ. Qual.* **1995**, submitted.
6. Coats, J. R. *Chemtech* **1993**, 23(3): 25-29.
7. Kruger, E. L.; Anhalt, J. C.; Anderson, T. A.; Coats, J. R. *Appl. Environ. Microbiol.* **1995**, submitted.
8. Kruger, E. L.; Somasundaram, L.; Kanwar, R. S.;. Coats, J. R. *Environ. Toxicol. Chem.* **1993**, 12: 1959-1967.
9. Kruger, E. L.; Rice, P. J.; Anhalt, J. C.; Anderson, T. A.; Coats, J. R. *J. Agric. Food Chem.* **1995**, submitted.
10. Kruger, E. L.; Zhu, B; Coats, J. R. *Environ. Toxicol. Chem.* **1995d**, in press.
11. Jayachandran, K.; Steinheimer, T. R.; Somasundaram, L.; Moorman, T. B.; Kanwar, R. S.; Coats, J. R. *J. Environ. Qual.* **1994**, 23, 311-319.
12. Blanchet, L.; Kanwar, R. S. *Proceedings of the American Society of Agricultural Engineers*, St. Joseph, MO, March 11-12, **1994**.
13. Mills, M. S.; Thurman, E. M. *Environ. Sci. Technol.* **1994**, 28, 600-605.
14. Sirons, G. J.; Frank, R.; Sawyer, T. *J. Agric. Food Chem.* **1973**, 21, 1016-1020.

Chapter 13

Transport of Nutrients and Postemergence-Applied Herbicides in Runoff from Corrugation Irrigation of Wheat

A. J. Cessna[1], J. A. Elliott[2], K. B. Best[2], R. Grover[3], and W. Nicholaichuk[2]

[1]Agriculture and Agri-Food Canada, Saskatoon Research Centre, Saskatoon, Saskatchewan S7N 0X2, Canada
[2]National Hydrology Research Institute, Saskatoon, Saskatchewan S7N 3H5, Canada
[3]Agriculture and Agri-Food Canada, Research Station, Regina, Saskatchewan S4P 3A2, Canada

Transport of plant nutrients and postemergence-applied herbicides was monitored in runoff water from four irrigations of a 12.6-ha wheat (*Triticum aestivum* L., *cv.* Owens) field. Cumulative loss of dissolved N was 8.25 kg and corresponded to 0.73% of that applied as fertilizer. Although no P was applied through fertilization, 2.75 kg of total P was also lost. Total losses of diclofop and bromoxynil from the experimental site were 0.098 and 0.035 kg, respectively, which were equivalent to approximately 1% of the amount of each herbicide applied to the wheat field. This relatively large loss was due to the short interval (7 h) between application and the first irrigation. Much less (0.07%) 2,4-D was transported when the corresponding interval was 7 d. The majority (70 - 80%) of each herbicide was transported in runoff water from the first irrigation after herbicide application. Maximum nutrient fluxes were 5.4 and 2.8 g ha^{-1} h^{-1} for N and P, respectively, with corresponding herbicide fluxes being 102, 37.1 and 7.3 mg ha^{-1} h^{-1} for diclofop, bromoxynil and 2,4-D.

The impact of agricultural practices on water quality in Canada with respect to plant nutrients and pesticides needs to be better defined to allow the development of effective environmental policies (1). The quantities of nutrients and herbicides entering surface waters through agricultural runoff can have important implications on future water use. Contaminated runoff waters may endanger freshwater aquatic life, be unsafe for human or animal consumption, or be unsuitable for downstream irrigation (2).

Agricultural runoff can originate from rainfall, snowmelt or surface irrigation. Transport of pesticides in agricultural runoff has been most extensively studied as a result of rainfall (3). In contrast, there is only a single report of pesticide transport in snowmelt runoff, that

0097–6156/96/0630–0151$15.00/0

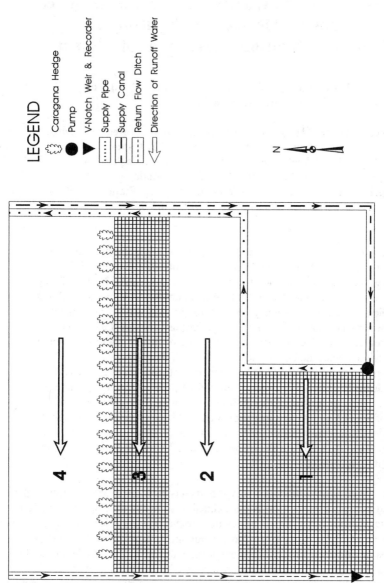

Fig. 1: Description of the study site indicating the sampling location (just upstream of the V-notch weir) and the order in which the four sections of the field were irrigated. Reproduced with permission from reference 9.

being a 6-yr study in which loss of 2,4-D [(2,4-dichlorophenoxy)acetic acid] fall-applied to wheat (*Triticum aestivum* L.) stubble was studied (4). In another 6-yr study, annual rainfall and snowmelt runoff losses of the plant nutrients N and P from corn and oat fields into which fertilizer had been spring-incorporated prior to seeding were reported (5). Losses of N and P in snowmelt runoff from summerfallow and stubble plots over a 6-yr period have also been compared (6).

Although herbicide concentrations in surface irrigation runoff in Alberta were recently reported (7), few reports which quantitate agrochemical transport in surface irrigation runoff are available. In a 5-yr study, Spencer and Cliath (8) determined transport of 20 pesticides, including six soil-applied herbicides, from several irrigated fields in the Imperial Valley in California. Seasonal losses, as percentages of amounts applied, were less than 1% for postemergence insecticides and ranged from 1 to 2% for the preemergence herbicides. More recently, the transport of three postemergence-applied herbicides in runoff from corrugation irrigation of wheat in Saskatchewan has been reported (9). With 6- to 15-d intervals between application and the first irrigation, seasonal herbicide losses were approximately 0.2% of amounts applied. Both of these studies showed that the factors which determined transport of pesticides in rainfall runoff (3) also affected losses in irrigation runoff. Loss of the plant nutrients N and P was also reported in irrigation runoff water from the corrugation irrigation of wheat (9). Transport of dissolved N and total P was 0.13 and 0.29%, respectively, of the amounts equivalent to those applied through fertilization.

In the present study, transport of two postemergence-applied herbicides [diclofop {(±)-2-[4-(2,4-dichlorophenoxy)phenoxy]propionic acid} and bromoxynil (3,5-dibromo-4-hydroxybenzonitrile)] applied only a few hours prior to the first irrigation was compared with that of another (2,4-D) where the interval between application and the first irrigation was several days. Associated transport of N and P was also determined even though only N was applied through preseeding/seeding fertilizer applications. Using a producer's field which was surface irrigated in accordance with normal irrigation practices, herbicide and nutrient concentrations in runoff from corrugation irrigation of wheat were monitored, and temporal trends in herbicide and nutrient fluxes and concentrations were examined with respect to runoff volume.

Materials and Methods

Field Operations. The study site (Figure 1), which consisted of a 12.6-ha field of loam soil (26% sand, 54% silt, 19% clay; 2.9% organic matter) on a producer farm near Outlook, Saskatchewan, has been described previously (9). On 1 May, 1988 the site was fertilized with anhydrous ammonia at a rate equivalent to 84 kg N ha^{-1} and then was seeded on 11 May to a soft white spring wheat (*cv.* Owens). Granular fertilizer (equivalent to 5.6 kg N ha^{-1}) was applied with the seed. No P was applied as fertilizer.

On 9 June, a tank mixture of diclofop plus bromoxynil [Hoe-Grass II, 310 g active ingredient (a.i.) L^{-1} (23:8) formulated as an emulsifiable concentrate of the methyl ester of diclofop and bromoxynil octanoate; Hoechst NOR-AM AgrEvo Inc.] was applied at 0.81 kg acid equivalent (a.e.) ha^{-1} and 0.28 kg phenol equivalent (p.e.) ha^{-1}, respectively, for grassy and broad-leaved weed control when the crop was at the 2- to 3-leaf stage. On 17 June, due to poor germination of the crop, the field was reseeded to wheat (cv Owens) by seeding directly into the first crop with no prior cultivation. On 6 July, 2,4-D (formulated as its dimethylamine salt) at 0.56 kg a.e. ha^{-1} was applied for broad-leaved weed control when the crop was at the 3- to 4-leaf stage.

Irrigation/Rainfall. The study site (Figure 1) was surface-irrigated in four sections by the corrugation method in accordance with normal practices for the region. With the exception of the irrigation of the first section, runoff originated from a minimum of two sections of the field. Corrugation irrigation, a type of surface irrigation, has been briefly described previously (9). The field was irrigated four times during the growing season at approximately 2 mm h^{-1} (Table I). The first irrigation commenced on 9 June, approximately 7 h after the application of the diclofop plus bromoxynil tank mixture. The second irrigation (22 June) followed re-seeding and re-corrugation. The third irrigation began 7 d (13 July) after the application of 2,4-D. The fourth irrigation was on 27 July.

Irrigation water inflow was determined by measuring the pumping rate to the supply pipe on the east side of the field using a Doppler flow meter. Runoff water, which flowed the length of field, was collected in a drainage ditch on the west side of the field which drained in a southerly direction. Runoff water outflow was determined by measuring the runoff rate in the drainage ditch using a V-shaped (1:2) sharp crested weir installed in the drainage ditch at the south end of the field and the head measured by a Stevens A-35 recorder. The runoff water, which was not reused for subsequent irrigation, was carried by the drainage ditch into a waste canal which in turn emptied into the South Saskatchewan River.

A totalizing tipping bucket rain gauge at the site indicated that no rainfall occurred during any of the four irrigations. Although rain fell during the intervals between irrigations, none of these events resulted in rainfall runoff. Soil moisture to 2.4 m was measured by gamma ray attenuation at four locations in the field. Measurements were made through June and July. After each irrigation, water lost through deep percolation was calculated as the change in water storage between 1.2 and 2.4 m.

Surface Runoff Water Sampling. Using two automated water samplers (Saskatchewan Research Council, Saskatoon, SK), hourly 1-L runoff water samples were collected just upstream of the sharp crested weir. Sampling and sample handling procedures were as described previously (9). One water sample was used to determine nutrient content, the second for herbicide residue analysis. The runoff water samples contained small and varying amounts of sediment.

During each irrigation, 1-L water samples were also collected from ports on the supply pipe to determine background concentrations of herbicides and nutrients in the irrigation water being applied to the experimental site. Two samples, one for each type of analysis, were collected for each irrigated section of the site.

Nutrient Analysis. Every second sample was analysed for nitrate/nitrite, ammonia, total P and orthophosphate using Environment Canada standard colorimetric methods (9). Unfiltered aliquots were used for total phosphorus and for ammonia analysis whereas filtered aliquots were used for nitrate/nitrite and orthophosphate analysis.

Total amounts of N and P transported off the experimental site in the runoff water were determined by summing the amounts of these nutrients lost each hour over the time period of outflow. Hourly losses were calculated as the product of the concentration of nutrient (mg L^{-1}) in the collected water sample and the volume of outflow (m^3) which occurred during that hour.

Herbicide Analysis.

Sample Extraction. As for the nutrient analyses, only every second sample was analysed for herbicide content. The unfiltered water samples were extracted and the extracts methylated according to procedures indicated previously (9). Following gas chromatographic quantitation of herbicide residues in the runoff water samples using electron-capture detection, total herbicide losses in the runoff outflow were determined in the same manner as for the nutrients.

Gas Chromatography. A Hewlett-Packard Model 5890 gas chromatograph, equipped with a ^{63}Ni electron-capture detector and on-column injector, was used with the Model 7673A autosampler set to inject 2 µL, and the Model 5895A data station. A 30-m x 0.53-mm i.d. HP-1 cross-linked methyl silicone fused silica column (Hewlett-Packard; film thickness, 0.88 µm) was used with the following operating conditions: a temperature program consisting of 70 C for 1 min, then 5 C min^{-1} until 270 C and hold for 5 min; carrier gas (helium) flow rate, 8 mL min^{-1}; detector make-up gas (nitrogen) flow rate, 70 mL min^{-1}; detector, 350 C. Under these conditions, the total run time was 46 min and the retention times were 18.6, 19.0, 33.1 and 34.0 min for bromoxynil methyl ether, 2,4-D methyl ester, bromoxynil octanoate and diclofop methyl ester, respectively.

Fortification Experiments. Recoveries were determined for 2,4-D, bromoxynil, bromoxynil octanoate and diclofop by extraction of water fortified at 1 µg L^{-1} with each herbicide. Because the analytical method would not permit differentiation between residues of diclofop and those of its methyl ester, fortification with diclofop methyl ester was not included in the recovery determinations. Fortification was by the addition of 0.5 µg of 2,4-D, bromoxynil, bromoxynil octanoate (in 1 mL of methanol) and diclofop (in 1 mL of acetone) to 500 mL of water. Respective percent recoveries, on the basis of 13 replicate samples and expressed as mean ± standard deviation, were 85.3 ± 10.3%, 78.2 ± 10.2%, 93.6 ± 13.8% and 105.8 ± 15.7%. Based on these recoveries, an operational limit of quantitation of the analytical method was considered to be 0.5 µg L^{-1} for each analyte.

Results and Discussion

Irrigation Efficiency. Although varying amounts of irrigation water were applied during the four irrigations, losses due to deep percolation were consistent at about 14% of the applied water over all irrigations (Table I). Efficiencies for the first, third and fourth irrigations were all around 65%. The poorest irrigation efficiency (52%) was found for the second irrigation when the least water was applied. Since runoff was consistently around 20 mm irrespective of the application amount, runoff was proportionally much greater for the second irrigation (34%). This occurred because the second irrigation was applied a few days after re-seeding and re-corrugation when the soil was hard-packed and there was only sparse crop to slow the passage of water. Because of hard-packed soil conditions over the growing season of the present study, percent runoff was greater for all irrigations than in the two irrigations made in the previous year at the same site (9) when even more water was applied. In that year, only 12 and 8% runoff occurred from the first (193 mm) and second (145 mm) irrigations, respectively.

Table I. Application amounts, losses of applied water, and efficiencies for the four irrigations

Irrigation	Applied	Runoff		Below Roots		Efficiency
	mm	mm	%	mm	%	%
1st	120	24	20	18	15	65
2nd	58	20	34	8	14	52
3rd	79	19	24	11	14	62
4th	95	18	19	13	14	67

Nutrient Runoff. Of the total dissolved inorganic N in the runoff water from the first three irrigations, NO_3-N accounted for at least 97% of dissolved N with the remainder being NH_3-N. In the fourth irrigation, NH_3-N concentrations were higher and accounted for 11% of dissolved N. Concentrations of NO_3-N and NH_3-N tended to be relatively constant from sample to sample over all irrigations. In contrast, there was much greater sample to sample variability in ortho-P and total P concentrations and in the proportion of ortho-P, especially in the first and second irrigations. The ortho-P content in the runoff water from the first and second irrigations accounted for approximately 40% of the total P whereas, for the third and fourth irrigations, about 65% was ortho-P. These observations may reflect greater soil erosion and availability of ortho-P as a result of the soil disturbance due to seeding and corrugation which preceded the first and second irrigations.

Dissolved NO_3-N concentrations (Table II) did not exceed the Canadian Drinking Water Quality Guideline of 10 mg L^{-1} (2). However, in all but the fourth irrigation, some samples exceeded the proposed multipurpose water quality objective for total inorganic N of 1 mg L^{-1} (10). Only during the second irrigation did the weighted concentration mean of dissolved N (1.13 mg L^{-1}) exceed the Saskatchewan objective. During all irrigations, weighted concentration means of total P exceeded the maximum desirable concentration in flowing water of 0.1 mg L^{-1} (11).

In a previous study (9), nutrient concentrations and fluxes were less in runoff water from the second irrigation than from the first irrigation. This was attributed to removal or leaching of available nutrients by runoff from the first irrigation, and uptake by the crop. The data in the present study followed the same trend with the exception of the second irrigation (Table II). The inefficiency of water use during the second irrigation resulted in the loss of more nutrients in runoff than in other irrigations and, as a consequence, the greatest concentrations and fluxes of dissolved N and total P were observed during the second irrigation. The soil disturbance accompanying the re-seeding and re-corrugation of

Table II. Nutrient concentration ranges, weighted concentration means, mean fluxes and amounts transported for the four irrigations

| Irrigation | Dissolved Inorganic N (NO$_3$ + NH$_3$) | | | | Total P | | | |
| | Conc. Range | Weighted Conc. Mean | Nutrient Flux | Amount Lost | Conc. Range | Weighted Conc. Mean | Nutrient Flux | Amount Lost |
	mg L^{-1}	mg L^{-1}	g ha^{-1} h^{-1}	g ha^{-1}	mg L^{-1}	mg L^{-1}	g ha^{-1} h^{-1}	g ha^{-1}
1st	0.27-1.97	0.85	3.7	207	0.27-1.00	0.26	1.6	63.4
2nd	0.32-4.00	1.13	5.4	225	0.03-1.12	0.56	2.8	111
3rd	0.37-1.19	0.65	2.9	126	0.08-0.42	0.12	0.6	23.3
4th	0.39-0.99	0.54	2.5	98.0	0.06-0.19	0.11	0.5	21.3

the field which took place a few days prior to the second irrigation may have also contributed to the increased nutrient losses during runoff.

With the exception of the second irrigation, nutrient concentration data in the present study (Table II) were comparable to those measured on the same site the previous year (9). Weighted concentration means determined from the second irrigation were more than double those of the previous year. The greater nutrient fluxes and total transport observed during the present study relative to the previous year reflect the inefficiencies of the irrigations in the second year and the greater percent runoff which occurred.

Because of the greater nutrient concentrations and fluxes in the runoff water from the second irrigation, the largest amounts of both nutrients were transported from the site during this irrigation (Table II). Amounts transported of either nutrient decreased with each subsequent irrigation. Over all four irrigations, a total of 8.25 kg of dissolved N was removed from the field which was equivalent to 0.73% of N applied as fertilizer. The cumulative loss of P was 2.75 kg from all four irrigations.

Figure 2 shows the temporal variability in NO_3-N concentration and runoff volume as irrigation water was applied to different sections of the field during the second irrigation. All flow during the first 8 h of runoff came from the first section to be irrigated after which the runoff consisted of contributions from more than one section of the field. In the first 8 h of runoff, there was a very clear inverse relationship between runoff flow rate and NO_3-N concentration. A similar relationship was noted previously (9) and was attributed to greater runoff-soil interaction during periods of low flow, and a dilution effect during periods of higher flow. After the first 8 h of runoff, the same effect can still be seen but with fluctuations in NO_3-N concentration which probably reflect the start and end of runoff from different sections of the field. Similar patterns were observed for the other irrigations although the inverse relationship between flow rate and concentration was weaker for the third and fourth irrigations.

Transport of Herbicides.

Residues of diclofop and bromoxynil were present in the runoff from all four irrigations whereas those for 2,4-D occurred only in the runoff from the third and fourth irrigations (Table III). The more hydrophobic bromoxynil octanoate was detected only in runoff from the first irrigation indicating that it was completely hydrolyzed prior to the second irrigation. Concentrations of bromoxynil octanoate in runoff from the first irrigation ranged from 2.6 µg L^{-1} (9 June) to < 0.5 µg L^{-1} (12 June). Because of the small amounts of sediment in the runoff water samples, none of the samples were filtered and the sediments analysed separately. Thus, the bromoxynil octanoate residue data do not indicate whether the octanoate was in solution or adsorbed to the sediments. Since bromoxynil octanoate was detected in runoff only from the first irrigation, bromoxynil transport (Table III) and concentration (Table IV) data for this irrigation were derived from the total of bromoxynil and its octanoate in the runoff water. Diclofop-methyl may also have been present in the runoff either in solution or adsorbed to sediments. However, because diazomethane was used for derivatization, the analytical method did not permit differentiation between diclofop-methyl and its hydrolysis product (diclofop).

Runoff transport of all three herbicides is presented both as amounts lost (g ha^{-1}) and as percent loss which normalizes loss with respect to application rate (Table III). The effects of time between herbicide application and the first irrigation, water solubility of the herbicide, and soil persistence of the herbicide are meaningful only in terms of normalized loss. Greater than 80% of the total transport of diclofop and bromoxynil occurred in runoff from the first irrigation. The short 7-h interval between the application of the tank mixture of diclofop plus bromoxynil and the first irrigation represents somewhat of a worst-case scenario for herbicide transport. Total amounts of diclofop and bromoxynil lost during four irrigations were in the same relative order of the amounts applied.. When normalized as a percent, losses of diclofop and bromoxynil were essentially the same, being ~ 1% of what was applied, even though they have marked differences in water solubility [diclofop: 122.7 g L^{-1} (M. Belyk, 1993, personal communication); bromoxynil: 0.1 g L^{-1} (12)] and in soil persistence (diclofop > bromoxynil; cf. 13,14). As a consequence of the short residence time on the soil surface prior to the first irrigation, there was insufficient time for expression of differences in rates of microbial degradation, photochemical degradation and formation of bound residues of these two herbicides.

Table III: Total and percent transport of diclofop, bromoxynil and 2,4-D in runoff water from four corrugation irrigations of the experimental site

Irrig	Total Transport			Percent Transport		
	Diclofop	Bromox[a]	2,4-D	Diclofop	Bromox[a]	2,4-D
	------------------ g ha^{-1}----------------			-------------------- % -----------------		
1st	6.4	2.3	-	0.79	0.83	-
2nd	1.0	0.34	-	0.12	0.12	-
3rd	0.27	0.06	0.29	0.03	0.02	0.05
4th	0.10	0.02	0.13	0.01	0.01	0.02

[a]Values were derived from the total of bromoxynil and the octanoate.

Greater than 95% of the total transport of the diclofop and bromoxynil occurred during the first two irrigations (Table III). Since the tank mixture was applied when the crop was at the 2- to 3-leaf stage, <50% of the application would have been intercepted by the weed/crop canopy (15) with the remainder depositing on the soil surface. Thus, transport of the two herbicides would have been in the order of 1-2% of what deposited on the soil surface. This magnitude of loss is similar to losses reported previously for soil-applied herbicides (8), and about an order of magnitude greater than losses of postemergence-applied herbicides when the interval between application and the first irrigation was much greater (9). The inefficiencies of the first two irrigations (Table I) relative to the previous report (9) may have also contributed to the relatively greater herbicide transport observed in the present study.

2,4-D was applied 7 d before the third irrigation when the re-seeded crop was at the 3- to 4-leaf stage. Due to the much longer interval between application and the first irrigation after application, much smaller amounts of 2,4-D were lost in the runoff water relative to

Table IV. Concentration ranges, weighted concentration means and average fluxes of diclofop, bromoxynil and 2,4-D in runoff water from four irrigations of the experimental site

Irrigation	Concentration Range			Weighted Concentration Mean			Average Flux		
	Diclofop	Bromox[a]	2,4-D	Diclofop	Bromox[a]	2,4-D	Diclofop	Bromox[a]	2,4-D
	-------- μg L^{-1} --------			-------- μg L^{-1} --------			-------- mg ha^{-1} h^{-1} --------		
1st	13.1-63.8	4.4-20.4	-	26.1	9.4	-	102	37.1	-
2nd	<0.5-10.7	<0.5-4.9	-	5.0	1.7	-	25.1	8.7	-
3rd	<0.5-4.8	<0.5-0.6	0.7-3.5	1.4	0.3	1.5	6.7	1.4	7.3
4th	<0.5-1.3	<0.5-0.2	0.5-1.2	0.6	0.1	0.7	2.6	0.6	3.2

[a]Values were derived from total of bromoxynil and the octanoate.

diclofop and bromoxynil (Table III). The cumulative amounts lost in the third and fourth irrigations were 3.7 and 1.6 g, respectively and, in total, corresponded to 0.07% of the amount applied. This much smaller loss of 2,4-D, relative to diclofop and bromoxynil, is similar to those reported previously for dicamba (3,6-dichloro-*o*-anisic acid), MCPA [(4-chloro-*o*-tolyl)oxy)acetic acid] and diclofop when a 6-d interval occurred between application and the first irrigation (9). The reduced loss of 2,4-D was most likely due to dissipation of the herbicide in the runoff-soil interaction zone during this interval. Moist surface soil due to a 23-mm rain on the day before spraying would have created optimal conditions for microbial degradation [soil half-life = <7 to 20 d (16)]. Although vapour losses would have been negligible because 2,4-D was applied as its dimethylamine salt, further dissipation of 2,4-D may have resulted due to photodegradation on the soil surface (17) or due to leaching out of the runoff-soil interaction zone.

The temporal pattern of herbicide loading in the runoff water was similar to that of the corresponding outflow hydrograph indicating that herbicide transport increased with runoff volume. As observed previously (8,9), even though herbicide concentrations in the runoff water showed an inverse relationship with outflow, herbicide fluxes increased with outflow. Diclofop concentration and flux patterns are illustrated in Figure 3 for the initial 16 h of the first irrigation when all runoff could be attributed to one section of the field. Similar patterns were observed for bromoxynil and 2,4-D. Average fluxes for diclofop, bromoxynil and 2,4-D during the first irrigation after application were 102, 37.1 and 7.3 mg ha^{-1} h^{-1}, respectively (Table IV). Fluxes of all three herbicides decreased significantly with each subsequent irrigation.

Relative to the first irrigation, percent transport of all three herbicides was greatly reduced but of the same order of magnitude during the second and subsequent irrigations (Table III). Reduced transport was also evident from average flux values, weighted concentration means and concentration ranges (Table IV) and was most likely due to herbicide removal from the runoff-soil interaction zone during each irrigation via the runoff water and/or leaching, and by microbial degradation during the interval between irrigations. Similarities in these losses of diclofop and bromoxynil are most likely explained by differences in their water solubilities and sorption coefficients; however, sorption coefficient values for these two herbicides have not yet been reported.

Concentrations of diclofop and bromoxynil, and thus their weighted concentration means, in the runoff water from the first irrigation exceeded Canadian Water Quality Guidelines (2) for drinking water, freshwater and irrigation (Table V), and thus would have been unsuitable for downstream irrigation. Diclofop concentrations and its weighted concentration mean also exceeded guidelines for livestock watering. During the second irrigation, some diclofop concentrations in the runoff water exceeded these guidelines but the weighted concentration means of both herbicides only exceeded irrigation guidelines for some crops. None of the concentrations or weighted concentration means of 2,4-D exceeded any of these guidelines.

Although the runoff water from the first two irrigations of the experimental site was unsuitable for downstream irrigation with respect to diclofop and bromoxynil concentrations, not all surface-irrigated fields within the Outlook Irrigation District were treated with these two herbicides. Thus, in general, there is potential for dilution of such herbicide concentrations when the runoff water via drainage ditches from individual fields enters the main waste canals that empty into the South Saskatchewan River. A study is currently in

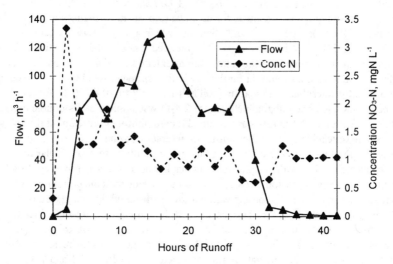

Fig. 2: Outflow (m³ h⁻¹) and NO₃-N concentration (mg L⁻¹) in the runoff water from the second irrigation.

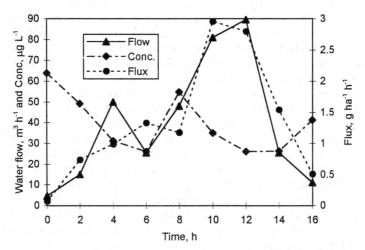

Fig. 3: Outflow (m³ h⁻¹), diclofop concentration (µg L⁻¹), and diclofop flux (mg ha⁻¹ h⁻¹) associated with the runoff water from the first irrigation of the first section of the experimental site.

progress to monitor herbicide concentrations in the effluent from the two main waste canals within the Outlook Irrigation District over a complete growing season to better determine the feasibility of downstream reuse and the magnitude of herbicide inputs into the South Saskatchewan River.

Table V: Canadian Water Quality Guidelines for diclofop, bromoxynil and 2,4-D

Herbi-cide	Raw water for water supply	Freshwater life	Agricultural uses	
			Irrigation	Livestock watering
		$\mu g\ L^{-1}$		
Diclofop	9	6.1	0.18 - 5.6	9
Bromox	5	5	0.35 - 7.4	11
2,4-D	100	4.0	100	100

Conclusions

Transport of dissolved N in runoff water from four irrigations was 0.73% of N applied as fertilizer. Because of soil disturbance due to re-seeding and re-corrugation, maximum transport of N and P occurred during the second irrigation. Maximum weighted concentration means were detected in runoff from the second irrigation and were 1.13 mg L^{-1} for dissolved N and 0.56 mg L^{-1} for total P. The value for dissolved N did not exceed Canadian Water Quality Guidelines, but that for total P exceeded the maximum desirable concentration in flowing water. With the very short interval (7 h) between application and the first irrigation, total runoff losses of diclofop and bromoxynil were relatively high at approximately 1% of what was applied. In contrast, losses of 2,4-D following a 7-d interval were only 0.07%. Approximately 80% of total herbicide transport occurred during the first irrigation producing maximum herbicide concentrations which were 63.8, 20.4 and 5.5 $\mu g\ L^{-1}$, respectively, for diclofop, bromoxynil and 2,4-D. Diclofop and bromoxynil weighted concentration means from runoff water of the first two irrigations exceeded guidelines for irrigation water and thus reuse of this runoff water for downstream irrigation would not be recommended without some dilution.

Acknowledgments

The authors thank Roger Pederson, Outlook, SK for permission to use his land for this study and for his very helpful co-operation in carrying out the study.

Literature Cited

1. Smith, W. In *Water resources in a global environment: the policy issues.* Milburn, P., Nicholaichuk, W., Topp, C., Eds.; Agricultural Impacts on Water Quality: Canadian Perspectives. Canadian Agricultural Research Council, Ottawa, ON, 1992, pp 2-8.

2. Canadian water quality guidelines. Task force on Water Quality Guidelines of the Canadian Council of Resource and Environment Ministers. Canadian Council of Resource and Environment Ministers. Water Quality Branch, Environment Canada, Ottawa, ON, 1987.

3. Wauchope, R. D. *J. Environ. Qual.* **1978,** 7, 459-472.

4. Nicholaichuk, W. and Grover, R. 1983. *J. Environ. Qual.* **1983,** 12, 412-414.

5. Burwell, R.E., Timmons, D.R., Holt, R.F. *Soil Sci. Soc. Amer. Proc.* **1975,** 39, 523-528.

6. Nicholaichuk, W. and Read, D.W.L. *J. Environ. Qual.* **1978,** 7, 542-544.

7. Miller, J.J., Foroud, N., Hill, B.D., Lindwall, C.W. *Can. J. Soil Sci.* **1995,** 75, 145-148.

8. Spencer, W.F., Cliath, M.M. *Environ. Sci. Res.* **1991,** 42 (Prot. Environ.), 277-289.

9. Cessna, A.J., Elliott, J.A., Kerr, L.A., Best, K.B., Nicholaichuk, W., Grover, R. *J. Environ. Qual.* **1994,** 23, 1038-1045.

10. Water quality objectives, January 1975. 3rd printing. Saskatchewan Ministry of the Environment. Water Pollution Control Branch, Regina, SK, 1983.

11. McNeely, R.N., Neimanis, V.P., Dwyer, L. *Water quality sourcebook: A guide in water quality parameters;* Inland Waters Directorate, Water Quality Branch, Environment Canada, Ottawa, ON, 1979.

12. Anonymous. 1989. Herbicide handbook of the Weed Science Society of America. 6th ed. WSSA, Champaign, IL.

13. Smith, A.E., R. Grover, R., Cessna, A.J., Shewchuk, S.R., Hunter, J.H. *J. Environ. Qual.* **1986,** 15, 234-238.

14. Grover, R., Smith, A.E., Cessna, A.J. *J. Environ. Qual.* **1994,** 23, 1304-1311.

15. Grover, R., Shewchuk, S.R., Cessna, A.J., Smith, A.E., Hunter, J.H. *J. Environ. Qual.* **1985,** 14, 203-210.

16. Smith, A.E., Hume, L. LaFond, G.P., Biederbeck, V.O. *Can J. Soil Sci.* **1991,** 71, 73-87.

17. Cessna, A.J., Muir, D.C.G. In *Photochemical transformations.* Grover, R. and Cessna A.J., Eds.; Environmental Chemistry of Herbicides; CRC Press, Boca Raton, FL, 1991, Vol 2; pp 199-263.

Chapter 14

Potential Movement of Certain Pesticides Following Application to Golf Courses

A. E. Smith and D. C. Bridges

College of Agricultural and Environmental Sciences, University of Georgia, Georgia Experiment Station, Griffin, GA 30223

The environmental fate and safety of pesticides used in the management of recreational facilities is a critical issue facing the turfgrass industry. The purpose of this research was to determine the potential movement of certain pesticides following application to golf course greens and fairways. Greenhouse and outside lysimeters were constructed and leachate, exiting the base of the lysimeters, was collected and analyzed for transport of treatment pesticides. Herbicide movement from simulated fairways was determined following treatment of small plots having a 5% slope and sodded with 'Tifway' bermudagrass. The data indicate that less than 1% of the applied pesticides were transported in the leachate from the base of the lysimeters. Highest concentrations of analytes in the leachate were less than 5.0 μg L^{-1}. However, higher concentrations (279 to 810 μg L^{-1}) of the treatment herbicides were transported in the runoff water from the treated simulated fairway plots.

Although agriculture is the largest user of pesticides in North America, turfgrass is typically the most intensively managed biotic system (*1*). The public demand for high quality and uniform turf often requires the use of intensive management strategies to maximize pest control and nutrient availability (*2*) resulting in increased public awareness of pesticide use. The enhanced interest is, in general, a response to the increased use of pesticides and fertilizer since the 1960s and advancements in technology allowing scientists to detect their presence at very low concentrations (*3*). The major concern about pesticides in the environment is their potential entrance into drinking water sources by movement in surface water and groundwater from the treated site (*4*).

Erosion and surface runoff processes in relation to water quality and environmental impacts have been examined by Anderson et al. (*5*), Leonard (*6*), and Stewart et al. (*7*). Information available on the mobility and potential for

0097–6156/96/0630–0165$15.00/0
© 1996 American Chemical Society

contaminating groundwater with pesticides used on turf is limited. Cohen et al. (*8*) sampled and analyzed water from 16 monitoring wells on golf courses and found chlorpyrifos (*O,O*-dimethyl *O*-3,5,6-trichloro-2-pyridyl phosphorothioate), 2,4-D [(2,4-dichlorophenoxy)acetic acid], dicamba (3,6-dichloro-2-methoxybenzoic acid), and isofenphos {*O*-[ethoxy-*N*-isopropylamino(thiophosphoryl)]salicylate} in at least one of the wells, chlorothalonil (2,4,5,6-tetrachloro-1,3-benzenedicarbonitrile) in 2 wells, DCPA (dimethyl 2,3,5,6-tetrachloro-1,4-benzenedicarboxylate) in 3 wells, heptachlor epoxide (1,4,5,6,7,8,8-heptachloro-3a,4,7,7a-tetrahydro-4,7methano-1*H*-indene) in 4 wells, and chlordane (1,2,4,5,6,7,8,8-octachloro-2,3,3a,4,7,7a-hexahydro-4,7-methaneoindene) in 7 wells. Eighty percent of the compounds were found in concentrations less than 0.5 µg L^{-1}.

Although conclusive evidence of health effects of long-term exposure to pesticides has yet to be established, there is intense public perception of risk concerning pesticides in drinking water (*4*). The EPA is currently establishing drinking water standards of reference doses for surface and groundwater (*9*). Standards will be based on the same toxicological research used to establish reference doses (formerly called Acceptable Daily Intake, or ADI) for food. These standards will be the maximum contaminant levels (MCLs) allowed for pesticide concentrations in potable water. The MCLs for only a few pesticides used on turfgrass have been recommended. The recommended MCL for 2,4-D is 70 µg L^{-1}. In addition to federal efforts to alleviate environmental quality concerns, state governments are also in the process of developing water quality regulations (*10,11*). State governments recognize the need to protect valuable surface and groundwater resources through both education and enforcement. Some states, such as California, New York, Nebraska, and Wisconsin, have selected a regulatory approach to water quality issues (*11*). Others, including Iowa, are legislating a combined approach of education, research, and demonstration (*12*).

Currently, there are over 14,000 golf courses in the United States and assuming an average size of 48.6 ha per course (*13*), there are over 0.68 million ha of turfgrass in the golf course industry. Additionally, the National Golf Foundation estimates that there are 21.7 million golfers in the United States, and by the year 2000, the number of players could easily exceed 30 million. To keep up with the demands of the rapidly increasing number of golfers, it is suggested that a new golf course must be opened every day over the next 10 years. Color, uniformity, and density of the turfgrass on these golf courses will be affected adversely by incursions of weeds, diseases, and insects. Turfgrass of high quality and uniform playing surface has become the expected necessity on golf courses and this condition often requires the use of intensive management to control pests.

Assuming that 2% of a golf course is managed as putting greens, there are 13.6 thousand ha of greens in the U.S. which are constructed for maximum infiltration and percolation of water through the rooting media. Root zone mixture composition generally includes at least 85% by volume (97% by weight) coarse sand allowing for rapid water percolation and having an extremely low cation exchange capacity. Additionally, soil sterilization is recommended during construction for weed and disease management (*14*). The sterilization ultimately influences the soil microbial decomposition of applied pesticides. These characteristics of the root zone mixture could result in rapid movement of pesticides through the rooting mixture allowing for a potential source of contamination of the effluent water from the greens. Additionally, the Ground Water Contamination Potential rating for use of the benzoic acid and phenoxyalkanoic acid herbicides on golf course greens, constructed according to U.S. Golf Association (USGA) specifications, would be RISKY (*15*). However, the tight thatch overlying the rooting media is very important in the retention and degradation of most pesticides. Studies by several investigators (*16-20*) showed that the chlorinated hydrocarbon insecticide chlordane and several organophosphorus pesticides, were retained in large amounts (>90%) by bluegrass thatch and that only small amounts (<10%) leached below the thatch zone. Gold et al. (*21*), in studies with dicamba and 2,4-D, only found 1.0 and 0.4 percent, respectively, of the total amounts applied to pass through the turf thatch.

Fairways compose approximately 98% of the golf courses and are typically intensively managed. The fairways are developed on soils typical for the region and in the Piedmont these soils have a high clay content allowing for a low water infilteration rate especially when crusted. As much as 70% of the rainfall will occur as runoff water from the sloped areas. This surface water can eventually terminate in potable water containments and it is estimated that 95% of the drinking water consumed by residents of the Piedmont comes from surface water containments resulting in a concern for surface water quality.

Materials and Methods

Measurement of Pesticide Movement from Simulated Greens. Thirty six lysimeters were constructed, in the greenhouse, by placing turfgrass growth boxes (40 x 40 x 15 cm deep) on top of bases (*22*). The bottom of the wooden growth boxes was perforated steel and at the inside-center of the growth boxes a 13-cm length of polyvinyl chloride (PVC) tube (15 cm diam.) was fastened to the bottom with acrylic caulk. The base of the lysimeter consisted of a 52.5 cm length of PVC tubing (15 cm diam.) capped at the bottom. The cap had a drain tube for the collection of aqueous effluent.

Prescribed rooting mixtures (sand:sphagnum peat moss) were based on the percolation rate for the sand. The proportions (85:15 and 80:20 v:v) were selected to give final percolation rates of 39 and 33 cm hr^{-1}, respectively. The rooting media were steam sterilized prior to use. The lysimeter bases were filled with sized gravel (10 cm), coarse sand (7.5 cm), and rooting mix (35 cm) in ascending sequence from the bottom simulating USGA specifications for greens construction. The bases of the lysimeters were enclosed and cooled by an air conditioner in order to maintain the soil temperature between 18-21°C. The lysimeters were housed in a greenhouse covered with LEXAN thermoclear sheet glazing. 'Penncross' creeping bentgrass (*Agrostis stolonifera* L.) was seeded into all boxes during September 1991. An automatic track-irrigation system was developed for controlling the rates and times for irrigation. The daily irrigation of 0.63 cm of water and a weekly rain event of 2.5 cm were controlled with an automatic timer. The conditions were chosen to simulate management practices and average rainfall events for golf course greens in central Georgia and the weekly water input was approximately 2X loss by open-pan evaporation. The treatments (pesticide/rate) for bentgrass (2,4-D/0.28; dicamba/0.07; and mecoprop/0.56 kg ha^{-1}) were applied to the growth boxes in 204 L ha^{-1} water diluent. A complete fertilizer was applied in water to an N rate of 2.44 g m^{-2}. The leachate in the sample bottles, at the base of the lysimeters, was collected and stored in a refrigerator maintained at 4°C. Collections were combined for weekly intervals and quantified for pesticide in the leachate.

The outside facility consisted of small greens subtended with lysimeters for directing the flow of water and pesticides into a collection area. The small greens were developed with similar rooting media as used in the greenhouse experiments. We used the 85:15 mix for the bentgrass green and 80:20 mix subtending the 'Tifdwarf' bermudagrass [*Cynodon dactylon* (L.) Pers. x *C. transvaalensis* Burtt-Davy] green. The interior diameter of each lysimeter is 55 cm and the depth is 52.5 cm allowing for layers of gravel, sand, and rooting media as developed in the bases of the greenhouse lysimeters. Bentgrass was seeded on October 15, 1991 and bermudagrass was sodded during March 1992. Pesticide treatments, for the respective species, are listed in Table I. A horizontal moving irrigation system was used to simulate the irrigation and rainfall events and an automatic moving rain shelter was constructed for movement over the greens during rain events. The simulated event intensities and frequencies were similar to the ones described for the greenhouse.

Table I. Pesticide treatments applied to 'Penncross' bentgrass and 'Tifdwarf' bermudagrass, in field lysimeter facility

Herbicide[1]	Rate	Quantity applied over each lysimeter
	kg ha^{-1}	mg
Bentgrass		
2,4-D DMA[2]	0.28	8.18
dicamba DMA	0.07	1.96
mecoprop DMA	0.56	16.36
control	----	0
Bermudagrass		
2,4-D DMA	1.12[3]	32.72
dicamba DMA	0.28	8.18
mecoprop DMA	1.40	40.91
control	----	0

[1]All pesticide mixes contained 0.5% X-77 surfactant (contains alkylarylopolyoxyethylene glycols, free fatty acids, isopropanol).
[2]DMA = dimethylamine salt formulation.
[3]Applied as a split application on 2-week interval.

Measurement of Pesticide Movement from Simulated Fairways. Twelve
individual plots (3.7 x 7.4 m), separated by landscape timbers, were developed in a
grid with a 5% slope from the back to the front. The subsoil was a clay loam and
the top soil was a sandy loam. A ditch was dug at the front of the plots to install a
trough for collecting the runoff water in a tipping-bucket sample collection
apparatus. The plots were sprigged with 'Tifway 419' bermudagrass on May 17,
1993 and the plots were completely covered by August 1, 1993. The WOBBLER
off-center rotary action sprinkler heads were mounted 7.4 m apart and 3.1 m above
the sod surface. When operated at 138 kPa, the system produced an even
distribution of simulated rainfall at an intensity of 3.3 cm hr^{-1}.

The plots were treated, in three replications, with the pesticides listed in Table
II between June 1 and November 1 during 1993 and 1994. Selection of the specific
treatment dates was based on meteorological forecasts that allowed for at least a
72-hr period with a low probability of rainfall. Rainfall was simulated at 24, 48,
96, and 192 hr after treatment (HAT). The subsamples were collected following
the rainfall event and stored at 4°C. Following the simulated rainfall period normal
rainfall events were monitored until herbicides in the runoff water were not
detected. The runoff water was quantified and subsamples were collected by the
tipping-bucket apparatus.

Table II. Pesticide applications made to simulated fairways

Pesticide		Rate
common name	chemical nomenclature	kg ha^{-1}
benefin	N-butyl-N-ethyl-2,6-dinitro-4-(trifluoromethyl) benzenamine	1.70
2,4-D DMA[1]	(2,4-dichlorophenoxy)acetic acid	2.24
dicamba DMA	3,6-dichloro-2-methoxybenzoic acid	0.56
dithiopyr	S,S-dimethyl 2-(difluoromethyl)-4-(2-methylpropyl)-6-(trifluoromethyl)-3,5-pyridinedicarbothioate	0.56
chlorothalonil	(2,4,5,6-tetrachloro-1,3-benzenedicarbonitrile)	9.50
chlorpyrifos	O,O-diethyl O-(e,5,6-trichloro-2-pyridyl)phosphorothioate	1.12
mecoprop DMA	(±)-2-(4-chloro-2-methylphenoxy)propanoic acid	1.68
pendimethalin	N-(1-ethylpropyl)-3,4-dimethyl-2,6-dinitrobenzenamine	1.70

[1]DMA = dimethylamine salt formulation.

Extraction and Analysis of Pesticides. Dicamba, 2,4-D, and mecoprop were
analyzed by procedures developed in our laboratory (Hong, M. K., Lee, A. S., and
Smith, A. E. *J. Agric. and Food Chem.*, in press; and Lee, A. S., Hong, M. K.,
Smith, A. E. *JAOAC Int.*, in press). Subsamples of 100 mL were transferred from
the storage bottle into a 250 ml beaker. An internal standard (2,4,5-T) was added
to the beaker and the mixed solution is acidified to pH 2 with 0.2M HCl (*23*).
The pesticides were extracted from the acidified solution by liquid-liquid

partitioning into 200 mL diethyl ether. The diethyl ether was evaporated and the analytes were esterified with triflouroethanol (TFE) and the esters were quantified by gas chromatography. The Hewlett Packard Model 5890 Series II was linked to a HP3365 ChemStation and equipped with an ECD.

Table III. GC-ECD operating conditions for the analytes analyzed

Analyte	Column[1]	Gas[2] flow (ml min⁻¹)	Inlet	Detector	Column initial/(min)	Column final/(min)	Program rate (°C min⁻¹)
benefin	RTX-1	12	270	300	160/3	192/5	30
pendimethlin	RTX-1	12	270	300	160/3	270/3	40
2,4-D	RTX-1	14	250	300	136/8	250/3	30
dicamba	RTX-1	14	250	300	136/8	250/3	30
mecoprop	RTX-1	14	250	300	136/8	250/3	30
chlorpyrifos	RTX-35	14	250	300	150/3	178/5	30
chlorothalonil	RTX-35	14	250	300	150/3	250/3	30

[1]RTX-1= Fused silica column (crossbond 100% dimethyl polysiloxane) 0.53 mm id; 1 micron thickness and 30 m long fit with a 5 m guard column.
RTX-35=Fused silica column (crossbond 65% dimethyl-35% diphenyl polysiloxane) 0.53 mm id; 0.5 micron thickness; 30 m long fit with 5 m guard column.
[2]Carrier gas=helium
Make-up gas=Ar/methane (5/95%)

Dithiopyr was extracted by solid phase extraction and analyzed by GC-ECD according to a method developed in our laboratory (*24*). Benefin and pendimethalin were extracted from the 50 mL aqueous subsamples by liquid-liquid partitioning into 150 mL dichloromethane. The dichloromethane was concentrated under vacuum to 1 mL and diluted with 1 mL toluene containing metribuzin as an internal standard. Analytes were quantified by GC-ECD. Chlorothalonil and chlorpyrifos were extracted by liquid-liquid partitioning into ethyl acetate. The ethyl acetate volume was reduced under vacuum to 1 mL and the hydroxy metabolites of chlorpyrifos and chlorothalonil were methylated using 1 mL diazomethane and 0.034 g silica gel. Following a 30 min reaction period and ether dry down, the solution containing the analytes and methylated metabolites was brought to volume with ethyl acetate containing dicamba-methyl ester as an internal standard. The analytes and metabolites were quantified by GC-ECD and data for the analytes are presented as the additive of the metabolites and analytes. Column and conditions for GC-ECD are listed in Table III. The helium carrier gas head pressure was adjusted until a 1 µL head space sampling from methylene chloride has a retention time of 2.5 min. Methane/argon (95/5) gas used for make-up and purge gas was maintained at flow rates of 12-14 and 50 mL min⁻¹, respectively. The extraction and quantification systems were established to give a minimum detectable concentration (MDC) of 1 μg L^{-1} in the aqueous effluent. This lower limit was determined as adequate considering that the concentration is 70-fold less than the MCL (70 μg L^{-1}) established for 2,4-D (*9*).

Results and Discussion

Measurement of Pesticide Movement Through Simulated Greens. Only small quantities of the pesticides were transported from the greenhouse lysimeters containing bentgrass during the 1992 experiments (Table IV). Less than 1.0% of the applied pesticides was transported over the 70 day experiment and the total fraction of applied pesticide transported over this period was very similar for the 3 pesticides. The highest concentration of 2,4-D, dicamba, and mecoprop in the leachate was 3.2, 0.1, and 0.6 μg L^{-1}, respectively. Similar data were obtained from 1993 experiment with 'Tifdwarf' bermudagrass on the two soil types. The percentages of total pesticides transported from the lysimeters, during 1993, were less than 1.0% of the applied and the highest concentrations for 2,4-D, dicamba, and mecoprop were 3.2, 3.6, and 3.8 μg L^{-1}, respectively (data not shown).

Table IV. Cumulative analyte (μg and % of applied) transported from greenhouse lysimeters containing bentgrass treated on July 3, 1992. Data are averages over both soil mixtures

| Days After Treatment | Herbicide | | | | | |
| | 2,4-D-DMA | | Dicamba | | Mecoprop | |
	μg	%	μg	%	μg	%
7	0	0	0.2	0.2	0.8	0.1
14	0	0	0.2	0.2	2.9	0.3
21	1.0	0.2	0.2	0.2	4.1	0.4
35	1.2	0.2	0.2	0.2	4.2	0.4
49	1.2	0.2	0.4	0.4	4.2	0.4
56	1.4	0.3	0.5	0.4	4.2	0.4
63	2.0	0.4	0.5	0.4	4.2	0.4
70	2.0	0.4	0.5	0.4	4.2	0.4

Data on analyte movement in the leachate from the field lysimeter experiments conducted during 1992 and 1993 for bentgrass and bermudagrass were not significantly different for years and species, therefore, the data over years and for species were combined (Figure 1). The data indicate that the concentrations of 2,4-D and mecoprop in the leachate did not exceed MDC for our methods. However, dicamba was detected in one weekly collection from these treatments. The highest analyte concentration for these collections was 1.9 μg L^{-1} accounting for transport of 0.4% of the applied herbicide.

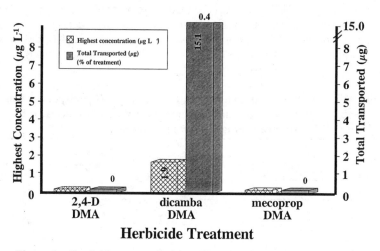

Figure 1. Pesticide transported in effluent water from field lysimeters.

Measurement of Pesticide Movement from Simulated Fairways. At 24, 48, 96 and 192 HAT the plots received only simulated rainfall events at averages of 5.0, 5.0, 2.4, and 2.5 cm, respectively. Only samples collected over the first 192 HAT contained concentrations of the pesticides above the MDC of 1 μg L^{-1}. The fraction of water leaving the plots at 24, 48, 96, and 192 HAT was 44.8 (580.6), 72.1 (448.5), 40.0 (248.8), and 35.5% (230.0 L), respectively. The highest concentrations of pesticides in the runoff water occurred during the first rainfall event applied at 24 HAT (Table V). The concentrations of 2,4-D, dicamba, and mecoprop in the runoff water were 810.7, 279.2, and 820.0 μg L^{-1}, respectively, and the analytes in the runoff water accounted for 7.26, 9.94, and 9.88% of the respective pesticides applied to the plots.

Table V. Analyte in runoff water from simulated rainfall 24 HAT, total analyte transported over 192 HAT (% of applied) and pKow

Analyte	Concentration 24 HAT	Total Transported	pK$_{ow}$
	(μg L^{-1})	(%)	
dicamba-DMA	279	14.4	-1.97
mecoprop-DMA	820	14.4	-1.77
2,4-D-DMA	810	9.6	-1.40
chlorothalonil	294	0.8	1.53
dithiopyr	22	1.9	4.70
chlorpyrifos	19	0.1	4.97
benefin	3	0.01	5.62
pendimethalin	9	0.01	5.11

The runoff plot system was developed in the Piedmont Region on a kaolinite-clay loam soil which resulted in at least 40% of the rainfall (3.3 cm hr^{-1}) to leave the plots as surface runoff. Additionally, this water contained moderately high concentrations of the treatment pesticides; 2,4-D, mecoprop, and dicamba; at 24 and 48 HAT. Concentrations of 2,4-D in the runoff water was a factor above the recommended MCL of 75 μg L^{-1} by the Environmental Protection Agency (Table V). Although, the concentration of chlorothalonil transported in the first simulated rainfall event was moderately high, it must be realized that much higher rates were applied compared to other pesticides. Only 0.7% of the applied chlorothalonil was transported from the plots (Table V).

The analytes were applied in a water carrier which evaporated soon after application allowing the analytes to reside in close contact with the wax cuticle of the turfgrass. The following simulated rainfall would result in a partitioning of the analytes from the lipophilic wax into the transporting water. Therefore, the log of the octanol:water partitioning coefficient (pK$_{ow}$) was determined from the equation presented by Dao et al. (*25*):

$$pK_{ow} = \frac{\log S_w - 4.184}{-0.922} \qquad S_w = \text{water solubility in ppm } (26)$$

The pKow's were plotted against the total % transported (Table V) and fit to linear and quadratic equations. The resulting R^2 values were 0.82 and 0.95, respectively. The better fit of the data to the quadratic equation might be due to the precipitation and crystallization, from application solution/suspension, of the analytes with low water solubility resulting in limited availability of analyte molecules for dissolving in runoff water.

It must be realized that the runoff water will probably be diluted many fold prior to reaching potable water systems. Additionally, the rainfall simulation probably represents extreme conditions compared to normal rainfall occurrences. The simulated rainfall is instantly turned on at maximum intensity as compared to a natural rainfall event which probably would not be at maximum intensity at the onset. A light shower prior to maximum intensity would allow the pesticides, having lower pKow's to be transported into the soil. However, these data would indicate that precautions must be exercised when applying the more water soluble pesticides to golf course fairways, in the Piedmont region, having a 5% slope.

Conclusion

Partitioning of the pesticides in the environment and potential loss of pesticides to groundwater and surface water is determined by innumerable interacting factors and conditions. The potential for pesticides to leach to ground water depends on the: 1) properties of the chemical, 2) properties of the soil, 3) application conditions, and 4) climatic conditions (27,28). Chemicals found most often in groundwater had many of the following characteristics: 1) highly mobile in soil leaching studies [high Rf values], 2) low retention by soil in adsorption studies [low K_{oc} values], 3) applied at moderate to high rates over large areas, and 4) moderate to long lived in the environment (half-lives of 30 days or longer) (7).

The data from the greenhouse and outside lysimeter experiments indicate that only small quantities of the applied pesticides are transported from the lysimeters. In our research, the analyte concentrations in the effluent water from the lysimeters were near the MDC established for our methods of analyte extraction and analyses (1 μg L^{-1}). Therefore if the analyte concentration is slightly above this level, the data are a real number. If the concentration is slightly below the MDC, the data becomes zero. This allows for a large error term that will mask small differences that might be attributed to the differences in rooting mixtures or grass species. Of the pesticides used in this research, only 2,4-D has a recommended MCL by EPA. The MCL of 70 μg L^{-1} was at least 10-fold above the concentrations in the of 2,4-D transported from the three lysimeter studies included in this report.

The Groundwater Loading Effects of Agricultural Management Systems (GLEAMS) model (28) was used to predict the potential for 2,4-D to be transported through the lysimeters subtending bermudagrass in the greenhouse. The prediction of transport was compared with data on 2,4-D transported during a 1991 greenhouse lysimeter experiment (29). The GLEAMS model, with the defined parameters of the 1991 study, overestimated the actual 2,4-D transported from the lysimeters through the two rooting-media profiles used in the experiments. The GLEAMS model data inferred that the increased level of sphagnum peat moss in the rooting medium should reduce the 2,4-D concentration in the aqueous effluent. Since the observed levels of 2,4-D were near the MDC, for that experiment, it was impossible to determine an influence of the increased organic matter content, in the rooting media, on 2,4-D transported in the aqueous effluent. The inclusion of the thatch layer in the GLEAMS model did not alter the predicted transport of 2,4-D in the aqueous effluent from the 85:15 rooting media compared to non-inclusion. The probable reason for no significant influence expressed by the thatch layer is due to the very thin thatch layer (1.40 cm) and due to the limited increase in organic matter used for this layer compared to the rooting media (5.80 vs 2.26%). Even though the GLEAMS model greatly overestimated the observed herbicide load, the maximum predicted concentration of 48 μg L^{-1} is less than the recommended MCL standard (70 μg L^{-1}) for 2,4-D.

Data from the runoff plots are probably the first data of this type to be obtained from the Piedmont region including turfgrass as a cover. Watschke et al. (*30*) described a series of plots established at Pennsylvania State University with slopes of 9 to 14% containing a turfgrass cover. Rainfall intensities as high as 15 cm hr^{-1} were needed to obtain runoff. Our system, developed in the Piedmont on a kaolinite-clay loam soil, resulted in at least 40% of the rainfall applied at 3.3 cm hr^{-1}, leaving the plots as surface runoff. Additionally, this water contained moderately high concentrations of the treatment pesticides 2,4-D, mecoprop, and dicamba at 24 and 48 HAT. The concentrations of 2,4-D in the runoff water were a factor above the recommended MCL. However, the runoff water from lawns and golf courses will probably be diluted several fold prior to reaching potable water systems. Additionally, the simulated rainfall was instantly turned on at maximum intensity as compared to a natural rainfall event which most probably would gradually attain maximum intensity over a period of time. However, these data would indicate that precautions must be exercised when applying these herbicides to golf course fairways, in the Piedmont region, having at least a 5% slope.

Acknowledgments

The author gratefully acknowledges the technical assistance of Mr. Nehru Mantripragada and Mr. Hal Peeler, the secretarial and editorial assistance of Ms. Mary Flynn, and the financial support furnished by the United States Golf Association.

Literature Cited

1. Walker, W. J.; Balogh, J. C.; Tietge, R. M.; Murphy, S. R. *Environmental issues related to golf course construction and management: A literature search and review*; Spectrum Research, Inc.: Duluth, MN, 1990.

2. Potter, D. A.; Cockfield, S. D.; Morris, T. A. In *Integrated pest management for turfgrass and ornamentals*; Leslie, A. R.; Metcalf, R. L., Eds.; Office of Pesticide Program, U.S. Environmental Protection Agency: Washington, DC, 1989, 1989-625-030; pp 33-44.

3. Lehr, J. H. *J. Prod. Agric.* **1991**, *4*, 282-290.

4. Pratt, P. F. 1985. *Cast Report No. 103*; Council for Agricultural Science and Technology, 1985, 62 pp.

5. Anderson, J. L.; Balogh, J. C.; Waggoner, M. *State Nutrient and Pest Management Standards and Specification Workshop*, St. Paul, MN; USDA Soil Conservation Service, 1989; *Section I & II*, 453 pp.

6. Leonard, R. A. In *Environmental chemistry of herbicides*; Grover, R., Ed.; CRC Press: Boca Raton, FL, 1988, Vol. 1; pp 45-87.

7. Stewart, B. A.; Woolhiser, D. A.; Wischmeier, W. H.; Caro, J. H.; Frere, M. H. *Control of water pollution from cropland: A manual for guideline development*; EPA 600/2-75-026a; U. S. Environmental Protection Agency and USDA Agricultural Research Service, 1975, Vol. I; 111 pp.

8. Cohen, S. Z.; Nickerson, S.; Maxey, R.; Dupay, Jr., A.; Senita, J. A. *Ground Water Monit. Rev.* **1990**, *10*, 160-173.

9. Office of Drinking Water, U. S. Environmental Protection Agency. *Drinking water regulations and health advisories*; U. S. Printing: Washington, DC, 1990; H04-IID-PESTI.

10. Fairchild, D. M. In *Ground water quality and agricultural practices*; Fairchild, D. M., Ed.; Lewis Publishing: Chelsea, MI, 1987; pp 273-294.

11. Morandi, L. In *Agricultural chemicals and groundwater protection: Emerging management and policy*; Proceedings of a Conference, Oct. 22-23, 1987, St. Paul, MN; Freshwater Foundation, 1988; pp 163-166.

12. Johnson, P. W. In *Agricultural chemicals and groundwater protection: Emerging management and policy*; Proceedings of a Conference, Oct. 22-23, 1987, St. Paul, MN; Freshwater Foundation, 1988; pp 167-170.

13. National Golf Foundation. *Golf Market Today* **1991**, *March/April*, 2-3.

14. Beard, J. B. *Turf management for golf courses;* Macmillan Publishing Co.: New York, NY, 1982.

15. Weber, J. B. *North Carolina Turfgrass* **1990**, *9(1)*, 24-29.

16. Braham, B. E.; Webner, D. J. *Agron. J.* **1985**, *77*, 101-104.

17. Goh, K. S.; Edminston, S.; Maddy, K. T.; Meinders, D. D.; Margetich, S. *Bull. Environ. Contam. Toxicol.* **1986**, *37*, 27-32.

18. Niemzcyk, H. D.; Filary, A.; Krueger, H. R. *Western Views Magazine* **1988**, *Jan./Feb.*, 7.

19. Sears, M. K.; Bowhey, C.; Braun, H.; Stevenson, G. R. *Pestic. Sci.* **1987**, *20*, 223-231.

20. Sears, M. K; Chapman, R. A. In *Advances in Turfgrass Entomology*; Niemzcyk, H. D.; Joyner, B. G., Eds.; Hammer Graphics, Inc.: Piqua, OH, 1984; pp 57-59.

21. Gold, A. J.; Morton, T. G.; Sullivan, W. M.; McClory, J. *Water Air Soil Poll.* **1988**, *37*, 121-129.

22. Smith, A. E.; Weldon, O.; Slaughter, W.; Peeler, H.; Mantripragada, N. *J. Environ. Qual.* **1993**, *22*, 864-867.

23. *EPA Methods*; Graves, R. L., Ed.; U.S. Environmental Protection Agency, 1989; pp 221-253.
24. Hong, S.; Lee, A. S,; Smith, A. E. *J. Chromatog. Sci.* **1992**, *32*, 1-3.
25. Dao, T. H.; Lavy, T. L.; Dragun, J. *Residue Rev.* **1983**, *87*, 91-104.
26. Wauchope, R. D.; Buttler, T. M.; Hornsby, A. G.; Augustijn-Beckers, P. W. M.; Burt, J. P. In *Reviews of Environmental Contamination and Toxicology*; Ware, G. W., Ed.; Springer-Verkag: New York, NY, 1992; pp 1-35.
27. Kimm, V. J.; Barles, R. In *Agricultural chemicals and groundwater protection: Emerging management and policy*; Proceedings of a Conference, Oct. 22-23, 1987, St. Paul, MN; Freshwater Foundation, 1988; pp 135-145.
28. Leonard, R. A.; Knisel, W. G.; Still, D. A. *Trans. ASAE* **1987**, *30*, 1403-14-18.
29. Smith, A. E.; Tillotson, W. R. In *Pesticides In Urban Environments*; Racke, K. D.; Leslie, A. R., Eds.; American Chemical Society: Washington, DC, 1993; pp 168-171.
30. Watschke, T. L.; Harrison, S.; Hamilton, G. W. *Green Section Record* **1989**, *27*, 5-8.

Chapter 15

Relation of Landscape Position and Irrigation to Concentrations of Alachlor, Atrazine, and Selected Degradates in Regolith in Northeastern Nebraska

Ingrid M. Verstraeten[1], D. T. Lewis[2], Dennis L. McCallister[2], Anne Parkhurst[3], and E. M. Thurman[1,4]

[1]Water Resources Division, U.S. Geological Survey, 406 Federal Building, 100 Centennial Mall North, Lincoln, NE 68508
[2]Department of Agronomy, University of Nebraska–Lincoln, Lincoln, NE 68588
[3]Department of Biometrics, University of Nebraska–Lincoln, Lincoln, NE 68588

Concentrations of alachlor, its ethanesulfonic acid degradate, atrazine and its degradates, deethylatrazine and deisopropylatrazine, in the upper regolith and associated shallow aquifers were determined in relation to landscape position (floodplains, terraces, and uplands) and irrigation (nonirrigated and irrigated corn cropland) in 1992. Irrigated and nonirrigated sites were located on each landscape position. Samples were collected from three depths. Canonical discriminant and multivariate analyses were used to interpret data. Herbicides and their degradation products tended to be present in soils with high percent organic matter, low pH, and low sand content. Atrazine was present more frequently on the floodplain at all depths than the other compounds. Atrazine (maximum 17.5 µg/kg) and ethanesulfonic acid (maximum 10 µg/kg) were associated with landscape position, but not with irrigation. Alachlor (maximum 24 µg/kg), deethylatrazine (maximum 1.5 µg/kg), and deisopropylatrazine (maximum 3.5 µg/kg) were not significantly associated with either landscape position or irrigation. Ground-water analytical results suggested that concentrations of these herbicides and degradates in ground water did not differ among landscape position or between irrigated and nonirrigated corn cropland.

[4]Current address: U.S. Geological Survey, 4821 Quail Crest Place, Lawrence, KS 66049

0097–6156/96/0630–0178$15.00/0
© 1996 American Chemical Society

Alachlor [2-chloro-2,6-diethyl-*N*-(methoxymethyl)acetanilide] and atrazine [6-chloro-*N*-ethyl-*N'*-(1-methylethyl)-1,3,5-triazine-2,4-diamine] are the herbicides most often present in ground and surface water in the midwestern United States (*1-8*). They have been detected at concentrations generally less than the U.S. Environmental Protection Agency Maximum Contaminant Levels for drinking water (2.0 µg/L for alachlor and 3.0 µg/L for atrazine) (*9*). These herbicides are mainly pre-emergent herbicides commonly applied to corn (*Zea mays* L.) and sorghum *(Sorghum bicolor L.)* to control annual broadleaf weeds and grasses, respectively. Atrazine in ground water has been associated with irrigated corn cropland in Nebraska (*8*). Alachlor is a nonionic herbicide with moderate vapor pressure and moderate solubility (*10*). Atrazine, a moderately persistent pesticide, is a weak base with low vapor pressure and moderate solubility (*11, 12*).

The degradation pathways of alachlor and atrazine include chemical, microbial, and photolytic degradation processes, which depend upon the chemical and physical conditions of the media and microbial populations (*13-15*). Numerous studies on the degradation pathways of alachlor and atrazine have been conducted (*13, 14, 16-20*). Baker and others (*21*) reported the presence of alachlor ethanesulfonic acid [2-[2,6-diethylphenyl(methoxymethyl)amino]-2-oxoethanesulfonic acid], a major alachlor degradate, in ground water. Other authors have identified this degradate in ground water by high performance liquid chromatography (HPLC) (*22, 23*) and by an enzyme-linked immunosorbent assay for alachlor (*21, 22, 24*). The half-life of atrazine is dependent upon water holding capacity, acidity, organic matter, the presence of microbial populations, and soil redox conditions (*25, 26*). Deethylatrazine and deisopropylatrazine mainly are formed through microbial degradation, a *N*-dealkylation process principally attributed to fungi (*27*). Microbial degradation of atrazine occurs frequently in the first meter of the soil profile and to a lesser extent at greater depth (*28*). Deethylation, rather than deisopropylation, is a more rapid degradation pathway (*14, 29*).

Agricultural chemicals are distributed to surface and ground water by transport as runoff and by leaching through the unsaturated zone. Agricultural chemicals also may volatilize to the atmosphere, degrade, and be adsorbed to and desorbed from mineral surfaces and organic matter. Water and solute transport are associated with the landscape position and intimately related to soil properties and water-balance relationships (*30*). Transport is controlled by soil biological, chemical, and physical properties, including antecedent soil-water content, the rate and amount of infiltrating water, and rate and amount of water applied during irrigation (*30, 31*). Hydraulic properties and ability of soil to retain organic chemicals vary across the landscape due to temporal and spatial changes in pedon characteristics, such as organic matter content and clay content, resulting in diverse patterns of water and solute distribution (*30*). Therefore, concentrations of herbicides and their degradates in the pedon and in ground water may be associated with landscape position and irrigation. The objective of this research was to determine whether: (1) concentrations of alachlor, atrazine, and selected degradates in and beneath soil on uplands, terraces, and floodplains differ across landscape positions, and (2) concentrations of alachlor, atrazine, and selected degradates in and beneath soils differ depending upon irrigation use.

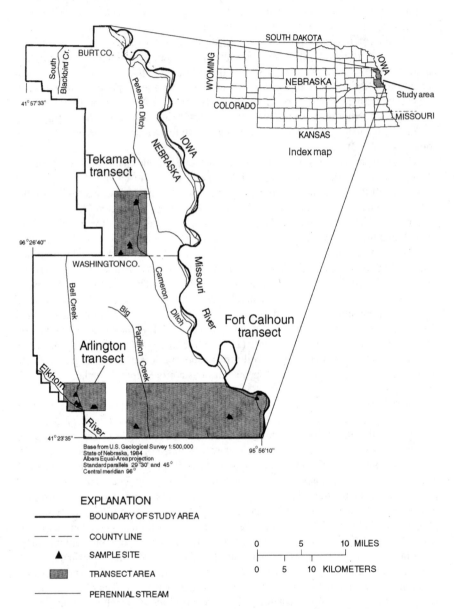

Figure 1. Study area and generalized locations of sample sites.

Materials and Methods

Site Selection. Sites were located along three transects that crossed the floodplain, terrace, and upland landforms located within a loess-capped glaciated region of eastern Nebraska (Fig. 1). Irrigated and nonirrigated sample sites were located along these transects on each landscape position.

Three sampling subsites within each major site were identified. Upper regolith samples (162 samples) were collected using a Giddings probe or spade in May 1992 from three depths: 15 to 31 cm (depth 1), 82 to 115 cm (depth 2), and 137 to 305 cm (depth 3). Soils at the sampling sites were within the fine-silty particle-size class. Samples were not collected from depths less than 15 cm to avoid the presence of herbicides from spring application. Land use was limited to cultivated land, with sprinkler irrigated and nonirrigated land in close proximity, and where alachlor and atrazine were applied within 5 years prior to sampling. In addition, 8 representative ground-water samples were collected from 7 irrigation wells and 1 domestic well completed at depths generally less than 200 feet in unconfined aquifers at or near the 18 upper regolith sample sites in accordance with U.S. Geological Survey ground-water sampling protocols (7).

Chemical and Physical Analytical Methods. The upper regolith samples collected for analyses of chemical and physical properties were air-dried and subsequently sieved to less than 2 mm. Particle size (Soil Survey Interim Report, No. 43, 1994), soil pH and excess lime (32), organic carbon (33), exchangeable cations and cation exchange capacity (34), and surface area (35) were determined for each sample. Atrazine, deethylatrazine, and deisopropylatrazine in upper regolith samples were determined using an automated solid-phase extraction, followed by automated GC/MS (gas chromatography/mass spectrometry) analysis of the eluates on a Hewlett-Packard Model 5890 gas chromatograph and a 5970A mass-selective detector (36). The atrazine extraction method was modified for extraction of alachlor and its ethanesulfonic acid through deletion of the anion-exchange procedure during the extraction process. The extract was blown to dryness using a Turbovap and subsequently dissolved in 5.0 mL organic-free water for analysis by an enzyme-linked immunosorbent assay (24).

To test this new extraction procedure for alachlor and its ethanesulfonic acid degradate, batch equilibrium experiments were conducted using three samples collected at three depths. Five grams of air dry sample were weighed into a teflon-lined screw-top test tube. Six mL of organic-free water with 0.005 M $CaSO_4 \cdot 2H_2O$ were added to the samples. Zero ng, 10 ng, 100 ng, 500 ng, and 1,000 ng alachlor or its ethanesulfonic acid were added to the samples in separate tubes. The solution was equilibrated for 24 hours utilizing a rotary mixer at 30 rpm. The slurry was centrifuged until the supernatant was clear (generally in excess of 15 min). The clear supernatant was removed with a pipette into a glass vial for evaporation and solid-phase extraction. Samples were analyzed with automated GC/MS for alachlor (37). Samples were analyzed with an enzyme-linked immunosorbent assay (ELISA) for ethanesulfonic acid (24). Percent recovery was 85-125% and reproducibility was about 15% (Diana Aga, U.S. Geological Survey, personal communication, 1994).

Water samples initially were tested qualitatively using enzyme-linked immunosorbent assay kits for triazines and alachlor (*37*). If atrazine or alachlor were detected with this technique, the samples were analyzed for alachlor, atrazine, deethylatrazine, and deisopropylatrazine using extraction procedures and gas-chromatographic separation as described by Thurman and others (*37*) and Meyer and others (*38*). Concentrations of ethanesulfonic acid were determined by automated solid-phase extraction and enzyme-linked immunosorbent assay by the U.S. Geological Survey Laboratory in Lawrence, Kansas (*24*).

Statistical Analytical Methods. Statistical analyses were performed with SAS Version 6.08 and 6.09, SAS Institute, Inc., Cary, North Carolina, 1990. Correlations among upper regolith properties and extracted concentrations of herbicides and selected degradates were calculated to suggest relationships. The data were determined not to be normally distributed. Several transformation were done to normalize the data. Neither log reciprocal, or quadratic transformations significantly improved the distribution of the data, with the exception of a log transformation of the sand variable. Subsequent parametric analyses therefore were done on untransformed data. Consequently the validity of many of the interpretations in the dicussion of results may depend upon the robustness of the statistical procedures. Multivariate analyses of variance, MANOVA, were used to compare several treatment means (*39*) with class variables land use, landscape position, and transect. MANOVA included interpretations of interactions and addition of contrasts. Depth was added to these MANOVA analyses as a repeated measure. Univariate analyses were conducted to test whether differences in mean concentrations existed for each depth. Herbicides and degradation products were detected in few samples. Therefore, canonical discriminant analyses were conducted to identify relationships between herbicide occurrence and chemical and physical properties (*39*), and to identify upper regolith factors that can discriminate among the landscape positions by depth. The assumption of equal covariance between discriminant groups was tested and failed. Subsequently, analyses for quadratic discriminant functions were done. As the error for quadratic functions was equal or greater than those for linear functions, linear functions were used as a tool to help understand the results.

Results and Discussion

Chemical and Physical Properties of Upper Regolith. In general, chemical and physical properties of upper regolith in floodplains differed from those properties found on terraces and uplands. Upper regolith properties had differences in chemical and physical properties at all depths and landscape position, e.g. sand, silt, clay, pH, exchangeable calcium, surface area, cation exchange capacity, and organic matter (*40*) (Table I). Differences in upper regolith properties by landscape position and depth were partially confirmed through statistical analyses. Physical and chemical characteristics of the upper regolith implied that (1) organic matter, fine silt, and clay probably contribute most of the herbicide retention sites, especially in the A and B horizons rich in organic matter or clay, (2) basic organic molecules tend to be adsorbed in A and B horizons, rich in organic matter and clay with relatively low pH, and (3) soil with deep sandy layers and relatively higher pH (especially common in the floodplain near the rivers) does not adsorb organic molecules as frequently (*40*).

Multivariate and univariate analyses. Multivariate and univariate analyses can suggest interactions and significant differences in chemical and physical properties (Table II). Interactions were absent among landscape positions, transects, and depths for all measured chemical and physical properties. A relationship existed between landscape position and the amounts of sand, pH, organic matter, and cation exchange capacity (Table II).

Table I.--Summary statistics of alachlor, atrazine, and selected metabolites in upper regolith sampled (N=54)

Variable	Units	Depth	Mean	Standard Deviation	Minimum	Maximum
Alachlor	ug/Kg	1	1.3	3.1	<0.5	17.7
		2	<0.5	0.5	<0.5	2.5
		3	<0.5	0.5	<0.5	3.5
Ethanesulfonic acid	ug/Kg	1	1.2	2.1	<0.5	10.1
		2	<0.5	0.5	<0.5	2.6
		3	<0.5	0.1	<0.5	0.8
Atrazine	ug/Kg	1	3.4	5.3	<0.5	23.9
		2	0.4	0.9	<0.5	3.7
		3	0.7	2	<0.5	10.2
Deethylatrazine	ug/Kg	1	<0.5	0.1	<0.5	0.7
		2	<0.5	0.2	<0.5	1.5
		3	<0.5	0.1	<0.5	0.5
Deisopropyl-atrazine	ug/Kg	1	0.1	0.6	<0.5	3.3
		2	<0.5	0.2	<0.5	0.9
		3	<0.5	0	<0.5	0

Amounts of organic matter were significantly different between floodplains and other landscape positions and between floodplains and terraces. The differences in soil properties between landscape positions may affect the amount of herbicides in the regolith at these landscape positions. Statistically significant differences did not exist

Table II. --Multivariate and univariate analytical p-values of upper regolith properties, alachlor, atrazine, and selected metabolites

Statistical analyses	Sand	Silt	Clay	pH	OM	CEC	SA	AL	ESA	AT	DEA	DIS
D*I*(FvsT+U)	0	0.4	0.49	0.5	0.1	0.06	0.5	0.5	0.64	0.89	0.54	0.53
D*I*(F+TvsU)	0.01	0.8	0.46	0.8	0.6	0.9	0.9	0.8	0.48	0.58	0.54	0.84
D*I*TvsU	0.32	1	0.69	0.5	1	0.13	0.9	0.5	0.24	0.6	0.54	1
D*I*FvsT	0.05	0.4	0.77	0.4	0.2	0.03	0.5	0.3	0.29	0.92	0.54	0.6
D*I	0.34	0.3	0.71	0.3	0	0.46	0.4	0.5	0.54	0.94	0.83	0.08
P*I	0.92	0	0	0.8	0	0	0	0.2	0.16	0.89	0.76	0.02
D*TR*P	0	0	0	0	0	0	0	0.6	0.25	0.75	0.22	0.01
D*P	0.07	0.3	0.59	0.1	0	0.02	0.5	0.3	0.14	0.05	0.44	0.25
D*(FvsT+U)	0.15	0.4	0.88	0.2	0.1	0.13	0.7	0.7	0.31	0.19	0.54	0.41
D*(F+TvsU)	0.27	0.7	0.67	1	0.1	0.06	0.6	0.4	0.27	0.13	0.54	0.41
D*(TvsU)	0.75	0.9	0.75	0.4	0.4	0.12	0.8	0.4	0.23	0.11	0.54	0.6
D*(FvsT)	0.28	0.5	1	0.1	0.2	0.57	0.9	0.8	0.26	0.16	0.54	0.6
D1: FvsT	0.02	0.6	0.02	0.2	1	0.31	0	0.8	0.86	0.05	0.29	0.32
D1: F+TvsU	0.3	0.7	0.02	0.9	0.3	0.07	0.1	0.2	0.07	0.65	0.52	0.74
D1: TvsU	0.51	0.6	0.29	0.5	0.4	0.21	0.7	0.2	0.09	0.39	1	0.81
D1: FvsT+U	0.02	0.8	0.01	0.3	0.6	0.09	0	0.6	0.35	0.06	0.23	0.31
D2: FvsT	0.21	0.2	0.12	0.1	0.1	0.39	0	0.2	0.07	0.14	1	0.29
D2: F+TvsU	0.48	0.2	0.17	0.9	0.1	0.47	0	0.8	0.24	0.67	0.23	0.52
D2: TvsU	0.94	0.7	0.67	0.3	0.4	0.83	0.3	0.4	0.07	0.26	0.29	1
D2: FvsT+U	0.16	0.1	0.06	0.2	0.1	0.28	0	0.3	0.24	0.25	0.52	0.23
D3: FvsT	0.26	0.2	0.32	0.8	0.2	0.36	0.2	0.5	0.29	0.09	0.29	--
D3: F+TvsU	0.47	0.3	0.22	0.9	0.4	0.31	0.2	0.4	0.52	0.43	0.52	--
D3: TvsU	0.98	0.7	0.52	0.8	1	0.65	0.5	0.7	0.29	0.77	0.29	--
D3: FvsT+U	0.2	0.1	0.16	0.9	0.1	0.21	0.1	0.3	0.52	0.08	0.52	--

Key: OM: organic matter content; CEC: cation exchange capacity; SA: surface area; AL: alachlor; ESA: ethanesulfonic acid; AT: atrazine; DEA: deethylatrazine; DIS: deisopropylatrazine; D: depth, I: land use, F: floodplain, T: terrace, U: upland, P: landscape position, TR: transect, D1: at depth 15-31 cm, D2: at depth 82-115 cm, D3: at depth 137-305 cm.

between chemical and physical properties at depth 3. Consequently, presence or lack of herbicides at this depth could be random or could be explained by other physico-chemical variables not considered in this study. Statistically significant differences existed in amounts of exchangeable calcium and sodium and percent organic matter by land use and depth.

These differences were not noted with the inclusion of contrasts in the statistical analyses and could not be demonstrated by univariate analyses by depth and land use. Differences in soil properties with land use may reflect the effect of leaching on these soil properties, whereas differences in soil properties by landscape position are believed to be related mainly to the parent material, biological activity, and depositional or erosional character of the soil, suggested by Buol and others (*41*).

Canonical Discriminant Analyses. The canonical discriminant analyses suggested a combination of chemical and physical properties that may have affected the presence and concentrations of herbicides in the upper regolith by landscape position (Table III). Canonical discriminant functions by land use and depth were insignificant, but canonical discriminant functions by landscape position and depth were significant. At depth 1, particle-size fractions characterize canonical discriminant function 1 and organic matter and pH characterize canonical discriminant function 2. At depths 2 and 3, particle-size fractions and organic matter characterize canonical discriminant function 1 and pH mainly characterizes canonical discriminant function 2. All functions discriminated between similar landscape positions for all depths (Fig. 2).

Table III.--Canonical discriminant functions for landscape position on upper regolith properties, Wilk's Lambda, chi-squared, degrees of freedom, and significance by depth

Depth of sample	Description of canonical discriminant functions	Wilk's lambda	Chi-squared	Degrees of freedom	Significance
15 to 31 cm	Particle-size fraction	0.11	101.19	22	0.0000
	Organic matter and pH	0.38	44.23	10	0.0000
82 to 115 cm	Particle size fraction and organic matter	0.12	96.72	22	0.0000
	pH	0.48	33.59	10	0.0002
137 to 305 cm	Particle size fraction and organic matter	0.16	84.68	22	0.0000
	pH	0.67	18.13	10	0.0529

Function 1 generally discriminated between the upper regolith on the floodplains and the upper regolith on terraces and uplands. Function 2 generally discriminated

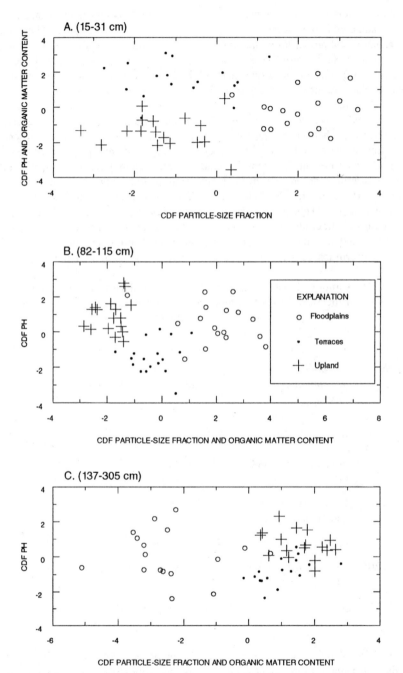

Figure 2. Discriminant function for landscape position by depth.

between the upper regolith on terraces and uplands. At shallow depths, particle size fractions represented canonical discriminant function 1 and organic matter and pH represented canonical discriminant function 2.

Herbicides and Selected Degradates in Upper Regolith. The majority of detections of the selected herbicides and degradates occurred in samples from depth 1 (Table I). Alachlor was present in 14 of 54 samples collected from depth 1, in 5 of 54 samples collected from depth 2, and in 3 of 54 samples collected from depth 3. Similarly, ethanesulfonic acid was detected in 23 of 54 samples collected from depth 1, in 6 of 54 samples collected from depth 2, and in 1 of 54 samples collected from depth 3. The numerous occurrences of ethanesulfonic acid suggest that it is a common degradate of alachlor that can occur at depths exceeding 270 cm. Whereas ethanesulfonic acid is transported to great depths and eventually can contaminate the ground water, the likelihood of alachlor contaminating the ground water is less, because alachlor, even though modrately water soluble, degrades readily. Atrazine was detected in 34 of 54 samples collected from depth 1 and in 12 of 54 samples collected from depths 2 and 3. Deethylatrazine, thought to be relatively persistent in soil (*27*) and more mobile than atrazine and deisopropylatrazine (*42*), was present in 1 of 54 samples collected from depth 1 and 1 of 54 samples collected from depths 2 and 3. Deisopropylatrazine, thought to be relatively nonpersistent (*27*) and less mobile than deethylatrazine and atrazine (*42*) in soil, was detected in 4 of 54 samples collected from depth 1 and 2 of 54 samples collected from depth 2 and suggests application of cyanazine.

The data indicated that (1) deethylation and deisopropylation reactions of atrazine exist and (2) deisopropylatrazine can be transported to depths exceeding 100 cm and deethylatrazine can be transported to depths exceeding 250 cm. The data also suggested that microbial populations that can degrade atrazine, deethylatrazine, and deisopropylatrazine were not as effective or present in sufficient numbers in the study area to degrade all atrazine applied. Therefore, atrazine, as well as both its degradation products, could affect ground water in the study area.

Maximum atrazine and alachlor concentrations were present in samples collected from depth 1 (17.5 and 24.0 µg/kg, respectively). Similarly, maximum ethanesulfonic acid and deisopropylatrazine concentrations occurred in samples collected from depth 1 (10.0 and 3.5 µg/kg, respectively). The highest concentration of deethylatrazine was present in a sample collected from depth 2. This suggests that microbial populations may be most effective in degrading atrazine at depth 1, characterized by soil with high organic matter and low pH. The highest mean alachlor concentrations were at depth 1 on floodplains and terraces. Highest mean atrazine concentrations were detected at all depths on the floodplains. Atrazine occurred more frequently at greater depths than alachlor. This suggests that atrazine is more persistent than alachlor as reported by Walker and others (*42*). Pothuluri and others (*43*) reported a longer residence time of alachlor in the vadose zone and ground water.

Correlations. The following selected relationships were established at an alpha level of 0.10 (Table IV): (1) alachlor was positively correlated with percent organic matter and cation exchange capacity; (2) alachlor was negatively correlated with pH;

(3) ethanesulfonic acid was positively correlated with percent organic matter, cation exchange capacity, and percent clay; (4) ethanesulfonic acid was negatively correlated with pH; (5) atrazine was positively correlated with percent organic matter and cation exchange capacity; (6) atrazine was negatively correlated with pH; (7) deisopropylatrazine was positively correlated with percent organic matter; and (8) deisopropylatrazine was negatively correlated with pH. Alachlor, atrazine, and deisopropylatrazine concentrations appear to be associated with higher percent organic matter and, to a smaller extent, higher cation exchange capacity and low pH.

Table IV.--Correlation matrix of selected upper regolith properties, alachlor, atrazine, and selected metabolites [Pearson correlation coefficients(PCC)/Prob>|R| under Ho: Rho=0/N=162]

Regolith property	Statistic	Alachlor	Ethane-sulfonic acid	Atrazine	Deisopropylatrazine
pH	PCC	-0.29	-0.44	-0.21	-0.18
	p-value	0.0002	0.0001	0.0083	0.0251
Organic matter content	PCC	0.29	0.47	0.37	0.15
	p-value	0.0002	0.0001	0.0001	0.0551
Cation exchange capacity	PCC	0.15	0.21	0.15	0.06
	p-value	0.0611	0.0089	0.0588	0.448

These relationships suggest that alachlor, atrazine, and deisopropylatrazine tend to be adsorbed mainly onto the organic matter and to a lesser extent onto clay minerals with greatest cation exchange capacity (smectites) at relatively low pH as was reported by Walker and others (*43*). The negative relationship of atrazine and pH could indirectly have been affected by the relationship of atrazine and organic matter or by the effect of pH on microbial activity. It also is possible that the relationship between the presence of ethanesulfonic acid with organic matter is caused by the relationship of the degradation rate of alachlor with organic matter content and microbial biomass. Numerous detections of ethanesulfonic acid at depths 1 and 2 and the detection of this product at more than 250 cm confirm the study of Baker and others (*21*), who postulated that this degradate is relatively mobile and persistent.

There are no significant correlations of deethylatrazine with any measured chemical and physical properties. These results may be caused by the low number of detections of deethylatrazine or could imply that the presence of deethylatrazine is not as readily related to these chemical and physical characteristics because this atrazine degradate tends to be more mobile than its parent herbicide and deisopropylatrazine. A positive correlation between alachlor and ethanesulfonic acid exists (PCC=0.54 with p-value=0.0001). This correlation shows that ethanesulfonic acid commonly occurs in areas where alachlor is applied to corn.

The data also suggest that ethanesulfonic acid is a common degradation product that may be relatively persistent and more common than deethylatrazine and deisopropylatrazine. The data also suggest that deethylatrazine may be more persistent than deisopropylatrazine, an explanation that is supported by the studies of Wolf and Martin (*45*) and Adams and Thurman (*14*).

Multivariate and Univariate Analyses. Multivariate analyses showed that interactions of landscape position, transect, and depth, and interactions of landscape position and land use only existed for deisopropylatrazine (Table II).   Atrazine concentrations and, to a lesser extent, ethanesulfonic acid concentrations were a function of landscape position and depth. Differences in atrazine concentrations exist on floodplains and terraces versus uplands by depth and to lesser extent on floodplains versus terraces.   Differences in atrazine concentrations between floodplains and terraces are further indicated through univariate analyses by depth. The data suggest that atrazine occurs more frequently on the floodplains at all depths than the other compounds, which could be a result of the amount of organic matter and cation exchange capacity present on the floodplains.

Some differences in ethanesulfonic acid concentrations at depth 1 were suggested by the univariate analyses of floodplains and terraces versus uplands. These differences in ethanesulfonic acid concentrations only were very weakly supported by the multivariate analyses. Ethanesulfonic acid concentrations tended to be greater in floodplains and terraces.   There were no differences in ethanesulfonic acid concentrations with regard to land use. Differences in atrazine concentrations between floodplains and terraces versus uplands may be related to differences in organic matter and cation exchange capacity among these landscape positions. Atrazine tends to be present in floodplains and terraces, which tend to have a higher organic matter content but lower cation exchange capacity, especially at depth.

Canonical Discriminant Analyses. Canonical discriminant analyses showed that the presence or absence of herbicides in the upper regolith were associated with landscape position and suggested the importance of soil properties in affecting the presence of these herbicides and degradates (Table V).   Significant canonical discriminant functions were derived for (1) alachlor in samples collected from depth 1, (2) ethanesulfonic acid in samples collected from depth 1, and (3) atrazine in samples collected from depth 3 (Fig. 3).

The discriminant function of alachlor for samples collected from depth 1 suggested that mainly percent organic matter, and, to a lesser extent, surface area and percent clay affect the presence or absence of alachlor. Soil pH plays only a minor role in determining presence or absence of alachlor at depth 1. However, in samples collected from depth 3, pH and percent clay play a much larger role in determining presence or absence of alachlor. Alachlor apparently tends to be adsorbed mainly to organic matter at depth 1, a process which apparently is not strongly pH dependent. On the other hand,

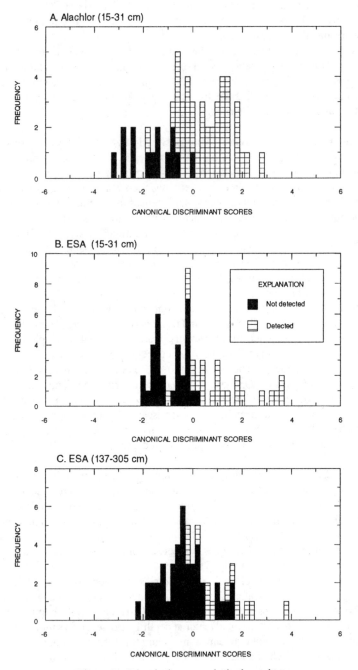

Figure 3. Discriminant analytical results.

at depth, adsorption is mainly related to the amount of clay present and appears to be more pH dependent, because the amount of organic matter and microbial activity generally is reduced.

The canonical discriminant function of ethanesulfonic acid suggested that the presence of this degradate in samples collected from depth 1 is related to percent organic matter, and, to a lesser extent, pH and cation exchange capacity. At all depths, the relationship of ethanesulfonic acid and organic matter was reduced. Soil pH played a moderate role in affecting the presence or absence of ethanesulfonic acid regardless of depth. It is postulated that ethanesulfonic acid primarily adsorbs to organic matter, a process that apparently is pH dependent. As organic matter content and, consequently, available adsorption sites decrease with depth, the relationship of organic matter and ethanesulfonic acid is reduced.

Table V.--Wilk's lambda, chi-squared, degrees of freedom, and significance of canonical functions of alachlor, atrazine, and selected metabolites in relation to upper regolith properties by depth

Parameter	Depth of sample	Wilk's lambda	Chi-squared	Degrees of freedom	Significance
Alachlor	15 to 31 cm	0.490	33.125	11	0.0005
	82 to 115 cm	0.712	15.778	11	0.1496
	137 to 305 cm	0.802	10.270	11	0.5062
Ethanesulfonic acid	15 to 31 cm	0.460	36.077	11	0.0002
	82 to 115 cm	0.772	12.058	11	0.3593
	137 to 305 cm	0.838	8.2209	11	0.6934
Atrazine	15 to 31 cm	0.898	5.0189	11	0.9302
	82 to 115 cm	0.816	9.4929	11	0.5765
	137 to 305 cm	0.684	17.690	11	0.0891
Deethylatrazine	15 to 31 cm	0.702	16.437	11	0.1257
	82 to 115 cm	0.840	8.1007	11	0.7043
	137 to 305 cm	0.743	13.811	11	0.2436

The relationship of ethanesulfonic acid with organic matter probably also indicates that the degradation rate of alachlor is greater at depth 1 and reduced at greater depth, a relationship that was established by Walker and others (*43*). In samples collected from depth 3, the relationship of atrazine with organic matter is important, whereas that of pH is minimal. The smaller effect of pH probably is related to the pKa of atrazine (1.68) (*11*), suggesting smaller effects of pH at a pH several units higher than its pKa.

The canonical discriminant analyses for deethylatrazine and deisopropylatrazine resulted in the correct classification of a high percentage of the data (>95%), even

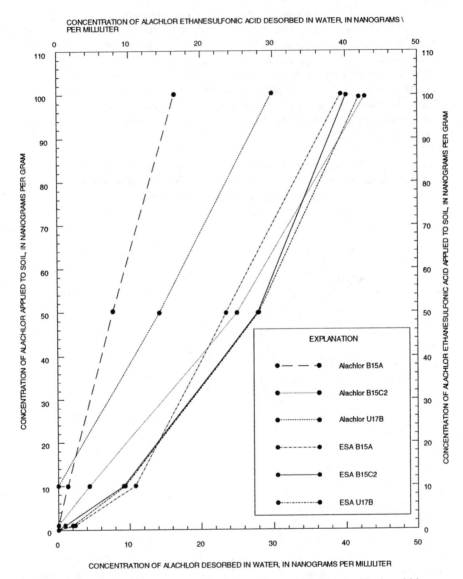

Figure 4. Batch equilibrium results of alachlor and its ethanesulfonic acid in three regolith samples.

though the canonical discriminant functions remained statistically not significant. The canonical discriminant functions suggested that percent sand and silt play a large role and pH and percent organic matter play a minor role in determining the presence of deethylatrazine in samples collected from depth 1. Previous research by Muir and Baker (*46*) and Shiavon (*42*) indicated that deethylatrazine is more mobile than atrazine and deisopropylatrazine, suggesting that deethylatrazine is not readily adsorbed by organic matter. This study confirms this lack of relationship between deethylatrazine and organic matter content. The relatively small dependence of presence of deethylatrazine with pH is more difficult to explain. It has been suggested in the literature (*28, 29*) that microbial degradation of atrazine is not only dependent on organic matter but also on pH, because microbial activity and populations are highly dependent on pH. Thus, overall differences in soil properties by landscape position appeared to be associated with the presence and changing concentrations of alachlor, atrazine, and selected degradates.

Batch Equilibration Experiments. Batch equilibration experiments for alachlor and its ethanesulfonic acid degradate support the canonical discriminant analytical results. Batch experiments on three upper regolith samples, which vary in organic matter and pH among other soil properties, illustrate the effect of organic matter and pH on the adsorption of alachlor and its ethanesulfonic acid degradate. A sample containing a larger amount of organic matter (2.68 %) and lower pH (5.76) than the other two samples tended to adsorb larger amounts of these organic products (Fig. 4). Both alachlor and its degradate are adsorbed least in the sample with the smallest percent organic matter and highest pH.

Preliminary equilibrium distribution coefficients (Kd) established through this research suggested that the Kd for alachlor varied from 6.2 to 7.7 in sample B15A, 2 to 2.3 in sample B15C2, and 3.3 to 3.6 in sample U17B. In addition, the Kd for ethanesulfonic acid varied to a smaller degree with soil sample: the Kd of sample B15A varied from 1.9 to 2.6, the Kd of sample B15C2 varied from 1.1 to 2.5, and the Kd of sample U17B varied from 1.0 to 2.4. The batch results illustrated that the tendency of ethanesulfonic acid of alachlor to adsorb to organic matter and mineral surfaces is smaller than the tendency of the parent herbicide, alachlor, to adsorb to soil organic matter and mineral surfaces. This is probably because this degradate is an acid, which in its anionic form (at high pH), will have a small tendency to adsorb.

Alachlor, Atrazine, and Selected Degradates in Ground Water. Alachlor, atrazine, and selected degradates were not detected in the ground-water samples. Only a small amount of deisopropylatrazine (0.08 µg/L) was detected in water collected beneath the upland of one transect. Based on the analyses of herbicides and degradates in the prior sections, higher concentrations of herbicides and degradates were expected in the ground water beneath the uplands and in the ground water of floodplain with soils low in organic matter and clay and high in pH. Therefore, on the basis of limited ground-water data collected, it appears that no relationship existed between irrigation and landscape position and the presence and concentrations of the herbicides and degradates studied in the ground water. Other factors, such as depth to water (generally less than 200 feet) and the low sample size may have affected the concentrations of the contaminants studied and the results presented herein. However, a study by Perry and others (Charles Perry, U.S. Geological Survey, written communication, 1995) appears to confirm the results of this study.

Conclusions

Analyses of soil properties suggested differences in sand content, pH, exchangeable calcium, organic matter, and cation exchange capacity by landscape position and depth. Two canonical discriminant functions for each depth were developed to discriminate between the landscape positions. These canonical discriminant functions demonstrated the relative importance of the soil properties in characterizing the landscape positions. Analyses for alachlor, its ethanesulfonic acid degradate, atrazine, deethylatrazine, and deisopropylatrazine in upper regolith suggested that the majority of detections of selected parent herbicides and degradates occurred at shallow depths. The number of ethanesulfonic acid detections and positive correlations of alachlor and ethanesulfonic acid verified that ethanesulfonic acid is a common, relatively persistent, degradation product of alachlor in the upper regolith. Alachlor, its ethanesulfonic acid degradate, atrazine, and deisopropylatrazine were present in higher concentrations in those samples characterized by a high percent of organic matter, high cation exchange capacity, and low pH.

Atrazine concentrations and, to a lesser extent, ethanesulfonic acid concentrations, were associated with landscape position and depth, but were not associated with irrigation. Deisopropylatrazine concentrations were not dependent upon landscape position, but could be associated with irrigation. Alachlor and deethylatrazine concentrations were not associated with landscape position or irrigation. Differences in soil properties by landscape position were associated with the presence and concentrations of alachlor, atrazine, and selected degradates. Canonical discriminant functions suggested that the presence of alachlor in samples collected from depth 1 mainly was associated with percent organic matter and, to a lesser extent, with exchangeable bases, surface area, and percent clay. The data also suggested a negative relationship of alachlor with pH and positive relationship of alachlor with percent clay. Ethanesulfonic acid in samples collected from depth 1 mainly was associated with percent organic matter and, to a lesser extent, pH and cation exchange capacity. The data suggested a negative relationship of this degradate with pH and positive relationship with organic matter and cation exchange capacity.

Batch experiments of alachlor and ethanesulfonic acid illustrated the effect of organic matter and pH on the adsorption of alachlor and its ethanesulfonic acid degradate. The batch results suggested that the tendency of ethanesulfonic acid to adsorb to organic matter and mineral surfaces was smaller than the tendency of the parent herbicide, alachlor, to adsorb to organic matter and mineral surfaces. The canonical discriminant functions also suggested that in samples collected from depth 3 the presence of atrazine was associated with organic matter, whereas the association of atrazine and pH appeared minimal.

Thus, concentrations of the alachlor ethanesulfonic acid degradate and atrazine in and beneath soil on uplands, terraces, and floodplains differed across landscape positions at selected depths. Finally, in this research, concentrations of alachlor, its ethanesulfonic acid degradate, atrazine, deethylatrazine, and deisopropylatrazine in and beneath soils did not differ within landscape positions depending upon the use of irrigation. Analyses of ground-water samples for herbicides provided no additional data to support the hypotheses that concentrations of selected herbicides and metabolites in the ground water differed among landscape position and with irrigation use.

Literature Cited

1. Spalding, R.F., Burbach, M.E., and Exner, M.E. *Ground Water Monitoring Rev.*, **1989**, Fall 1989, pp. 126-133.
2. Exner, M.E. *Ground Water Monitoring Rev.*, **1990**, Winter 1990, pp. 148-159.
3. Bushway, R.J., Hurst, H.L., Perkins, L.B., Tian, L., Cabanillas, L.G., Young, B.E., Ferguson, B.S., and Jennings, H.S. *Bull. Environ. Contam. Toxicol.*, **1992**, *vol.* 49, pp. 1-9.
4. Goolsby, D.A., Thurman, E.M., and Kolpin, D.W. *Irrigation and Drainage Proc. 1991*, **1991**, pp. 17-23.
5. Thurman, E.M., Goolsby, D.A., Meyer, M.T., and Kolpin, D.W. *Environ. Sci. Technol.*, **1991**, *vol. 25*, pp. 1794-1796.
6. Goolsby, D.A., Boyer, L.L., Mallard, G.E. *U.S. Geological Survey Open-File Report 93-418*, **1993**, 89 pp.
7. Verstraeten, I.M., and Ellis, M.J. *U.S. Geological Survey Water-Resources Investigations Report 94-4197*, **1995**, 88 pp.
8. Verstraeten, I.M., Sibray, S., Cannia, J.C., Tanner, D.Q. *U.S. Geological Survey Water-Resources Investigations Report 94-4057*, **1995**, pp. 114.
9. U.S. Environmental Protection Agency in *U.S. Code of Federal Regulations*, Title 40, parts 100 to 149, revised as of July 1, 1991, pp. 585-588.
10. Tiedje, J.M., and Hagedorn, M.L *J. Agric. Food Chem.*, **1975**, *vol. 23 (1)*, pp. 77-81.
11. McGlamery, M.D., and Slife, F.W. *J. Weeds*, **1966**, *vol. 14*, pp. 237-239.
12. Kearney, P.C. *J. Residue Rev.*, **1970**, *vol.* 32, pp. 391-399.
13. Gamble, D.S., and Khan, S.U. *J. Agric. Food Chem.*, **1990**, *vol.* 38, pp. 297-308.
14. Adams, C.E., and Thurman, E.M. *J. Environ. Qual.*, **1991**, *vol.* 20 (3), p. 540-547.
15. Schnoor, J.L. *Fate of pesticides and chemicals in the environment*; John-Wiley & Sons, Inc., New York, NY, 1992, 436 pp.
16. Kaufman, D.D., and Kearney, P.C. *J. Residue Rev.*, **1970**, *vol. 32*, pp. 235-265.
17. Sethi, R.K., and Chopra, S.L. *J. Indian Soc. Soil Sci.*, **1975**, *vol. 23*, pp. 184-194.
18. Fang, C.H. *J. Chinese Agric. Chem. Soc.*, **1979**, *vol.* 17, pp. 47-53.
19. Giardina, M.C., Giardi, M.T., and Filacchioni, G. *J. Agric. Bio. Chem.*, **1985**, *vol.* 46 (6), pp. 1439-1445.
20. Lee, J.K. *J. Korean Agric. Chem. Soc.*, **1986**, *vol. 29*, pp. 182-189.
21. Baker, D.B., Bushway, R.J., Adams, S.A., Macomber, C. *Environ. Sci. Technol.*, **1993**, *vol.* 27, pp. 562-564.

22. Macomber, Carol, Bushway, R.J., Perkins, L.B., Baker David, Fan, T.S., and Ferguson, B.S.   *J. Agric. Food Chem.*, **1992**, *vol. 40*, pp. 1450-1452.

23. Kolpin, D.W., Burkart, M.R., and Thurman, E.M., *U.S. Geological Survey Water-Supply Paper 2413*, **1991**, 34 pp.

24. Aga, D.S., Thurman, E.M., and Pomes, M.L.   *J. Anal. Chem.*, **1994**, *vol. 66*, pp. 1495-1499.

25. Roeth, F.W.   *Atrazine degradation in two soil profiles*; M.S. Thesis, Univ. of Nebraska-Lincoln, Lincoln, NE, **1967**, 45 pp.

26. Khan, S.U.   *Pesticides in the soil environment: Fundamental Aspects of Pollution Control and Environmental Science 5*; Elsevier Co., New York, NY, **1980**, 240 pp.

27. Couch, R.W., Gramlich, J.V., Davis, D.E., Funderburk, Jr., H.H.   *Proc. South. Weed Sci. Soc.*, **1965**, *vol. 18*, pp. 623-31.

28. Konopka, Allan, and Turco, Ronald   *J. Appl. Environ. Microbiol.*, **1991**, *vol. 57*, pp. 2260-2268.

29. Mills, M.S.   *Field dissipation of encapsulated herbicides: Geochemistry and degradation;* M.S. Thesis, Univ. of Kansas, KS, **1991**, 98 pp.

30. Bathke, G.R., Cassel, D.K., and McDaniel, P.A.   *J. Environ. Qual.*, **1992**, *vol. 21*, pp. 469-475.

31. Germann, P.F., Edwards, W.M., and Owens, L.B.   *J. Soil Sci. Soc. Am.*, **1984**, *vol. 48*, pp. 237-244.

32. Eckert, D.J.   *North Dakota agricultural experiment station, North Dakota State University, Fargo*, **1988**, pp. 6-8.

33. Schulte, E.E, in *Recommended chemical soil test procedures for the north central region;* Dahnke, W.C., Ed., : North Dakota Agric. Exp. Sta., North Dakota State Univ., Fargo, ND, **1988**, pp. 29-32.

34. Holmgren, G.S., Juve, R.L., and Geschwender, R.C   *Soil Sci. Soc. Am. J.*, **1975**, *vol. 41*, pp. 1207-1208.

35. Heilman, M.D., Carter, D.L., and Gonzales, C.   *J. Soil Sci.*, **1965**, *vol. 100*, pp. 409-413.

36. Mills, M.S., and Thurman, E.M.   *J. Anal. Chem.*, **1992**, *vol. 64 (17)*, pp. 1985-1990.

37. Thurman, E.M., Meyer, M.T., Pomes, M., Perry, C.A., and Schwab, A.P.   *J. Anal. Chem.*, **1990**, *vol. 62 (18)*, pp. 2043-2048.

38. Meyer, M.T., Mills, M.S., and Thurman, E.M., *J. Chromatography*, **1993**, vol. 629, pp. 55-59.

39. Johnson, R.A., and Wichern, D.W. *Applied multivariate analysis;* Prentice Hall; Englewood Cliffs, NJ, 1988, 605 pp.

40. Verstraeten, I.M.   *Influence of landscape position and irrigation on alachlor, atrazine, and selected degradates in selected upper regolith and associated shallow aquifers in northeastern Nebraska;* University of Nebraska-Lincoln, Lincoln, NE, Ph.D. dissertation, **1994**, 258 pp.

41. Buol, S.W., Hole, F.D., and McCracken, R.J.   *Ames, Iowa State University Press*, **1984**, 439 pp.

42. Schiavon, M.   *J. Ecotoxicol. Environ. Safety*, **1988**, *vol. 15*, pp. 46-54.

43. Walker, A., Moon, Young-Hee, and Welch, S.J. *J. Pesticide Sci.*, **1992**, *vol. 35*, pp. 109-116.
44. Pothuluri, J.V., Moorman, T.B., Obenhuber, D.C., and Wauchope, R.D. *J. Environ. Qual.*, **1990**, *vol. 19*, pp. 525-530.
45. Wolf, D.C., and Martin, J.P. *J. Environ. Qual.*, **1975**, *vol. 4 (1)*, pp. 134-139.
46. Muir, D.C.G., and Baker, B.E. *J. Agric. Food Chem.*, **1976**, *vol. 24*, pp. 122-125.

WATER QUALITY STUDIES

Chapter 16

The Environmental Impact of Pesticide Degradates in Groundwater

Michael R. Barrett

Office of Pesticide Programs, U.S. Environmental Protection Agency, 401 M Street, Southwest, Washington, DC 20460

Pesticide environmental fate and ground-water monitoring studies often only analyze for residues of parent. Yet available environmental fate data for many pesticide degradates indicate a higher propensity to leach in soil than the respective parent compound. Even in cases where no mobile degradate has been found to accumulate in soil to any great extent (generally true with s-triazine and acetanilide herbicides, for example), the existence of a chemically stable molecular "core" is an indicator that some degradates could reach ground water. Recent studies with atrazine and alachlor have demonstrated that quantities of degradates in ground water may exceed (sometimes by a great amount) the quantities of the parent compound. Ground-water monitoring for newer, low-rate pesticides such as sulfonylurea herbicides and their degradates is almost nonexistent. These low-rate herbicides have environmental fate properties indicating they may be more mobile in soil than some pesticides that have already been found to impact ground water at numerous locations. In cases where degradates are found to be of toxicological significance, use limitations might be needed to mitigate impact of the degradates on ground water.

Until recently, little research has been available on residues of pesticide degradates in ground water. This has limited our understanding of the scope of ground-water contamination by pesticides. Most pesticides have degradates that have little apparent biological activity on target pests but these degradates often do have environmental fate characteristics such that the potential for leaching to ground water appears to be significant. This is particularly true for the majority of the major soil-applied herbicides currently registered in the United States. This paper examines the available data on the occurrence of some of these degradates in ground water and estimates from basic environmental fate data the likely potential for leaching to ground-water of degradates of selected pesticides. The data are also briefly compared to toxicological endpoints, including established or proposed lifetime Health Advisories (HAs) or Maximum Contaminant Levels (MCLs) in drinking water.

Overview Of Environmental Concerns For Pesticide Degradates

Toxicological Concerns. Except for propesticides (pesticides designed to be converted after application to a new compound with pesticidal activity), activity of degradates on target pests is generally significantly lower than the parent compound activity. Indeed, a major selection criterion for prospective pesticides is toxicity to pests at low doses which do not adversely affect crops or other non-target organisms. These factors may reduce, but do not preclude the possibility of significant activity of degradates on various nontarget species. With respect to occurrence in water that may be used for drinking, there is also the human health issue. Regulations are designed to ensure safety to humans even with the many unknown factors about the effects of relatively low-level (but often long-term) exposure to pesticides. Often the toxicological endpoints upon which drinking water standards are based may not be related to the mode of action of the pesticide on the target organism. The mode of action of herbicides, for example, is usually not applicable to mammalian species.

Some pesticides are readily converted to other compounds which are toxic to the target organisms. Organophosphate and organosulfur insecticides commonly have initial degradation products with well-established insecticidal activity, often of greater potency than the compound originally applied. In some cases, rapidly formed degradates may be relied on to provide the bulk of the pest control for which the chemical is applied. Maloxon is such a product formed by oxidative desulfuration of the thione moiety (a common activation pathway for organophosphates) of malathion. A very common reaction observed in many sulfide-containing pesticides is oxidation to sulfoxides and sulfones which are usually quite active on a spectrum of pests similar to the parent compound. Formation of aldicarb sulfoxide and sulfone is an example of this. Toxicologically active degradates of insecticides tend to be more mobile in soils than the respective parent compound (2). However, the efficacy and toxicity of insecticides in the form applied tends to be inversely correlated with water solubility (3). Even though few insecticides have previously been found to extensively impact ground water, the potential for persistent and mobile degradates to impact ground water may be significant.

Ground-Water Residues. Overall, monitoring data for pesticide degradates are relatively sparse. The U.S. EPA Pesticides in Ground Water Database (1) includes reports on analyses of 29 degradates of currently used pesticides, with, except for a handful of compounds, very little evidence of a significant impact of these compounds on ground water (Table I). Exceptions to the general lack of research emphasis on degradates have been for a few pesticides which rapidly transform into compounds that are intended to be biologically active on target pests. For example, extensive

Table I. Detections of metabolites included in EPA Pesticides in Ground Water Database (1)

Compound[†]	No. of wells	No. with detections	Range of detections ug L^{-1}	MCL or HA[‡]
2,6-diethylaniline (alachlor, etc.)	305	0	---	2P
3-hydroxy-carbofuran	22314	42	0 - 10.0	40P
3-keto-carbofuran	839	3	0.03 - 0.03	40P
3,5-dichlorobenzoic acid (pronamide)	126	0	---	50P
3-hydroxy-dicamba	87	0	---	200P
aldicarb sulfone	37652	5070	0 - 153	2
aldicarb sulfoxide	37593	4991	0 - 1030	4
carbofuran phenol	126	0	___	40P
DCPA acid	118	59	0.2 - 431	4000P
demeton sulfone	188	0	---	None
deethyl atrazine	689	27	0.05 - 2.9	3P
deisopropyl atrazine (= deethyl simazine)	689	24	0.1 - 3.5	3P
endosulfan sulfate	1969	6	0.05 - 1.4	None
ethylene thiourea (maneb, mancozeb, etc.)	183	1	0.7	None
fenamiphos sulfone	180	0	---	2P
fenamiphos sulfoxide	180	0	---	2P
malaoxon (malathion)	1	0	---	200P
methyl paraoxon (methyl parathion)	125	0	---	2P
molinate sulfoxide	196	1	0.8	None
phorate sulfone	12	0	---	None
phorate sulfoxide	12	0	---	None
phoratoxon (phorate)	9	0	---	None
phoratoxon sulfone (phorate)	9	0	---	None
phoratoxon sulfoxide (phorate)	9	0	---	None
phosmet oxygen analog	3	0	---	None
pirimicarb sulfone	1	0	---	None
rotenolone (rotenone)	4	0	---	None
terbufos sulfone	13	0	---	0.9
thiobencarb sulfoxide	157	0	---	None

[†]Full chemical names available from the author upon request.

[‡]Values followed by a P are for parent compound only; no MCL or (lifetime) HA exists for the degradate.

monitoring has been conducted for aldicarb and its sulfoxide and sulfone degradates by the registrant and some state agencies (1). The need for inclusion of these degradates in the analyses for aldicarb was obvious because all three forms (parent, sulfoxide, and sulfone) are known to be highly active on target pests and toxic to many nontarget organisms. In recent years, studies of several herbicide degradates (particularly of atrazine) have begun to provide insight into the impact that the consideration of degradates may have on pesticide regulation, and are discussed in detail below.

As monitoring efforts have increased and analytical methods have become more sensitive, there have been many more detections of pesticides in ground water and more public concern about the possible health effects of these residues. Concerns will increase further if it turns out that some of the most environmentally significant degradates of pesticides have adverse ecological effects or toxicological effects similar to those upon which the MCL for the parent molecule is based.

Evaluation of the Leaching Potential of Pesticide Degradates

A comprehensive discussion of all registered pesticides and their degradation products is beyond the scope of this paper. The remainder of this paper discusses three families of herbicides with the following distinctive characteristics:

▶ High volume use with relatively high soil persistence (the *s*-triazine herbicides, with specific discussion of atrazine and simazine) and moderate soil mobility.
▶ High volume use with low to moderate soil persistence but potentially persistent degradates (the acetanilide herbicides with specific discussion of acetochlor, alachlor and metolachlor) and moderate parent compound soil mobility.
▶ Very low use rates and variable soil persistence, but high soil mobility (the sulfonylurea family of herbicides).

Numerous ground-water monitoring studies over the last several years have included the *s*-triazine and acetanilide herbicides as analytes (except for acetochlor, which was not registered in the United States until 1994). Monitoring studies for sulfonylurea herbicides, which are much more difficult to analyze for, have not yet been published. Acetanilide and *s*-triazine herbicide parent compounds have been found to occur in ground water (1), and, more recently, some of their degradates have been analyzed for and found in ground water (discussed below). It is no accident that these compounds are herbicides: With the lone exception of aldicarb (for which the registered uses have already been substantially reduced) a large majority of ground-water detections of currently used pesticides have been of soil-applied herbicides. The potential contribution of the degradates to the total residues occurring in ground water is discussed directly from ground-water monitoring data, if available, and also by inference from soil degradation and adsorption studies. Full chemical names as well as abbreviated names used in the text for the pesticides and degradates discussed in the following sections are given in Table II, or; for the sulfonylurea herbicides discussed, chemical structures of parent compounds and degradates are given in Figures 1 and 2, respectively.

Table II. Chemical Names of Pesticide Parent Compounds and Respective Degradates Discussed in Detail in the Text

Common & Abbreviated Names (Parent in **bold**)	Chemical Name
acetochlor	2-chloro-N-ethoxymethyl-2'-ethyl-6'methylacetanilide
acetochlor sulfonic acid	N-ethoxymethyl-2'-ethyl-6'-methyl-2-sulfoacetanilide
acetochlor sulfinyl acetic acid	{[N-ethoxymethyl-N-(2'-ethyl-6'-methyl)phenyl]-2-amino-2-oxoethyl} sulfinylacetic acid
acetochlor EOM-oxanilic acid	N-ethoxymethyl-2'-ethyl-6'-methyloxanilic acid
alachlor	2-chloro-2',6'-diethyl-N-(methoxy-methyl)-acetanilide
alachlor sulfinyl acetic acid	[N-methoxymethyl-N-(2,6-diethylphenyl)-2-amino-2-oxoethyl] sulfinylacetic acid
alachlor oxanilic acid	2',6'-diethyloxanilic acid
alachlor sulfonic acid (ESA)	N-methoxymethyl-2',6'-diethyl-2-sulfoacetanilide
atrazine	2-chloro-4-ethylamino-6-isopropylamino-s-triazine
deisopropyl atrazine (DEIS)[†]	2-amino-4-chloro-6-ethylamino-s-triazine
deethyl atrazine (DEET)	2-amino-4-chloro-6-isopropylamino-s-triazine
diaminochloro triazine (DIAM)[†]	2,4-diamino-6-chloro-s-triazine
hydroxy atrazine (HYAT)	2-ethylamino-4-hydroxy-6-isopropylamino-s-triazine
metolachlor	2-chloro-N-(2-ethyl-6-methylphenyl)-N-(2-methoxy-1-methylethyl) acetamide
metolachlor acid	(N-(2'-methoxy-1'-methylethyl)-2-ethyl-6-methyl-oxalic acid) anilide
(de-MOME) metolachlor acid	(2-ethyl-6-methyl-N-oxalic acid) anilide
(de-MOME) hydroxy acetamide	N-(2-ethyl-6-methylphenyl)-2-hydroxy-acetamide
(demethyl) hydroxy metolachlor	2-chloro-N-(2-ethyl-6-methylphenyl)-N-(2-hydroxy-1-methylethyl)acetamide
methyl morpholin	4-(2-methyl-6-ethylphenyl)-5-methylmorpholin
simazine	2-chloro-4,6-bis(ethylamino)-s-triazine
hydroxy simazine (HYSI)	2-hydroxy-4,6-bis(ethylamino)-s-triazine

[†] These compounds are degradates of both atrazine and simazine.

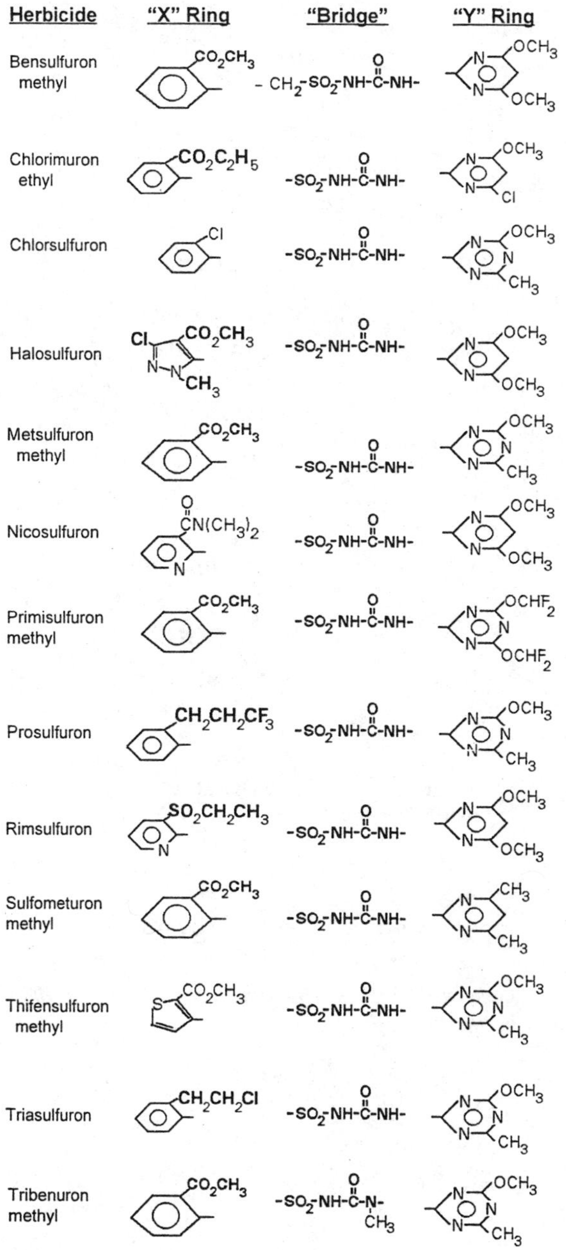

Figure 1. Chemical structures of sulfonylurea herbicides registered in the United States.

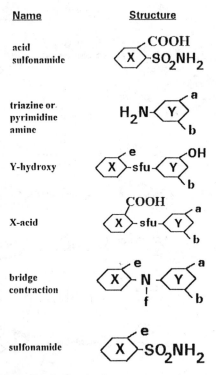

Figure 2.   Generalized chemical structures of major degradates of sulfonylurea herbicides.

Leaching Potential of s-Triazine Herbicides. Some recent investigations have focused on residues of these degradates in ground water, particularly of atrazine, which is one of the most heavily used pesticides in the United States (4) and one of the most commonly detected pesticides in ground water (1). Each of the degradates (except for hydroxy atrazine) analyzed for to date has consistently been found to constitute a significant proportion of the total residues in ground water (Table III).

The extent of the impact of atrazine degradates on ground water was not necessarily evident from earlier studies that examined the formation of atrazine degradates in soil. The soil and ground-water residue data are examined more closely here, with an emphasis on the more general implications regarding the possible occurrence of pesticide degradates in ground water.

Soil Residue Studies. Most studies have detected modest accumulations of primarily three degradates: Hydroxy atrazine, deethyl atrazine, and deisopropyl atrazine (Table IV). Only hydroxy atrazine has been shown to commonly accumulate at more than a few percent of the amount of atrazine applied to the soil. The accumulation of hydroxy atrazine may also have been underestimated in some studies because hydroxy atrazine is much more strongly adsorbed in most soils than the other atrazine degradates (Table V) and therefore may not have been efficiently extracted from soil.

Table V summarizes the best available data on the relative leaching potential of atrazine, simazine, and their degradates. Of the degradates listed, deisopropyl atrazine and diaminochloro triazine are degradates of both atrazine and simazine. Other degradates of these herbicides have been identified, but there are insufficient data on these to provide a credible evaluation of their leaching potential. From Table V it is clear that even with best-case assumptions about the persistence and mobility of the dealkylated degradates of atrazine and simazine that there is still a potential for leaching to ground water. All studies to date indicate these compounds are more weakly adsorbed in soils than parent atrazine or simazine. Data on the soil persistence of the dealkylated degradates are more equivocal (Table IV), but these compounds may be significantly more persistent in most ground water than in soils. Hydroxy atrazine is much more strongly adsorbed than atrazine or its dealkylated degradates in most soils and is not expected to be as likely to leach (Table V). Limited data indicates hydroxy simazine may generally be even more strongly adsorbed than hydroxy atrazine, and therefore this compound also appears to be relatively unlikely to leach. However, of several studies of hydroxy atrazine, some measurements of the K_{oc} are low enough (the lowest in the literature are around 300) that, if hydroxy atrazine persists for a very long time (i.e., half-life of at least several months) this compound potentially could leach in vulnerable soils. Clay and Koskinen (25) found that there is less adsorption of hydroxy atrazine at higher pH (more than a two-fold difference in K_{oc} over 1.6 to 2.1 pH units) in two soils that they studied.

The amounts of the dealkylated degradates detected in soil have often been so low as for the study authors to conclude that dealkylation is a minor initial degradation pathway in soil; and to conclude that, in spite of somewhat greater soil mobility, these residues will not impact ground water. However, Capriel et al. (14) found that there can be a great difference between the persistence of residues extractable by conventional methods and those by more exhaustive methods. They found that these unextractable

Table III. Relative Amounts of Atrazine Parent and Degradates Recorded in Ground-Water Studies

Study Reference/Location and study scale (large-scale studies in bold)	Number of samples[†]	Mean atrazine ug L[-1]	Degradate/parent atrazine ratios[‡]				
			DEIS	DEET	DIAM	HYAT	combined
Adams & Thurman (5), KS field study.	1	2.30	0.20	1.13	----	----	1.33
Cai et al. (6). NE field study.	6	2.05	----	----	----	0.02	----[§]
Denver & Sandstrom (7). DE watershed.	20	0.41	0.34	1.76	----	----	2.10
Frank et al. (8), subset of data used. Ontario field study.	7	0.34	----	2.99	----	----	2.99
Isenee et al. (9), values approximated from data subset. MD field study.	many	~0.7	----	~2.0	----	----	~2.0
Kross et al.(10), IA statewide.	55	0.57	0.49	0.57	----	---	1.06
Pionke and Glotfelty (11), PA watershed.	34	0.95	0.63	1.50	----	----	2.14
Squillace et al. (12). Alluvial wells, Cedar R., IA.	132	0.22	----	0.73	----	----	0.73
Dep. Agric., Trade, & C.P. (13), WI statewide.	256	0.95	0.27	0.71	0.90	----	1.89

[†]For IA and WI studies, this is the number of wells with detections of one or more of the listed compounds (many more wells were sampled). Although the KS data presented are from a single ground water sample, other soil water samples in this study show similar trends (i.e., DEET ≥ atrazine > DEIS) after residues had leached from the soil surface. If no value is given, the amount of the degradate was not reported.

[‡]See Table II for interpretation of abbreviations. Combined is the ratio of the sum of all atrazine degradate residues analyzed for to the atrazine concentration.

[§]This study analyzed for the first time the level of hydroxyatrazine residues in ground water - since other degradates which are more important ground-water contaminants were not analyzed for, a degradate/ parent ratio could not be calculated.

Table IV. Extent of Formation of Atrazine degradates in Soil

Study identification	Last sample, DAT[†]	% of applied residue at final sampling time[‡]				Maximum residues as % of applied[‡]		
		Atra-zine	DEIS	DEET	HYAT	DEIS	DEET	HYAT
Adams & Thurman (5)	150	6	1.5	<0.5	ND	----	----	----
Capriel et al. (14)	3300	atraz., HYAT > DEIS, DEET				----	----	----
Khan et al. (15)	365	9	-----	-----	16	----	----	32
Khan & Marriage (16)	1275[§]	<1	9.0	<1	4.2	9.0	<1	10
Kruger et al. (17)[#]	180	4	<1	1.6	1.8	<1	2.4	1.8
Kruger et al. (18)	105	9	1.2	4.2	<1	----	----	----
Mills & Thurman (19)	~150	3	<1	2	----	3.1	7.0	----
Sorenson et al. (20)	485	21	1.0	5.3	26	1.6	7.3	26
Winkelmann & Klaine (21)	180	1	<1	<1	6	<1	3	9

[†] DAT = days after treatment.

[‡] Refer to the Table II for full common and chemical names for each degradate. For some studies maximum residues are not given because residues were analyzed for at only one sampling interval.

[§] Days after last of several annual applications; percent remaining for each compound calculated on the basis of the rate of the last application made.

[#] Several other degradates were analyzed for. DIAM constituted 4.1 % of residues 60 days after treatment, but was much less than 1 % of applied by 180 days posttreatment.

Table V. Relative Leaching Potential of Atrazine, Simazine, and Degradates as Estimated from K_{oc} and Soil Degradation Half-Lives[†] using the GUS Index [‡]

Compound	Half-life days	K_{oc} L kg^{-1}	GUS Score[§]	Leacher?
atrazine	100	89	4.1 (2.9-4.6)	yes
HYAT	150E	1668	1.7 (1.4-3.9)	no
DEET	100E[¶]	35	4.9 (2.7-6.2)	yes
DEIS	100E[¶]	51	4.6 (2.5-5.8)	yes
DIAM	100X	56	4.5 (3.4-5.8)	yes
simazine	100	118	3.9 (2.5-4.6)	yes
HYSI	200E	4906	0.7 (0.2-2.9)	no

[†] Values are direct measurements except when followed by an E (estimated from laboratory or field studies) or by an X (no specific data, estimated from data for similar compounds).

[‡]The GUS Index provides a comparison of soil leaching potential of pesticides based solely on measurements of K_{oc} and soil degradation half-life for each pesticide using the formula: GUS score = $\log(t\frac{1}{2}) * [p - \log(K_{oc})]$ where p = 4 (22). A GUS of greater than 2.8 corresponds to pesticides that have been found to occur in vulnerable ground water, a GUS of 1.8 to 2.8 is associated with pesticides that have been less consistently found to impact ground water.

[§]Values in parenthesis are the range of scores calculated from a range of values for K_{oc} and soil half-life taken from EPA Office of Pesticide Programs' registration files plus published values (22-27)

[¶]Some published values of only a few weeks derived from direct application of these degradates to soil exist (28, 29).

residues may remain bound in the soil as parent compound or one of the initial degradates for many years. Several other authors have found that total ^{14}C-residues persisted much longer than extractable residues (20, 21, 28-30), and, while rarely are the "bound" residues identified, some of these authors have speculated on the potential of some of these residues persisting as parent atrazine or initial degradates.

In trying to anticipate or understand pesticide impact on ground water, it is important to consider the change in the behavior of these compounds over time and as they move through the profile. With s-triazine herbicides, as well as many other pesticides, there is an increase in the amount of bound residues and the apparent adsorption coefficient over time. This seems to result in detoxification of herbicides since there is very little evidence of phytotoxicity of degradates accumulating in the soil after many years of use (31). However, there is no direct evidence that once s-triazines or other pesticides have been bound to soil the pesticide becomes permanently transformed into non-toxic compounds. The possibility of some sort of longer-term environmental impact of *s*-triazines on plants or other organisms cannot be discounted because the available evidence indicates that the *s*-triazine ring probably persists for many years in most soil environments (14, 21). Only in rare instances are soil microbes present which can readily completely degrade *s*-triazine compounds, including mineralization of the s-triazine ring (32, 33).

Another factor which increases the potential of parent and degradates to reach ground water is an often dramatic decrease in the rate of metabolism of pesticides and their degradates as the chemicals move below the A horizons of soils into subsoils with much lower organic matter content and microbial activity. For example, in a recent study (34) the degradation half-lives of atrazine increased from an average of about 70 days in surface soil at two locations to an average of about 300 days in soil at a 150 cm depth at the same locations. Similar or greater decreases in degradation rate with depth were observed for alachlor, metolachlor, and metribuzin (an asymmetrical triazine herbicide). Baluch et al. (28) found that the half-life of deethyl atrazine averaged 36 days in the surface layer of 5 soils and 111 days in soil taken from about a 78 cm depth at the same locations. Average half-lives of hydroxy atrazine for the same soils were 100 and 162 days for the surface and 78 cm-depth soil, respectively. At a site where atrazine appeared to be readily mineralized by soil microorganisms in surface soil, the atrazine half-life ranged from 11 days in surface soil to 248 days in deep soil (32). Residues may be even more persistent in ground water: The half-life of atrazine may be several years or more, and the dealkylated degradates are also quite persistent in ground water (35).

Ground-Water Residue Studies. Until recently, degradates of *s*-triazine herbicides had rarely been analyzed for in ground water. The advent of lifetime Health Advisories and/or Maximum Contaminant Levels in the last few years for parent atrazine, cyanazine, and simazine (all under 3 μg L^{-1}) has brought to the attention of many the possibility of low levels of these pesticides and perhaps some of their degradates in ground water affecting human health.

All of the studies available to date confirm that atrazine degradates can contribute significantly to the total triazine residues in ground water arising from atrazine use (Table III). Although some other *s*-triazine herbicides have degradates in common with atrazine (e.g., simazine), the primary source of the residues observed was probably atrazine because the use of atrazine has been much more extensive in most of the areas where ground water was sampled (see, e.g., 36). In most of these studies both deisopropyl atrazine and deethyl atrazine were analyzed. In all cases deethyl atrazine was present, on average, at higher concentrations than deisopropyl atrazine, although in one of the large-scale studies (10), the difference in concentrations was slight. Analyses from Wisconsin domestic wells statewide (13) indicate that the three dealkylated degradates on average account for approximately double the residue of parent atrazine (this is the only study available so far in which the diaminochloro triazine degradate was analyzed). All of the wells sampled from in this Wisconsin study had yielded samples with atrazine concentrations greater than 0.35 μg L^{-1} in a previous study. Therefore, it is possible that samples from other wells would have less atrazine and a higher ratio of degradates to parent. The state of Wisconsin regulates residues of all chloro-triazine compounds in ground water, including the three dealkylated degrades (Jeff Postle, Wisconsin Department of Agriculture, Trade, and Consumer Protection, Madison, personal communication, 1993). Therefore, in Wisconsin, the contribution of the degrades to the total residues can significantly increase the number of wells out of compliance with the state standards. No studies have yet been published concerning residues of the secondary degrades of atrazine or for degrades of other triazine herbicides.

With still only a limited number of studies focusing primarily on two degrades of atrazine, it is already clear that degrades are a major contributor to the total residues in ground water. A single study on a third degrade, diaminochloro triazine, indicates this compound also is a major component of the residues. Existing studies indicate that total ground-water residues of these three degrades plus atrazine may constitute, on average, two or three times the residues of atrazine parent alone. There may also be other *s*-triazine compounds (for example, hydroxydeethyl atrazine) occurring in ground water that have not been analyzed for yet. Although no other degrades have been confirmed to occur in significant quantities in soil, the persistence of the *s*-triazine moiety in soil and the long residence time of water in most aquifers indicates that there is a potential for secondary metabolites to occur. Hydroxy atrazine and other hydroxylated degrades have been presumed to not impact ground water because of their high propensity to be adsorbed in most soils. For hydroxy atrazine, so far only results from analysis of a few wells in Nebraska known to have atrazine residues have been published: On average, the amount of hydroxy atrazine was only about 2 % of the amount of atrazine parent. Even if hydroxy degrades do not leach as is, they might be formed in ground water, and results of additional studies will be needed to determine if there are any circumstances under which it becomes an important component of the total ground water residue. However, hydroxy metabolites lacking the chlorine atom do not appear to form in significant amounts in humans (37) or laboratory animals (38) and may not elicit the same sort of toxicological effects as chloro-triazine herbicides; therefore if hydroxy degrades occur in drinking water, these residues will not necessarily be regulated in the same way as chloro-triazine compounds.

There are differences in the toxicology, structure, and degradation pathways of *s*-triazine herbicides other than atrazine and simazine that will lead to differences in the concern for degrades of each compound. However, in each case, the *s*-triazine ring appears to be quite persistent in soil under most field conditions. Therefore, it is likely that degrades of compounds such as cyanazine and prometon will also be found in ground water, once they are looked for in the right places. Supporting this hypothesis is a study in which it was demonstrated that the dealkylated degrades of each of three *s*-triazine herbicides (atrazine, propazine, and simazine) leached at similar concentrations in a Eudora silt loam in Kansas (19).

Leaching Potential of Acetanilide Herbicides - Soil and Ground-Water Residue Data. Alachlor and metolachlor are the most heavily used of this class of herbicides (4) and acetochlor is expected by the registrant to reach a similar use level; in its very first year of use (1994), acetochlor was already used on 7 % of the United States corn acreage (39). The environmental fate of these compounds is better understood than for other herbicides in this family, but few studies have directly focused on the fate of the degradates. Monitoring results for degradates in ground water are also very sparse, but some very intriguing results have been reported recently, and are discussed below. Table VI provides an estimate of the relative leaching potential of these herbicides and selected degradates of each from the best available data on soil adsorption coefficients and degradation rates. Of the parent compounds, metolachlor appears to have the highest intrinsic leaching potential, since it is generally more persistent and only a little less mobile than acetochlor and alachlor. However, the extent of impact on ground water is also of course dependent on the extent of use; and all three of these compounds are (or presumably will be soon for acetochlor) among the most heavily used pesticides in the United States. There are numerous ground-water detections for both alachlor and metolachlor (1, 40), but both compounds are generally less persistent than most *s*-triazine herbicides and also strongly enough adsorbed in most soils that movement to ground water should be less frequent than for *s*-triazine herbicides. This has been confirmed in studies in areas where alachlor, atrazine, and metolachlor have all been heavily used (10, 40). Acetochlor was only registered in the United States in 1994, but extensive ground-water monitoring is in progress, as this was a requirement for its registration.

Only a limited amount of indirect information on the persistence of alachlor and metolachlor degradates in soil is available. The data are sufficient to demonstrate that most of the degradates of alachlor and metolachlor are considerably more mobile than the respective parent compound and of at least moderate persistence (Table VI). For each of these herbicides, several other degradates have been identified or at least postulated in studies to support their pesticide registration. Clearly, at the very least, this summary of the available data for a few degradates demonstrates that more attention needs to be paid to the possibility of residues of these compounds reaching ground water. The degradation of these herbicides is quite complex and it is not possible at the current time to predict the relative impact of each degradate on ground water.

The Potential for Widespread Occurrence of Acetanilide Herbicide Degradates in Ground Water: The Case of the Alachlor Sulfonic Acid Degradate. Recent reports on the occurrence of a degradate of alachlor in ground water clearly confirm that it has impacted ground water at much higher concentrations than alachlor. The chain of events that led to the discovery of the occurrence of this degradate in drinking well water provides important insight as to how our vision of ground-water contamination by pesticides is limited when samples are only analyzed for parent compounds.

Large amounts of the sulfonic acid degradate of alachlor (also commonly referred to as the ethanesulfonic acid degradate; hereafter referred to as ESA) have been recently found to occur in private, rural (mostly drinking) water well samples (41). This was discovered after investigation of a high false positive rate for immunoassay of ground-water samples for alachlor. In the late 1980s' researchers began to utilize immunoassay methods in ground-water monitoring programs for alachlor. Immunoassays using polyclonal antibodies may exhibit significant cross-reactivity to various compounds chemically related to the target analyte. The types of compounds for which significant cross-reactivity occurs can be controlled or predicted to some extent by selection of the hapten molecule, carrier protein, and procedure for preparation of the immunogenic

Table VI. Relative Leaching Potential of Alachlor, Metolachlor, Acetochlor, and Degradates as Estimated with the GUS Index from K_{oc} and Soil Degradation Half-Lives. Refer to Table V for further explanation

Compound	Half-life days	K_{oc} ml g^{-1}	GUS Score[†]	Leacher?
alachlor	15	177	**2.1** (1.6-2.8)	marginal
al. sulfinyl acetic acid	90E	50X	**4.5** (3.0-6.8)	yes
al. oxanilic acid	150E	50X	**5.0** (3.4-7.4)	yes
al. sulfonic acid	250X	50X	**4.8** (3.4-6.2)	yes
metolachlor	67	273	**2.9** (2.4-3.8)	yes
metolachlor acid	275E	7	**7.7** (4.1-9.7)	yes
(de-MOME) meto-lachlor acid	275E	20E	**6.6** (4.1-8.2)	yes
(de-MOME) hydroxy acetamide	60E	20E	**4.8** (3.4-6.2)	yes
(demethyl) hydroxy metolachlor	60E	50E	**3.6** (2.5-5.6)	yes
methyl morpholin	90E	50E	**4.5** (2.8-6.1)	yes
acetochlor	12	165	**1.9** (1.1-3.4)	marginal
ac. sulfonic acid	250	49	**6.6** (5.1-8.4)	yes
ac. sulfinylacetic acid	40E	18	**4.4** (2.9-5.7)	yes
ac. EOM-oxanilic acid	400	27	**5.5** (4.2-7.0)	yes

[†] Values in parenthesis are the range of scores calculated from a range of values for K_{oc} and soil half-life. Data sources include EPA Office of Pesticide Programs' registration files and the Pesticides Properties Database (27). For some of the degradates, selection of the extreme K_{oc} and half-life values to calculate this range was arbitrary due to the paucity of available measurements.

conjugate (antigen). Once antibodies are made and antiserum developed for the immunoassay, cross-reactivities must be individually tested with each compound suspected of potentially causing a reaction in the analysis of environmental samples. However, if a compound is not anticipated to occur or is not available in the form of analytical standard material, then it is of course not tested for. This was originally the case for ESA.

Commercially developed kits were developed for the immunoassay of alachlor parent and have been increasingly used by researchers to screen ground-water samples for alachlor, with follow-up confirmation by compound-specific analytical methods. Baker et al. (41), in a well water quality sampling program primarily covering Indiana, Kentucky, and Ohio; found that a high rate of false positives for alachlor could be attributed almost entirely to the reaction of ESA with the immunoassay test kits that were used to screen for alachlor. In a resampling of wells from which the original samples had a detection of alachlor by immunoassay, 76% of the samples (103 of 136) which were still positive by the immunoassay were found not to contain any alachlor parent by gas chromatography - mass spectrometry (GC-MS). Reanalysis of 30 of these samples for ESA by liquid chromatography-MS-MS confirmed that all false positives by the immunoassay could clearly be attributed to the presence of ESA (in spite of the immunoassay being about 50 times less reactive with ESA than with alachlor). Furthermore, the median value of ESA in these samples was 14 μg L^{-1}, close to 50 times the average amount of alachlor parent determined in all of the samples analyzed by GC-MS (David B. Baker, Heidelberg College, Tiffin, Ohio, July 9, 1993; personal communication).

In the 8137 rural well samples analyzed by immunoassay, Baker and his colleagues reported a 5% positive ratio for "alachlor"; from the resampling program it is apparent that likely only about 1 to 1.5% of the well samples actually contained alachlor, the remainder nearly all contained ESA at greater than 5.0 μg L^{-1} (the effective minimum detection limit of the immunoassay for ESA). Undoubtedly, the rate of occurrence of ESA would have been found to be higher if the immunoassay had been as sensitive for ESA as it was for alachlor. More recently, Kolpin et al. (42), in a survey of near-surface aquifers in the midwestern United States, detected ESA in 45% of the samples analyzed for at 0.1 μg L^{-1} or greater whereas alachlor parent was detected in only 5% of the wells at 0.002 μg L^{-1} or greater (50 times greater sensitivity than for ESA). Clearly, this study confirms that ESA is much more prone to leach to ground water than alachlor. Of more than 50 analytes, ESA was the most commonly detected, in spite of the methods for all of the other pesticides or their degradates being more sensitive.

Studies submitted to EPA to support registration imply that alachlor degradates other than ESA may be formed in as great or greater quantities. In fact, for acetochlor, alachlor, and metolachlor there are several degradates (the ones that so far appear to be the most important are given in Table VI) which may or may not accumulate in soil in large quantities (i.e., greater than 10 or 20% of the applied parent compound). Most or all of these compounds appear to be more mobile in soils than the respective parent pesticide, and often much more mobile. Therefore, the story of ESA detection by immunoassay should provide an impetus for greater emphasis on developing methods and sampling programs for these degradates. Acetochlor, alachlor and metolachlor appear to have a "core" of the molecule that is highly resistant to degradation in most soil environments. For alachlor the persistent moiety is 2,6-diethyl-acetanilide, whereas for acetochlor and metolachlor the persistent moiety is 2-ethyl-6-methyl-acetanilide (however, different degradates predominate for each). Most, but not all, of the identified initial degradates also retain the N-ether moiety that each has (ethoxymethyl for acetochlor, methoxymethyl for metolachlor, and methoxypropyl for metolachlor). Immunoassays can be developed which are sensitive to various molecules containing one of these moieties, and therefore provide a method for general screening of degradates of alachlor or of metolachlor.

The Leaching Potential of Sulfonylurea Herbicides. Most of the herbicides registered for use in recent years have very different characteristics from previous generations of herbicides. Most are applied at very low rates (as low as a few grams per acre) and are quite mobile in soil. Very little research has been published on the subsurface movement and impact on ground water of these herbicides. There are three main reasons for the lack of published research on the environmental fate of these compounds: (1.) Most have been registered for use only in the last few years and undoubtedly many have not yet reached their potential for market share; (2.) the assumption that these compounds are of very low potential for environmental effects on non-target organisms; and (3.) the significant technical challenges in developing sufficiently sensitive analytical methods to track the fate of these compounds in the environment. The data discussed below come from studies to support pesticide registrations and are being compiled in a publicly available data base by the USEPA (43).

Sulfonylurea herbicides are acids with pKa values generally ranging from 3.0 to 5.5. All compounds registered so far are relatively mobile in soil, with average Koc values determined from batch equilibrium adsorption studies between 5 and 100 ml g^{-1} for 11 of the 13 compounds (43). Some evidence indicates that halosulfuron and bensulfuron methyl may be generally more strongly adsorbed than the other sulfonylureas (43, 44). However, the strength of adsorption of sulfonylurea herbicides to soil organic matter and clay is often highly dependent on pH, with weaker adsorption in neutral to alkaline soils than in acidic soils. Hydrolytic stability also increases at higher pH, therefore sulfonylurea herbicides are more likely to be persistent in alkaline soils. Degradation half-lives determined in laboratory aerobic soil studies range from about 5 to 140 days, with only halosulfuron, rimsulfuron, thifensulfuron methyl, and tribenuron methyl having average half-lives shorter than three weeks. For some compounds, calculated field dissipation half-lives were significantly shorter than laboratory metabolism half-lives, possibly because of more rapid degradation under field conditions (but it is not specifically known how much of the dissipation was due to leaching and volatilization rather than degradation). Overall, most sulfonylurea herbicides have a high intrinsic leaching potential -- only thifensulfuron methyl and tribenuron methyl have been demonstrated to commonly have GUS scores below 2.8, indicating moderate to low leaching potential (22).

Application rates of sulfonylurea herbicides are generally very low (Table VII), ranging from about 5 to 50 times lower than older generation herbicides that are often applied at rates close to 1 lb ai A^{-1}. Reference doses {"an estimate, with an uncertainty of perhaps an order of magnitude, of a daily exposure that is likely to be without appreciable risk of deleterious health effects in the human population (including sensitive subgroups) over a lifetime" (45)}, for sulfonylureas range from 0.006 to 1.25 mg per kilogram of body weight per day. Since these values are in the same range or higher (indicating less potent toxicity) than for longer-established herbicide families like *s*-triazines and acetanilides, and since sulfonylureas are applied at lower rates, this means that sulfonylureas are generally much less toxic at the dose applied to humans and animals. Beyer et al. reviewed acute and chronic toxicity data for several sulfonylurea herbicides showing them to be toxic to a wide variety of organisms only at concentrations exceeding a few hundred parts per million in the diet or, for aquatic organisms, in water (44). However, even if no direct effects on animals occur at environmental concentrations, there remains a potential for toxicity of sulfonylureas to sensitive plants at very low doses.

The 13 sulfonylurea herbicides currently registered in the United States have in common a sulfonylurea bridge between two ring structures, one of which (at the amine end of the bridge) is always either a pyrimidine or *s*-triazine ring (Figure 1, Table VIII). This ring is substituted with a methoxy or methyl moiety (halogenated in the case of primisulfuron methyl) at the number four and six positions (except for chlorimuron ethyl, which has a chlorine atom at one of these positions). The ring attached to the sulfonyl end of the bridge was a phenyl ring in the earliest-registered compounds, but compounds have now also been registered with either pyrazole, pyridine, or thiophene rings.

The degradates of sulfonylurea herbicides are generally very similar. A summary of the major categories of degradates for sulfonylurea herbicides is provided in Figure

Table VII.   Current maximum label rates for agricultural uses of sulfonyl-urea herbicides [†]

Common Chemical Name	Maximum Rate, lb ai/A
sulfometuron methyl	0.3750
halosulfuron	0.1250
nicosulfuron	0.0625
bensulfuron methyl	0.0625
primisulfuron methyl	0.0356
prosulfuron	0.0356
rimsulfuron	0.0313
triasulfuron	0.0263
chlorsulfuron	0.0156
tribenuron methyl	0.0156
chlorimuron ethyl	0.0117
metsulfuron methyl	0.0113
thifensulfuron methyl	0.0039

[†]Some compounds have higher rates permitted for applications to noncrop land to control vegetation. The halosulfuron and triasulfuron rates are for split applications. Maximum rates are for row crops except for sulfometuron methyl (forestry) and metsulfuron methyl (pasture).   Some compounds are more commonly applied in combination with other herbicides and under these conditions the maximum rate may be lower.

Table VIII. Sulfonylurea herbicide nomenclature used in Figures 1 and 2

Symbol	Explanation
X	The ring structure (phenyl, pyrazole, pyridine, or thiophene) at the sulfonyl end of the sulfonylurea bridge.
Y	The ring structure (s-triazine or pyrimidine) at the urea end of the sulfonylurea bridge.
e	The moiety attached to the X ring adjacent to the position of attachment with the sulfonyl end of the sulfonylurea bridge. This is most commonly a carboxylic ester.
f	Rearranged fraction of the sulfonylurea bridge, commonly an aminocarbonyl moiety.
a and b	The moieties substituted on the pyrimidine or s-triazine ring; usually a methoxy or methyl group (can be halogenated).
sfu or bridge	The sulfonylurea moiety that connects the ring structures in a sulfonylurea herbicide.
SFU	Abbreviation for the entire herbicide molecule.

2. Sometimes degradates with the bridge intact form in significant concentrations, these are most commonly a "Y-hydroxy" or "X-acid" derivative. The Y-hydroxy degradate has one of the methyl or methoxy moieties on the "Y" ring (i.e., s-triazine or pyrimidine) replaced by a hydroxyl moiety. The X-acid degradate is the carboxylic acid derivative from hydrolysis of the ester that is located on "X" ring of several of the sulfonylurea herbicides (ortho to the attachment of the ring to the sulfonylurea bridge).

The sulfonylurea bridge typically cleaves early in the degradation process with the sulfonylamine moiety retained by the X ring and the remaining amine moiety retained by the Y ring. The resultant degradates are a triazine or pyrimidine amine from the "Y" end of the molecule and a sulfonamide from the "X" end of the molecule. In the case of rimsulfuron, which uniquely has a sulfonyl ester on the X ring, bridge contraction often occurs before bridge cleavage, with only a nitrogen atom connecting the pyridine and pyrimidine rings of rimsulfuron.

The vast majority of sulfonylurea herbicide degradates have a high potential to leach to ground water (Table IX). Degradates with the bridge intact can be moderately persistent and quite mobile. The sulfonamide degradates vary in persistence and to some extent sorptivity depending on the structure of the X ring. Phenyl and thiophene sulfonamides usually persist for only a few months in surface soils, but pyrazole and especially pyridine sulfonamides can persist for months or years. Triazine and pyrimidine amines vary widely in their sorptivity. Pyrimidine amines especially, have a strong affinity for organic matter and clays, with K_{oc} values ranging up to several

Table IX. **Relative Leaching Potential of Degradates of Sulfonylurea Herbicides, Summarized by Class of Degradate and estimated from K_{oc} values and Soil Degradation Half-Lives[†] using the GUS Index**

Compound class	Applicable compounds[‡]	Half-life range, days	K_{oc}, range, ml g[-1]	Maximum % applied	GUS Score[§]	Leachers?
Intact bridge:						
Y-hydroxy SFU	possibly all	50-300	25-100	8-30	3.4-6.5	yes
X-acid SFU	be cm mt pm sm tf tb	15-150	5-20	10-50	3.2-7.2	yes
Altered bridge:						
contracted bridge	rm	40-600	100-300	9-80	2.4-5.6	probably
X ring degradates:						
phenyl sulfonamide	be cm mt pm sm tb ts	30-200	25-140	4-60	2.7-6.0	probably
phenyl acid sulfonamide	be cm mt pm sm tb ts	30-200	1-5	7-30	4.9-9.2	yes
thiophene sulfonamide	tf	30-120	10-50	10-30	3.4-6.2	yes
pyrazole sulfonamide	ha	125-500	0-350	10-40	3.0-10.8	yes
pyridine sulfonamide	nc rm	500-2000	10-40	12-93	6.5-9.9	yes
Y ring degradates						
triazine amine	cs mt po tf ts tb	150-2000	5-200	30-97	3.7-10.9	yes
pyrimidine amine	be cm ha nc pm sm rm	200-4500	50-800	27-89	2.5-8.4	probably

[†] Values are taken from studies submitted to support registration (43).

[‡]Abbreviations: be = bensulfuron methyl, cm= chlorimuron ethyl, cs = chlorsulfuron, ha = halosulfuron, mt = metsulfuron methyl, nc = nicosulfuron, pm = primisulfuron, po = prosulfuron, sm = sulfometuron methyl, rm = rimsulfuron, tf= thifensulfuron methyl, ts = triasulfuron, tb = tribenuron methyl.

[§] The GUS score is explained in Table V. The range is based on the extreme values of degradation half-lives and organic carbon partition coefficients measured for compounds in this class.

hundred ml g^{-1}. Both triazine and pyrimidine amines tend to be the most persistent of all of the degradates of sulfonylureas.

GUS scores range well above 2.8 (indicating a high propensity for leaching) for all classes of sulfonylurea degradates considered in this paper. However, the leaching potential of phenyl sulfonamides, pyrimidine amines, and contracted bridge degradates appears to be marginal in some cases with the lower range of GUS scores extending below 2.8. All classes of degradates listed in Table IX have been shown to form in significant concentrations at times, increasing the possibility that these compounds will leach to ground water.

There are no known environmental effects of the degradates *per se*. Even though, for example, the triazine amine degradates of sulfonylureas appear very likely to leach to and persist in ground water at some locations, the concentrations would be relatively low because of the low application rates for these herbicides. Applications of *s*-triazine herbicides at rates 10 to 100 times greater than the rates of *s*-triazine-containing sulfonylurea herbicides are likely to result in much more contamination of ground water. Some pesticides or degradates have been shown to be much more persistent in ground water than they generally are in surface soil (35) and so there is some possibility of a cumulative impact on ground water from long-term use of pesticides. With respect to sulfonylurea herbicides, the only documented environmental effects are on sensitive plants and crops. Carryover of residues representing a small fraction of the applied dose can adversely affect susceptible rotational crops (46); current labels for many of these herbicides have rotational crop restrictions to prevent this. The potential for some sulfonylurea herbicides to leach to vulnerable ground water has been demonstrated (47) at levels that are close to soil concentrations that affect sensitive crops.

Conclusions

The key initial consideration in determining whether degradates of a particular pesticide might have a substantial impact on ground water is whether there is a key moiety or molecular skeleton which itself is particularly resistant to degradation in the environment, even though transformation from one degradate to another containing the resistant moiety may occur relatively rapidly. Such a recalcitrant moiety clearly exists for two of the three families of pesticides (*s*-triazine herbicides and acetanilide herbicides) discussed in detail in this paper. Many other pesticide families also have a molecular core that is highly resistant to degradation under most environmental conditions. The toxicological properties of degradates usually (but not always) differ substantially from the parent compound, and when this is the case, there may be less concern for potential environmental effects of degradates. Therefore, regulation of degradates often will not end up to be as stringent as for the respective parent compound. Residues of degradates may be considered for inclusion in federal or state regulatory standards when either (1.) the degradate has activity on some or all of the target pests or (2.) the degradate might contribute to toxicity observed in animal or plant effect studies.

Many additional pesticide degradates are likely to be found in ground water as monitoring studies begin to focus more on the total residues including degradates. If any of these degradates comes to be included in regulatory standards such as MCLs, this could make it necessary to further manage their use to limit the impact on ground water.

If xenobiotics are persisting for a long time in soil and water, residues should at least be monitored for to continually evaluate whether any unforeseen environmental effects may occur over the long-term. Such long-term effects occurred as a result of the heavy, indiscriminate use of persistent chlorinated hydrocarbon insecticides in the 1950s' and 1960s', and monitoring for modern pesticides and their degradates is needed to anticipate if their use might someday result in unforeseen effects.

For many pesticides, immunoassays may provide the ideal method for relatively low-cost screening of a sample for a large number of possible degradates at one time. Unlike gas chromatography - mass spectrometric (GC-MS) and liquid chromatography (LC)-MS methods, immunoassays can be usually easily designed to detect a combination of structurally related compounds with a single analysis. Immunoassays are quite reliable in that false negatives, i.e., an immunoassay determining a compound is absent when it actually is present in a sample; are extremely rare. Immunoassays which are sensitive to more than one compound cannot be used to determine actual concentrations of these compounds, but they can provide a preliminary screen to determine whether additional (more expensive) compound-specific analysis of the sample is warranted.

For each parent compound, there may be a multitude of degradates that might occur in soil and water. Full consideration of these compounds and their environmental impact adds substantially to the cost of regulation. The study of pesticide degradation processes in soil along with the subsurface movement of residues can be quite tedious, but is necessary to ensure that unforeseen environmental effects do not occur from the degradates. That is why the pesticide registration process in the United States requires extensive study of pesticide degradation. The case studies presented in this paper illustrate that researchers need to devote more effort to the study of the subsurface environmental fate of pesticides including degradation products.

Disclaimer

The views expressed in this paper are entirely the author's, and do not represent or reflect the policy of the Environmental Protection Agency or any other entity of the United States government.

Literature Cited

1.   Jacoby, H.; Hoheisel, C.; Karrie, J.; Lees, S.; Davies-Hilliard, L.; Hannon, P.; Bingham, R.; Behl, E.; Wells, D.; Waldman, E. *Pesticides in Ground Water Data Base, A compilation of monitoring studies: 1971 - 1991, national summary.* Office of Pesticide Programs, EPA, Washington, DC, 1992. EPA 734-12-92-001.

2.   Miles, C. J. Degradation products of sulfur-containing pesticides in soil and water. In *Pesticide transformation products: Fate and significance in the environment*; L. Somasundaram and J. R. Coats, editors. Symposium Series no. 459, Amer. Chem. Soc.: Washington, DC, 1991, pp. 61-74.

3.   Felsot, A. S.; Pederson, W. L. Pesticidal activity of degradation products. In *Pesticide transformation products: Fate and significance in the environment*; L. Somasundaram and J. R. Coats, editors. ACS Symposium Series 459, Amer. Chem. Soc.: Washington, DC, 1991, pp. 172-187.

4.   Aspelin, A.L. *Pesticide industry sales and usage: 1992 and 1993 market estimates*. Biological and Economic Analysis Division, Office of Pesticide Programs, United States Environmental Protection Agency: Washington, DC, 1994; 33 pp.

5.   Adams, C.D.; Thurman, E.M. Formation and transport of deethylatrazine in the soil and vadose zone. *J. Environ. Qual.* **1991**, 20:540-547.

6.   Cai, Z.; Sadagopa Ramanujam, V. M.; Gross, M. L.; Monson, S. J.; Cassada, D. A.; Spalding, R. F. Liquid-solid extraction and fast atom bombardment high-resolution mass spectrometry for the determination of hydroxyatrazine in water at low-ppt levels. *Analyt. Chem.* **1994**, 66:4202-4209.

7.   Denver, J. M.; Sandstrom, M. W. Distribution of dissolved atrazine and two metabolites in the unconfined aquifer, southeastern Delaware. In *U.S. Geological Survey Toxic Substances Hydrology Program -- Proceedings of the technical meeting, March 11-15, 1991, Monterey, CA.* U.S. Geological Survey Water Resources Investigations Report 91-4034.

8.   Frank, R.; Clegg, B. S.; Patni, N. K. Dissipation of atrazine on a clay loam soil, Ontario, Canada, 1986-1990. *Arch. of Environ. Contam. Toxicol.* **1991**, 21:41-50.

9.   Isenee, A. R.; Nash, R. G.; Helling, C. S. Effect of conventional vs. no-tillage on pesticide leaching to shallow groundwater. *J. Environ. Qual.* **1990**, 19:434-440.

10.   Kross, B. C.; Halberg, G. R.; Bruner, D. R.; Libra, R. D.; Rex, K. D.; Weih, L. M. B.; Vermace, M. E.; Burmeister, L. F.; Hall, N. H.; Cherryholmes, K. L.; Johnson, J. K.; Selim, M. I.; Nations, B. K.; Seigley, L. S.; Quade, D. J.; Dudler, A. G.; Sesker, K. D.; Culp, M. A.; Lynch, C. F.; Nicholson, H. F.; Hughes, J. P. *The Iowa State-Wide Rural Well-Water Survey water quality data: Initial analysis.* Iowa Dep. of Natural Resources, Technical Information Series no. 19, Des Moines, IA, 1990.

11.   Pionke, H. B., Glotfelty. D. W. Contamination of groundwater by atrazine and selected metabolites. *Chemosphere* **1990**, 21:813-822.

12.   Squillace, P. J.; Thurman, E. M.; Furlong, E. T. Groundwater as a nonpoint source of atrazine and deethylatrazine in a river during base flow conditions. *Water Resour. Res.* **1993**, 29:1719-1729.

13.   Wisconsin Department of Agriculture, Trade, and Consumer Protection. *Residues of atrazine and chlorotriazine metabolites in follow-up well water samples from a rural well water sampling program.* Wisconsin D.A.T.C.P., Madison, 1992.

14.   Capriel, P.; Haisch, A.; Khan, S. U. Distribution and nature of bound (nonextractable) residues of atrazine in a mineral soil nine years after the herbicide application. *J. Agric. Food Chem.* **1985**, 28:1096-1098.

15.   Khan, S. U.; Saidak, W. J. Residues of atrazine and its metabolites after prolonged usage. *Weed Res.* **1981**, 21:9-12.

16.   Khan, S. U.; Marriage, P. B.   Residues of atrazine and its metabolites in an orchard soil and their uptake by oat plants. *J. Agric. Food Chem.* **1977**, 25:1408-1413.

17.   Krugar, E. L.; Somasundaram, L.; Kanwar, R. S.; Coats, J. R.   Movement and degradation of [^{14}C]atrazine in undisturbed soil columns. *Environ. Toxicol. Chem.* **1993**, 12:1969-1975.

18.   Krugar, E. L.; Somasundaram, L.; Kanwar, R. S.; Coats, J. R.   Persistence and degradation of [^{14}C]atrazine and [^{14}C]deisopropylatrazine as affected by soil depth and moisture conditions. *Environ. Toxicol. Chem.* **1993**, 12:1959-1967.

19.   Mills, M. S.; Thurman, E. M.   Preferential dealkylation reactions of *s*-triazine herbicides in the unsaturated zone. *Environ. Sci. Technol.* **1994**, 28:600-605.

20.   Sorenson, B. A., Wyse, D. L.; Koskinen, W. C.; Buhler, D. D.; Lueschen, W. E.; Jorgenson M. D.   Formation and movement of ^{14}C-atrazine degradation products in a sandy loam soil under field conditions. *Weed Sci.* **1993**, 41:239-245.

21.   Winkelmann, D. A.; Klaine, S. J. Degradation and bound residue formation of atrazine in a western Tennessee soil. *Environ. Toxicol. Chemistry* **1991**, 10:335-345.

22.   Gustafson, D. I. Groundwater ubiquity score: A simple method for assessing pesticide leachability. *Environ. Toxicol. Chem.* **1989**, 8:339-357.

23.   Armstrong, D. E.; Chesters, G. Adsorption catalyzed chemical hydrolysis of atrazine; *Environ. Sci. and Technol.* **1968**, 9:683-689.

24.   Brouwer, W. W. M., Boesten, J. J. T. I.; Siegers, W. G. Adsorption of transformation products of atrazine by soil. *Weed Res.* **1990**, 30:123-128.

25.   Clay, S. A.; Koskinen, W.C. Adsorption and desorption of atrazine, hydroxyatrazine, and *s*-glutathione atrazine on two soils. *Weed Sci.* **1990**, 38:262-266.

26.   Robinson, R. C., Dunham, R. J. The uptake of soil-applied chlorotriazines by seedlings and its prediction. *Weed Res.* **1982**, 22:223-236.

27.   Wauchope, R. D.; Buttler, T. M.; Hornsby, A. G.; Augustin-Beckers, P. W. M.; Burt, J. P. The SCS/ARS/CES pesticide properties database for environmental decision-making. *J. Rev. Environ. Contam. Toxicol.*, Springer-Verlag: New York, 1992, Vol. 123; 164 pp.

28.   Baluch, H. U.; Somasundaram, L.; Kanwar, R. S.; Coats, J. R. Fate of major degradation products of atrazine in Iowa soils; *J. Environ. Sci. Health B* **1993**, 28:127-149.

29.   Winkelmann, D.A. and S.J. Klaine. Degradation and bound residue formation of four atrazine metabolites, deethylatrazine, deisopropylatrazine, dealkylatrazine, and hydroxyatrazine, in a western Tennessee soil. *Environ. Toxicol. Chem.* **1991**, 10:347-354.

30.   Schiavon, M. Studies of the movement and formation of bound residues of atrazine, of its chlorinated derivatives, and of hydroxyatrazine in soil using ^{14}C ring-labeled compounds under outdoor conditions. *Ecotoxicol. Environ. Safety* **1988**, 15:55-61.

31.   Duke, S. O.; Moorman, T. B.; Bryson, C. T. Phytotoxicity of pesticide degradation products. In *Pesticide transformation products: fate and significance in the environment*; L. Somasundaram and J. R. Coats, editors.

Symposium Series no. 459, Amer. Chem. Soc.: Washington, DC, 1991, pp. 188-204.

32.   Stolpe, N. B.; Shea, P. J. Alachlor and atrazine degradation in a Nebraska soil and underlying sediments. *Soil Sci.* **1995**, 180:359-370.

33.   Tyess, D. L.; Shea, P. J.; Comfort, S. D.; Stolpe, N. B. Atrazine degradation as influenced by microbial adaptation and availability of residues in soil. *WSSA Abstracts* **1995**, 35:94.

34.   Lavy, T. L.; Mattice, J.; Massey, J.; Skulman, B.; Senseman, S.; Gbur, E. *Minimizing the potential for ground water contamination due to pesticides.* Report submitted to Office of Pesticide Programs, United States Environmental Protection Agency. EPA contract no. CRT-815154-01-0. Washington, DC. 1993.

35.   Perry, C.A. *Source, extent, and degradation of herbicides in a shallow aquifer near Hesston, Kansas.* United States Geological Survey, Water-Resources Investigations Report 90-4019, Lawrence, KS, 1990.

36.   USDA. *Agricultural chemical usage - 1990 field crops summary.* National Agricultural Statistics Service and Economic Research Service, United States Department of Agriculture. Washington, DC, 1991. 154 pp.

37.   Catenacci, G.; Barbieri, F.; Bersani, M.; Feroli, A.; Cottica, D.; Maroni, M. Biological monitoring of human exposure to atrazine. *Toxicol. Letters* **1993**, 69:217-222.

38.   Ikonen, R.; Kangas, J.; Savolainen, H. Urinary atrazine metabolites as indicators for rat and human exposure to atrazine. *Toxicol. Letters* **1988**, 44:109-112.

39.   USDA. *Agricultural chemical usage - 1994 field crops summary.* National Agricultural Statistics Service and Economic Research Service, United States Department of Agriculture. Washington, DC, 1995. 106 pp.

40.   Holden, L. R.; Graham, J. A.; Whitmore, R. W.; Alexander, W. J.; Pratt, R. W.; Liddle, S. K.; Piper, L. L. Results of the National Alachlor Well Water Survey. *Environ. Sci. Technol.* **1992**, 26:935-943.

41.   Baker, D. B.; Bushway, R. J.; Adams, S. A.; Macomber, C. Immunoassay screens for alachlor in rural wells: False positives and an alachlor soil metabolite. *Environ. Sci. and Technol.* **1993**, 27:562-564.

42.   Kolpin, D. W.; Goolsby, D. A.; Thurman, E. M. Pesticides in near-surface aquifers: An assessment using highly sensitive analytical methods and tritium. *J. Environ. Qual.* **1995**, 24:1125-1132.

43.   USEPA. *Pesticide environmental fate "one-line" summary.* Environmental Fate and Ground Water Branch, Office of Pesticide Programs, United States Environmental Protection Agency. Washington, DC, 1996. (Current version in preparation).

44.   Beyer, E. M.; Duffy, M. J.; Hay, J. V.; Schlueter, D. D. Sulfonylureas. In *Herbicides: Chemistry, degradation, and mode of action*, P. C. Kearney, D. D. Kaufman, editors; Marcel Dekker, Inc.: New York, 1987, Vol. 3; pp. 117-189.

45.   USEPA. *Drinking water health advisory: Pesticides (United States Environmental Protection Agency Office of Drinking Water Health Advisories).* Lewis Publishers, Inc.: Chelsea, MI, 1989, pp. xi-xii.

46. Friesen, G. H.; Wall, D. A.; Residual effects of CGA-131036 and chlorsulfuron on spring-sown rotational crops; *Weed Sci.* **1991**, 39:280-283.

47. Barrett, M. R. Using PRZM modeling to compare leaching potential of wheat herbicides. In *Agrochemical environmental fate: State of the art*; M. L. Leng, E. M. K. Leovey, and P. L. Zubkoff, editors; Lewis Publishers: Boca Raton, FL; 1995; pp. 335-342.

Chapter 17

Herbicide Mobility and Variation in Agricultural Runoff in the Beaver Creek Watershed in Nebraska

Li Ma and Roy F. Spalding

Water Sciences Laboratory, University of Nebraska–Lincoln, 103 Natural Resources Hall, Lincoln, NE 68583–0844

Herbicide mobility and variation in agricultural runoff were studied with respect to the site specific conditions, herbicide solubilities and degradations in a small watershed near York, Nebraska. The Beaver Creek watershed includes 3,327 ha of primarily row-cropped, heavily irrigated farmland. The estimated average runoff is 1.48×10^6 m^3 yr^{-1} under a precipitation norm of 635 mm yr^{-1}. Maximum atrazine concentrations (24.23 μg L^{-1} at station A and 18.93 μg L^{-1} at station B) were observed in the first post-application runoff event on June 22, 1994. Both degradates, deethylatrazine (DEA) and deisopropylatrazine (DIA), were also present in runoff samples. High cyanazine and metolachlor concentrations were detected in the first post-application runoff at station B; in contrast, only trace concentrations of these compounds were detected at station A. The DEA to atrazine molar ratio (DAR) decreased rapidly over a period of several hours in the first post-application runoff suggesting that DEA was more mobile than atrazine, and atrazine degradation occurred predominantly in the top soil. Average DARs in the first post-application runoff were low compared to pre-application runoff, which is consistent with inputs from recently atrazine treated soil. Higher average DARs in later runoff events indicated an increasing DEA production as soil residence time increased.

Widespread degradation of water quality (*1, 2*) and herbicide residues in surface water (*3, 4*) are of major concern in the Midwestern United States. Dispersion of herbicide residues into the environment by surface runoff from agricultural lands has received major attention for about the last 20 years (*5, 6*) because large amounts of relatively water soluble herbicides, such as atrazine, cyanazine, alachlor and metolachlor, are applied each year in the Midwestern United States as pre- and post-emergent weed-control agents on corn, grain sorghum and soybeans (*7*). The

0097–6156/96/0630–0226$15.00/0

potential for herbicide transport by surface runoff is governed by the persistence and mobility properties of herbicides, the timing of the rainfall/runoff events with respect to herbicide application, duration and intensity of the rainfall events, and texture, slope, surface roughness, antecedent water content and organic matter content of the soils (*8*).

The rates of herbicide transformation in soil environments are dependent upon a variety of factors including the physical/chemical properties of the herbicide, temperature, moisture, pH, light, and the presence and abundance of organic matter in the surrounding environment (*9*). Atrazine is the most prevalent herbicide residue detected in surface water in the Midwest (*10-12*). Atrazine can be degraded by both abiotic (isomerization, hydrolysis, photolysis) and biotic mechanisms (Figure 1). Deethylatrazine (DEA) and deisopropylatrazine (DIA) are two common microbial degradation products of atrazine formed in soil (*13*). Atrazine deethylation reaction is more rapid than the deisopropylation reaction resulting in preferential DEA production over DIA (*14*). Small concentrations of DIA associated with the degradation of atrazine may be due to a rapid degradation of DIA in the unsaturated zone (*14, 15*). Hydrolysis of atrazine to hydroxyatrazine (Figure 1) is the major abiotic degradative pathway (*16*). Hydroxyatrazine is strongly bound to the soil, and is almost immobile in soil (*17*).

The molar ratios of DEA to atrazine (DAR) and DIA to DEA (D^2R) are indicators of soil residence times (*18, 19*). High DAR (>1) indicates soil bacteria and fungi convert significant quantities of atrazine to DEA (*18*). Since the removal of the cyanomethylethyl group from cyanazine can also produce DIA (Figure 1), the D^2R also reflects the contribution from cyanazine degradation (*20*).

The objectives of this research were to 1) characterize the variation of herbicides and degradates in agricultural runoff in a small watershed and 2) study the fate of herbicides during the growing season in response to runoff events.

Methodology

Site Description. The study area includes a 3,327 ha agricultural watershed on a tributary of Beaver Creek located in York county, Nebraska (Figure 2). The Beaver Creek watershed drainage area is highly productive land used mostly for the production of irrigated (88%) row crops. Corn comprised the greatest land use component of this area (2,062 ha). Grain sorghum was grown on 302 ha, soybeans on 284 ha and conserving use crops on 165 ha. Non-cropland (pastures, roads, right-of-ways and farmsteads) represented 514 ha. The watershed is relatively flat with about 94% of the area having 6% slope or less. The predominant soil type of the area is Hastings silt loam, representing 81% of the area. The Hastings series consists of deep, well-drained, nearly level to steeply sloping soils on uplands. Most of the area (87%) is not considered highly erodible (*21*). The estimated average runoff is 1.48×10^6 m^3 yr^{-1} under a precipitation norm of 635 mm yr^{-1} (*21*).

Herbicide applications in the watershed follow the guide for herbicide use in Nebraska (*22*). Percentages of acreage of herbicide usage, application rates, total annual amount of herbicide applied to corn, grain sorghum and soybean acres are shown in Table I. Atrazine was most commonly used in the watershed followed by

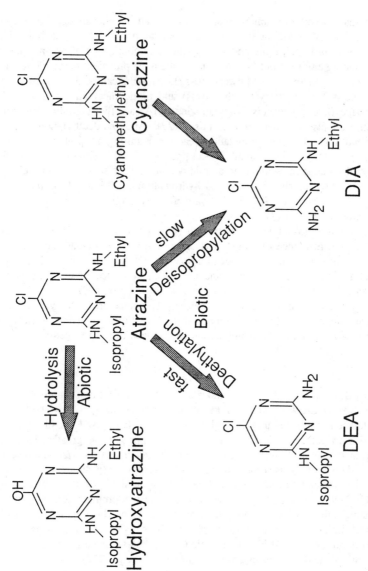

Figure 1. Degradation pathways of atrazine to hydroxyatrazine, deethylatrazine (DEA), and deisopropylatrazine (DIA) and of cyanazine to DIA.

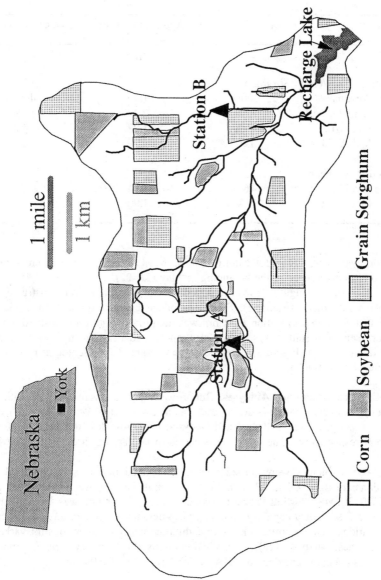

Figure 2. Watershed location map and autosampler stations.

Table I. **Percentage of Acreage of Herbicide Usage, Application Rates, Total Annual Application on Corn, Grain Sorghum and Soybean Acres in the Watershed**

Herbicide	Percent of Acreage	Application Rate	Annual Application
		kg a.i. ha^{-1}	kg a.i.
Applied to Corn/Grain Sorghum			
Atrazine	100%	1.79	4000
Alachlor	36%	2.34	1921
Cyanazine	14%	1.23	392
Metolachlor	12%	1.79	633
Applied to Soybean			
Alachlor	46%	2.34	293
Metolachlor	7%	1.90	32

alachlor, cyanazine, and metolachlor (Table I). Most of the herbicides were applied prior to crop emergence at the beginning of May (*21, 22*).

Two steam gauging stations equipped with autosamplers were constructed on the ephemeral streams (Figure 2). Station A collected surface runoff from a relative larger area compared to station B. Corn was the predominant crop in the drainage area represented by station A, soybeans also represented a significant amount of land in the drainage area (Figure 2). Dry land grain sorghum represented a major crop contributed to station B.

Sampling Procedures and Analyses. Runoff water samples were collected with the autosamplers every two hours during each runoff event. In 1994, a total of 31 and 30 runoff samples were collected and analyzed for herbicides from station A and station B respectively. The sampling occurred in late spring through the growing season.

The herbicides were sorbed to a C-18 bonded porous silica solid phase extraction cartridge. A 1 μL aliquot of the concentrated extract was autoinjected into a HP5890 GC/MS (Hewlett Packard, Avondale, PA). Separation was accomplished by using a fused silica capillary column with programmed temperature. Herbicide concentrations were determined by isotope dilution, which compensates for variable yields of these analytes (*23*). Quantitation limits (QL) are 0.05 μg L^{-1} for all compounds except cyanazine and DIA, which have QLs of 0.09 μg L^{-1}.

Results and Discussion

Herbicide in First Post-Application Runoff. Maximum herbicide inputs to streams and rivers draining agricultural watersheds usually occur in May and June, immediately after preplant disking and herbicide application (*2, 24, 25*). After herbicide application in 1994, the first major precipitation event (67.6 mm) occurred on June 22 and induced a 16-20 h runoff event (Table II). At the beginning of

Table II. Herbicide Concentrations in the First Post-Application Runoff (June 22, 1994) from the Two Gauging Stations

Hours	DIA	DEA	Atra-zine	Alac-hlor	Cyan-azine	Metol-achlor	DAR	D^2R
				$\mu g\ L^{-1}$				
Station A								
0	2.27	5.34	18.64	3.45	0.04	0.05	0.33	0.50
2	2.14	5.09	17.91	3.04	0.12	0.09	0.33	0.50
4	2.11	3.92	21.84	4.04	2.66	0.11	0.21	0.64
6	1.60	2.92	24.20	4.98	0.88	0.69	0.14	0.65
8	1.55	2.83	25.29	5.62	0.66	0.87	0.13	0.65
10	1.64	3.07	27.52	6.25	0.87	1.23	0.13	0.63
12	1.66	2.97	27.62	6.24	0.67	0.93	0.12	0.66
14	1.70	3.02	27.01	5.44	0.68	0.68	0.13	0.67
16	1.71	3.09	28.03	5.61	0.65	0.69	0.13	0.65
SD[a]	0.27	0.98	3.91	1.18	0.76	0.42	0.09	0.07
Station B								
0	5.35	11.90	10.65	5.39	32.01	54.89	1.28	0.53
2	1.77	2.47	20.09	3.80	13.63	4.27	0.14	0.85
4	1.77	2.60	20.34	3.83	12.35	4.17	0.15	0.81
6	1.45	2.17	17.55	2.91	11.40	3.59	0.14	0.79
8	1.49	2.28	21.19	2.76	10.42	3.81	0.12	0.77
10	1.81	2.49	19.63	2.88	12.35	4.19	0.15	0.86
12	1.44	2.27	20.46	2.46	11.38	3.98	0.13	0.75
14	1.74	2.59	20.18	2.33	10.83	4.59	0.15	0.79
16	1.87	2.67	19.56	2.77	15.19	5.13	0.16	0.83
18	2.00	2.85	20.06	2.52	12.72	5.64	0.16	0.83
20	1.79	2.53	18.50	2.23	12.66	5.54	0.16	0.84
SD[a]	1.11	2.84	2.92	0.93	6.09	15.21	0.34	0.09

[a]Standard Deviation.

runoff, a higher concentration of atrazine was observed at station A compared to that at station B (Table II); however, both DEA and DIA concentrations were higher at station B than at station A indicating extended degradation occurred in the field upgradient of station B. At both stations, lower atrazine and higher DEA and DIA concentrations were observed during the first two hours of runoff compared to the late flush (Table II) suggesting that atrazine degradation occurred predominately in top soil (15) where the compounds were easily flushed. DEA is more mobile (26) than atrazine and, therefore, was most pronounced in the early stages of the runoff event.

The DAR value in the first runoff sample at station B (1.28) was much higher than at station A (0.33) suggesting an earlier application time with a corresponding longer soil residence time for the triazines associated with drainage to station B than A. During the first runoff event on June 22, DAR values decreased rapidly at both stations within a period of several hours (Table II), also suggesting that the more mobile DEA (26) is being flushed preferentially compared to atrazine. DARs stabilized at ~0.13 after six hours at station A in comparison to the DARs which stabilized within the first two hours at station B (Table II).

Both cyanazine and metolachlor remained in trace concentrations during the first post-application runoff event at station A (Table II). High metolachlor (54.89 μg L^{-1}) and cyanazine (32.01 μg L^{-1}) concentrations were observed at station B at the beginning of the runoff. These concentrations may be the result of high metolachlor and Extrazine (cyanazine and atrazine combination at 3:1 ratio) applications in corn and dry land sorghum. Metolachlor concentration at station B declined from 54.89 to 4.27 μg L^{-1} during the first two hours of runoff. Cyanazine also decreased rapidly during the first two hours of runoff (Table II). The rapid declines of these two herbicides at station B may be the result of a smaller drainage area, relative high solubilities, and less persistence of the compounds (27).

The D^{2}Rs increased at both stations during the first runoff event indicating that DIA had a lower mobility compared to DEA (Table II). Since cyanazine concentrations were much higher at station B than station A (Table II), it was not surprising to find higher D^{2}Rs (Table II) associated with drainage to B than that flowing to A. It appears that cyanazine and metolachlor usages were more prevalent in drainage upgradient from collector B than collector A. The cyanazine contribution to DIA concentration was negligible due to its low concentration in the watershed represented by station A.

Variation of Herbicides in Seasonal Runoff Events. Average atrazine concentrations in pre-application runoff (April 14, 1994) were low compared to the post-application runoff events (Table III). Average atrazine concentrations in the first major runoff were 24.23 ± 3.91 and 18.93 ± 2.92 μg L^{-1} at station A and B respectively (Table III). At station B, average concentrations of cyanazine (14.08 ± 6.09 μg L^{-1}) and metolachlor (9.07 ± 15.21 μg L^{-1}) were much higher than at station A (Table III). The samples also contained 4.96 ± 1.18 and 3.08 ± 0.93 μg L^{-1} alachlor at station A and B respectively (Table III); these concentrations are consistent with the observations of Thurman at el. (28).

Table III. Average Herbicide Concentrations Per Runoff Event During 1994 Growing Season at the Two Gauging Stations

Runoff events	DIA	DEA	Atra-zine	Alac-hlor	Cyan-azine	Metol-achlor	DAR	D^2R
				μg L^{-1}				
Station A								
4/14/94	0.01	0.12	0.42	0.07	0.06	0.13	0.30	0.05
6/24/94	1.82	3.58	24.23	4.96	0.80	0.60	0.18	0.62
7/5/94	1.66	2.47	12.46	2.66	0.77	2.39	0.24	0.82
7/18/94	0.96	1.27	4.00	0.29	0.17	0.10	0.40	0.91
Station B								
4/14/94	0.01	0.12	0.43	0.07	0.07	0.14	0.31	0.05
6/24/94	2.04	3.35	18.93	3.08	14.08	9.07	0.25	0.79
7/13/94	1.01	2.10	6.56	0.83	0.38	6.87	0.37	0.58
7/18/94	0.82	1.42	5.13	1.19	0.97	3.19	0.32	0.71

With the exception of metolachlor at station A, herbicide concentrations decreased in the three subsequent post-application runoff events. The average cyanazine concentration in runoff declined rapidly from 14.08 ± 6.09 μg L^{-1} on June 22 to 0.38 ± 0.34 μg L^{-1} on July 13 indicating that most of the surface mobile cyanazine was flushed during the first runoff. This is consistent with its moderate solubility (160 μg mL^{-1}) and short half-life (~14 d) in soil (27, 29). Why average metolachlor concentration was highest (6.87 ± 5.62 μg L^{-1}) at station A in the third runoff on July 13, 1994 (Table III) is not presently understood.

Average DARs in runoff samples prior to herbicide application were high (Table III). The low average DARs (Table III) in the first post-application runoff reflected a higher proportion of the parent compound than the degradate which resulted from the flushing of a recently applied non-degraded parent compound (2, 4). As atrazine is microbially degraded, the ratio of DEA to atrazine in soil increases and is reflected, in turn, in the increasing DARs (Table III) in progressive runoff events (12).

The D^2R values were low (~0.05) in pre-application runoff samples collected at both stations (Table III). DIA is less persistent than either atrazine or DEA in soil environment (14, 15). Since DIA dealkylation is much faster than DEA dealkylation, low D^2R are commonly reported (14). The D^2R values increased to 0.62 at station A and 0.79 at station B in the first post-application runoff samples which was about a month and half after the spring herbicide applications. Thurman et al. (28) observed that D^2R was approximately 0.6 on a regional scale (Midwestern US), and this ratio increased when the basin contained significant contributions of cyanazine in surface water. The high D^2R in the first post-application runoff at station B may be associated with both atrazine and cyanazine degradation.

Herbicide loss in runoff was estimated by dividing the accumulated herbicide in the impoundment from runoff events during the growing-season by the herbicide applied to the watershed (Ma and Spalding, 1995, submitted to *J. Environ. Qual.*). About 0.19% (7.41 kg a.i.) of the total atrazine applied in the watershed was lost from the runoff events during the growing season. These results compare favorably with the 0.2-1.9% seasonal atrazine losses to runoff reported by Leonard et al. (*30*). Hall et al. (*31*) reported that average atrazine loss to runoff for all application rates was 2.4% of that applied to Pennsylvania soil. About 0.08% of the applied alachlor and 0.14% of the applied cyanazine were lost to the runoff. The shorter half-lives (15 d for alachlor and 14 d for cyanazine) in soil (*27*) are likely responsible for their rapid disappearances in the soil environments. About 0.25% of total applied metolachlor was lost to the runoff. The percentages of herbicides lost to runoff in this study were low compared to other studies (*30, 31*) due to the relatively flat watershed, low application rates, and the long period of time between herbicide application and major runoff.

Summary

Variation of herbicide concentrations in the first post-application runoff are associated with herbicide applications and watershed conditions. Lower atrazine concentrations and higher DEA concentrations (high DAR) at the beginning of the runoff suggests the DEA are more mobile, and atrazine degradation primarily occurred in top soil from where the compounds were first flushed. In contrast to station A, high cyanazine and metolachlor concentrations were observed at station B. Cyanazine and metolachlor concentrations decreased rapidly during the first several hours of runoff due to their high solubility. Cyanazine degradation to DIA at station B appears to contribute DIA concentration resulting in higher D^2R compared to station A.

Herbicide concentrations decreased in later runoff events except metolachlor at station A. Cyanazine concentration decreased rapidly at station B due to its moderate solubility and less persistence. DAR values progressively increased in later runoff events suggesting the accumulation of more DEA and increased soil residence time. Relative low percentages of herbicides lost to runoff were observed in the study.

Acknowledgements

The authors are grateful for the sampling assistance of Russell Callan and Rod DeBuhr and for the background information provided by Jay Bitner of the Upper Big Blue Natural Resources District. We thank Dan Snow and Dave Cassada of the Water Sciences Laboratory for overseeing the herbicide analyses and Lorraine Moon for her editorial assistance. We are also grateful for the helpful comments and suggestions of Dr. Michael Meyer and two anonymous reviewers. The project was funded by the Upper Big Blue Natural Resources District and Ciba-Geigy.

Literature Cited

1. Humenik, F.J.; Smolen, M.D.; Dressing, S.A. *Environ. Sci. Technol.* **1987,** *21,* 737-742.
2. Thurman, E.M.; Goolsby, D.A.; Meyer, M.T.; Kolpin, D.W. *Environ. Sci. Technol.* **1991,** *25,* 1794-1796.
3. Baker, J.L.; Johnson, H.P. *Trans. ASAE.* **1979,** *22,* 554-559.
4. Spalding, R.F.; Snow, D.D.; Cassada, D.A.; Burbach, M.E. *J. Environ. Qual.* **1994,** *23,* 571-578.
5. Wauchope, R.D. *J. Environ. Qual.* **1978,** *7,* 459-472.
6. Leonard, R.A. In *Environmental chemistry of herbicides;* Grover, R., Ed.; CRC Press: Boca Raton, FL, 1988; pp 45-88.
7. Gianessi, L.P. *A National Pesticide Usage Data Base, Resources for the Future;* U.S. Geological Survey: Washington, DC, 1986; pp 1-14.
8. Severn, D.J.; Ballard, G. In *Pesticides in the soil environment - an overview;* Cheng, H.H., Ed.; SSSA Book Ser. 2; SSSA: Madison, WI., 1990, Chap. 13. pp 467-492.
9. Day, K.E.; In *Pesticide transformation products. Fate and significance in the environment;* Somasundaram, L.; Coats, J.R., eds.; ACS Symposium Ser. 459. ACS, Washington, D.C., 1991, Chap. 16. pp 217-241.
10. Spalding, R.F.; Burbach, M.E.; Exner, M.E. *Ground Water Monit. Rev.* **1989,** *9,* 126-133.
11. Goolsby, D.A.; Coupe, R.C.; Markovchick, D.J. *USGS Water Resources Investigations Rep. 91-4163.* **1991,** USGS, Denver, CO.
12. Schottler, S.P.; Eisenreich, S.J.; Capel, P.D. *Environ. Sci. Technol.* **1994,** *28,* 1079-1089.
13. Behki, R.M.; Khan, S.U. *J. Agric. Food Chem.* **1986,** *34,* 746-749.
14. Mills, S.M.; Thurman, E.M. *Environ. Sci. Technol.* **1994,** *28,* 600-605.
15. Kruger, E.L.; Somasundaram, L.; Kanwar, R.S.; Coats, J.R. *Environ Toxicol Chem.* **1993,** *12,* 1959-1967.
16. Obien, S.R.; Green, R.E. *Weed Sci.* **1969,** *17,* 509-515.
17. Russell, J.D.; Cruz, M.; White, J.L. *Science.* **1968,** *160,* 1340-1342.
18. Adams, C.D.; Thurman, E.M. *J. Environ. Qual.* **1991,** *20,* 540-547.
19. Jayachandran, K.; Steinheimer, T.R.; Somasundaram, L.; Moorman, T.B.; Kanwar, R.S.; Coats, J.R. *J. Environ. Qual.* **1994,** *23,* 311-319.
20. Meyer. M.T. Ph.D. Thesis. University of Kansas, 1994.
21. Zoubek, G.L. *A baseline study to access farm operators practices and attitudes in the recharge lake basin drainage area.* Nebraska Cooperative Extension. Univ. of Nebraska-Lincoln, Lincoln, NE., 1993.
22. *A 1995 guide for herbicide use in Nebraska.* EC 95-130-D, Nebraska Cooperative Extension, University of Nebraska-Lincoln, Lincoln, NE., 1995.
23. Cassada, D.A.; Spalding, R.F.; Cai, Z.; Gross, M.L. *Anal. Chim. Acta.* **1994,** *287,* 7-15.
24. Ritter, W.R.; Johnson, H.P.; Lovely, W.G.; Molnau, M. *Environ. Sci. Technol.* **1974,** *8,* 38-42.

25.   Snow, D.D.; Spalding, R.F. *Agricultural Impacts on Ground Water - a Conference, Des Moines, IA. 21-23 Mar. 1988.* National Water Well Association, Des Moines, IA, 1988, pp 211-223.
26.   Widmer, S.K.; Spalding, R.F. *J. Environ. Qual.* **1995,** *24,* 445-453.
27.   *Water quality workshop: Integrating water quality and quantity into conservation planning.* Soil Conservation Service, SCS National Technical Center, Ft, Worth, TX., 1980.
28.   Thurman, E.M.; Meyer, M.T.; Mills, M.S.; Zimmerman, L.R. Perry, C.A. *Environ. Sci. Technol.* **1994,** *28,* 2267-2277.
29.   Hall, J.K.; Hartwig, N.L.; Hoffman, L.D. *J. Environ. Qual.* **1984,** *13,* 105-110.
29.   Leonard, R.A.; Langdale, G.W.; Fleming, W.G. *J. Environ. Qual.* **1979,** *8,* 223-229.
30.   Hall, J.K.; Pawlus, M.; Higgins, E.R. *J. Environ. Qual.* **1972,** *1,* 172-176.

Chapter 18

Monitoring Pesticides and Metabolites in Surface Water and Groundwater in Spain

D. Barceló[1], S. Chiron[1], A. Fernandez-Alba[2], A. Valverde[2], and M. F. Alpendurada[3]

[1]Department of Environmental Chemistry, CID-CSIC,
c/Jordi Girona 18–26, 08034 Barcelona, Spain
[2]Department of Analytical Chemistry, Faculty of Sciences,
Almeria Campus, 04071 Almeria, Spain
[3]Laboratory of Hydrology, Faculty of Pharmacy, University of Porto,
R. Anibal Cunha, 164.4000 Porto, Portugal

Automated on-line solid-phase extraction (SPE) followed by liquid chromatographic techniques [(LC)-Diode array, LC-post-column fluorescence detection (EPA Method 531.1) for carbamate pesticides) and LC-thermospray mass spectrometry (TSP-MS)] were used for the monitoring of various pesticides and metabolites in surface (river) water and ground water. The Ebro river estuary (Tarragona, North East) and a representative aquifer in Almeria (South East), both in Spain, were monitored during one year periods from March 92 to 93 and March 93 to 94, respectively. Alachlor, metolachlor, atrazine, simazine, molinate, propanil, bentazone, MCPA and the herbicide metabolites deethylatrazine, deisopropylatrazine and 8-hydroxybentazone were found to be the major pollutants in surface waters. Carbofuran, methiocarb, methomyl and the pesticide metabolites 3-hydroxycarbofuran and methiocarb sulfone were detected in ground waters . The contamination levels varied from non detectable (below 10 ng/L) up to 3.0 µg/L. The seasonal variation studies showed that pesticide pollution was conservative for triazines and their metabolites, alachlor and metolachlor, which were detected through-out the year at 0.1-0.3 µg/L levels, whereas the other pesticides exhibited a sporadic occurrence related to agricultural and irrigation practices.

The assessment of non-point sources of pollution from pesticides is an important water quality issue (1). These sources are difficult to identify and quantify , occur in pulses during spring or early summer precipitation events and are more complex to control than point sources of pollution (2). Pesticides applied to crops are transported to surface waters by various mechanisms such as run off, ground water discharge, and atmospheric deposition. Triazines and their metabolites deethylatrazine (DEA) and deisopropylatrazine (DIA) alachlor and metolachlor can be regarded as the most frequently encountered modern pesticides in surface waters (2-7). Atrazine and

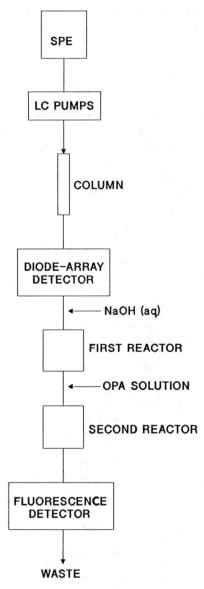

Figure 1. Genenal scheme of the analytical procedure

alachlor are mobilized and transported quickly during precipitation events and maximum concentrations are reached before rivers obtain maximum flow (generally May and June). Study shows that a portion of the herbicide to the soil is immediately available to be transported by runoff during precipitation events. DEA was present throughout the year, while DIA was infrequently detected and its formation was related to cyanazine (3). Metolachlor was fairly stable in natural water, with less than 4 % loss from sterile or nonsterile natural water after 100 days (8).

Pesticide use associated with rice production has aroused the suspicion of regulators and concerned citizens (9). Bentazone, MCPA, molinate and propanil are typical rice herbicides. Molinate was detected in the lower Mississippi river (10) and in Mediterranean estuarine waters (4,5) at concentration level -up to 1.75 µg/L during pilot survey studies in 1991. Recently (11) bentazone, 8-hydroxybentazone and MCPA were determined in Ebro estuarine waters (Tarragona, Spain) at levels varying from 0.1-0.22 µg/L (11).

Limited information is available on the contamination of ground water by the carbamate insecticides. A review article reported levels of carbamate pesticides such as aldicarb, carbaryl and carbofuran in ground water within the US at up to 50 µg/L (12). The National Pesticide Survey (NPS) reported detectable levels of aldicarb, carbaryl, carbofuran and their metabolites 3-hydroxycarbofuran, 3-ketocarbofuran and aldicarb sulfone in drinking water wells (13). Among the carbamates recently reported in ground water from Nebraska (14) methiocarb and propoxur (added to the previous NPS list) were found at concentrations below 0.5 µg/L (14). More than 500 tons of carbaryl per annum is used in the European Community; however only few pilot monitoring surveys took place within Europe for either carbaryl or carbamates (carbaryl levels ranged from non detectable up to 38 ng/L in rivers of Northern Greece) (15). Obviously, there is a need for better understanding of how carbamates behave in soil and whether they are transported to ground waters after hydrolysis, photolysis or microbial degradation . The metabolites of carbamate pesticides are generally more toxic than the parent compounds, (for example 1-naphthol has an EC_{50} lower than carbaryl in the Microtox system) It seems evident that there is a need to include the main polar carbamate metabolites in monitoring programs as recommended by the NPS (13).

Most of the methods for carrying out monitoring studies of pesticides in water samples are based on well established techniques, such as those used by the US EPA and /or the NPS. These methods are usually based on dichloromethane liquid-liquid extraction (LLE) or solid-phase extraction (SPE) using either cartridges or Empore disks (16). There are no examples of long-term monitoring survey studies of pesticides using on-line SPE coupled to liquid chromatographic techniques. The purpose of this work is to use the knowledge of our group in this field (11,17) and to apply currently developed on-line SPE methods followed by liquid chromatographic techniques with UV/diode array, post-column fluorescence and/or mass spectrometry detection for the monitoring of pesticides and metabolites in surface water and ground water from Spain (the analytical scheme is shown in Fig. 1). The comparison of automated on-line SPE with LLE has been carried out during interlaboratory studies, thus indicating the feasibility of automated methods for aquatic monitoring studies of pesticides (17).

The main interest of the present study lies in two case studies of contamination by different pesticide compounds, in the regions of the Ebro river (Tarragona) and in an aquifer in Almeria. The aim of this work is : (a) to use an automated on-line SPE LC method involving mass spectrometry confirmation of the samples in order to establish a preliminary assessment of agrochemical pollution by pesticides and metabolites in two Spanish areas; (b) to investigate the seasonal and spatial variation

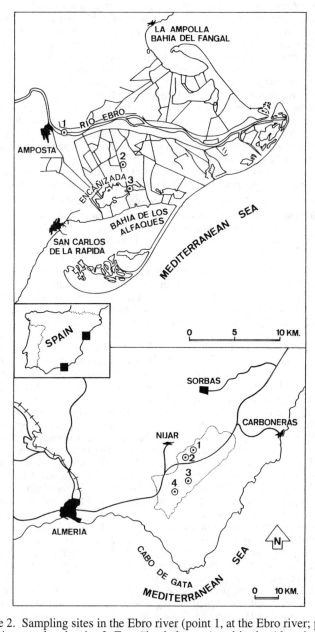

Figure 2. Sampling sites in the Ebro river (point 1, at the Ebro river; point 2, irrigation canal and point 3, Encañizada lagoon) and in the Almeria aquifer (sampling points 1-4, correspond to wells 1-4, respectively)

of the different pesticide families in both areas during a one year period (c) to assess the ability of carbamate metabolites to leach into ground water and (d) to publish the first long-term data levels on the occurrence of a wide variety of pesticides and metabolites monitored in Spanish surface and ground waters.

Experimental

Areas   of study The Ebro river is the third largest river flowing into the Mediterranean sea. Its average annual flow is 550 m^3/s and it is exceeded only by the Rhône (1,712 m^3/s) and the Po (1,559 m^3/s). The Ebro river flow represents ca. 8 % of the Mississippi river flow and the large agricultural basin comprises 84x10^3 km^2 and involves wheat, corn and vineyard cultivations. The Ebro river delta (Northeast of Spain) corresponds to a typical rice cultivation area which lays over 350 Km^2. Triazine   and chloroacetanilide herbicides (mainly alachlor and metolachlor) can reach the river mouth after transport in the dissolved water phase along the river basin. The estimation of triazine and related corn herbicides used in the whole Ebro river basin and the rice herbicides (MCPA, bentazone, molinate and propanil) used in the Ebro estuary are reported in Table I. Other typical rice herbicides,. (e.g.,thiobencarb and bensulfuron) are only sprayed out in small amounts (< 1 ton/year) and were not considered in the monitoring program.

Almeria is a Spanish province located on the southeast of the Mediterranean coast. With 36,460 hectares of cultivated fields (mainly greenhouses), this region has become one of the major suppliers of green vegetables to most of the EEC countries. The major crops cultivated include peppers, tomatoes, melons and cucumbers, and the total production reached 1,384,240 tons during the 92-93 season. With a hot and dry climate (average maximal temperature: 24.2°C, and an average annual precipitation of 300 mm) large amounts of insecticides, and to a lesser extent fungicides, are sprayed to improve crop yields (see Table I). One of the major factors affecting leaching is the water recharge rate (rainfall plus irrigation minus evapotranspiration). Extensive use of irrigation has been one of the reasons for the movement of agrochemicals in the soil. In the area of study, the irrigation flow was similar through-out all the year except for the period from April to June when the irrigation flow was approximately twice usual, and during July and August when no cultivation took place in the area. The irrigation practices were related to the high temperatures in this area of study from April to August.

Sample collection Three sampling points were selected in the Ebro estuary and are shown in Figure 2. Sampling point 1 (at the Ebro river), Sampling point 2 (at an irrigation ditch which receives water from many surrounding fields) and Sampling point 3 (at Encañizada lagoon which receives irrigation channels from the rice fields). Water samples (1L volume) were collected in jars below the surface layer and at water depth of 15-40 cm monthly in the spring and summer and every two months in the fall and winter , from March-1992 to March 1993. Samples were acidified to pH=2 with sulfuric acid to inhibit biological activity and were filtered through fiber glass filters to remove sand and debris (Millipore Corp., Bedford, MA, USA of 0.45 μm). Samples were stored at 4°C in clean glass bottles until analysis.

Water samples were collected from Aquifer *Campo de Nijar*, (Almeria), which covers an area of 157 Km^2. Water depth ranges from 80 to 200 m and a high pesticide load may be expected from normal cultural practices of 1500 Ha of greenhouses. This aquifer can be regarded as representative of the whole area. The four sampling sites are shown in Figure 2 and a summary of hydrologic characteristics is given in Table II. The analysis of the basic data (pH, conductivity) showed discrepancy between the four sampling points despite their proximity.

Table I: Physico-chemical properties and annual usage of studied pesticides

Pesticides	water solubility (mg/L)	soil half life (days)	Annual Usage AI* (ton/year)
			Ebro river basin
alachlor	240	15	58
atrazine	30	60	130
cyanazine	170	14	6.5
metolachlor	530	90	37
simazine	5	80	13
terbutylazine	8	114	6.5
			Ebro river estuary
bentazone	500	47	7
MCPA	825	15	1.5
molinate	880	21	70
propanil	225	1	50
			Almeria area
carbofuran	120	50	0.3
methiocarb	30	30	2
methomyl	58000	30	2

*AI: active ingredients. Ref. 18 and 19.

The aquifer flow is northeast to southeast. Two sampling points were chosen upstream (well 1,2) and two downstream (well 3,4). Sampling sites are located in the middle of the cultivated field and aquifer, far from point source of pollution like pesticide storage. Four duplicate samples were collected monthly from March 1993 to February 1994 (except in August and October 1993) after running the well pump for 2hr to ensure that a fresh water sample was received. Water samples (1L) were passed through a 0.45 μm membrane filter   (Millipore, Bedford MA) to remove sand and debris, acidified to pH = 2 with sulfuric acid to inhibit biological activity and sent to Barcelona by airmail express. Samples were stored at 4°C in clean glass bottles until analysis.

Table II :The four sampling points with the hydrologic parameters (Almeria)

Sampling point	Depth (meters)	pH*	Conductivity* (μS / cm)	NO_3^-* (mg/L)
Well 1	150	7.6	2400	6.8
Well 2	130	7.2	1600	12.6
Well 3	80	7.4	2600	9.7
Well 4	90	7.9	400	15.6

* Average value calculated over one year period.

Sample handling and analysis Samples were analyzed within four days of collection due to the rapid degradation of some carbamate insecticide metabolites, like methiocarb sulfone, methiocarb sulfoxide and 3- hydroxycarbofuran. Extractions were performed by the automated on-line SPE method described earlier (17,20). All analytes of interest gave recovery rates > 90 % except for deisopropylatrazine with a recovery of 65 % (20). The on-line SPE method combined with LC-DAD was validated by using dichloromethane LLE and the results obtained were equivalent, thus indicating the suitability of the on-line SPE for carrying out monitoring studies. The extract was then analyzed either by LC-DAD or post-column fluorescence detection according to EPA method 531.1 for *N*-methyl and *O*-(methyl carbomyl) oxime carbamates. Water volumes of 150 and 10 mL were required for diode array (DAD) and post-column fluorescence detection, respectively, in order to achieve a limit of detection (L.O.D.) below 0.1 μg/L in compliance with the EEC Drinking Water Directive (16). All data concerning reproducibility and L.O.Ds of the analytes have already been discussed and are adequate for environmental analyses, with L.O.Ds in most of the cases below 0.1 μg/l level when only 150 mL of sample is used. When on-line SPE-LC followed by post-column reaction and fluorescence detection is used, LODs for carbamates are in the range of 0.01 μg/L and require the use of only 10 mL of water sample (16, 18). Analytes monitored included 19 pesticides and 12 metabolites: aldicarb sulfoxide, aldicarb sulfone, oxamyl, methomyl, 3-hydroxy-7-phenol-carbofuran, deisopropylatrazine, 3-hydroxycarbofuran, methiocarb sulfoxide, deethylatrazine, methiocarb sulfone, 3-ketocarbofuranphenol, butocarboxim, aldicarb, 3-ketocarbofuran, 8-hydroxybentazone, cyanazine, simazine, baygon, carbofuran, carbaryl, chlortoluron, bentazone, MCPA, 1-naphthol, atrazine, isoproturon, propanil, methiocarb, molinate, alachlor and metolachlor.   Quantitative analyses were performed by LC-DAD at 220 nm or by post-column fluorescence detection at excitation and emission wavelengths of 330 and 465 nm, respectively. Calibration graphs were constructed by analyzing spiked aqueous solution at concentrations encompassing the range of interest and were linear over the range 0.02-1μg/L. All the results presented were corrected for recoveries.

   To avoid false positives, all compounds were confirmed by an on-line SPE LC-MS equipped with a thermospray (TSP) interface. For this purpose, a Hewlett-Packard (Palo Alto, Ca, U.S.A) Model 5988A thermospray LC-MS quadrupole mass spectrometer and a Hewlett Packard Model 35741B instrument for data acquisition and processing were employed. The temperatures of TSP were as follows: 90, 200, and 270ºC for the stem, tip and source, respectively. Chromatograms were recorded under positive ionization (PI) or negative ionization (NI) conditions depending on the compound, and in selected ion monitoring mode; each compound was identified by two main ions (19,20).

Results and Discussion

Surface Water Atrazine, DEA, metolachlor and alachlor were detected through-out the whole year, while DIA was not detected from river delta waters between September and January. The total concentrations measured were up to 0.3-0.4μg/L, following spring application and diminished significantly in fall, down to 0.1 μg/L. Atrazine had the least seasonal variation of all the compounds, exhibiting stable concentration of 0.1μg/L during the whole year. The global pollution behavior is not constant through-out   the whole delta. Water pollution by triazines and chloroacetanilides is highest in the river (see Figure 3, sampling point 1) showing that many of these compounds are transported significant distances from their application

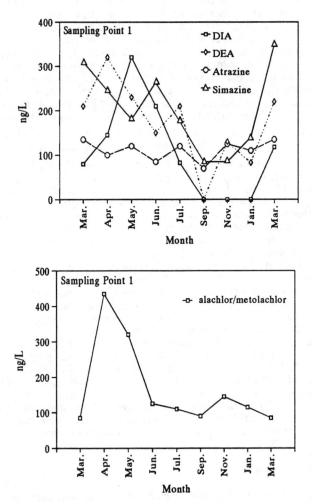

Figure 3. Seasonal variation of concentration of corn and rice herbicides and selected metabolites at sampling points 1-3 from the Ebro estuary. March 1992-March 93.

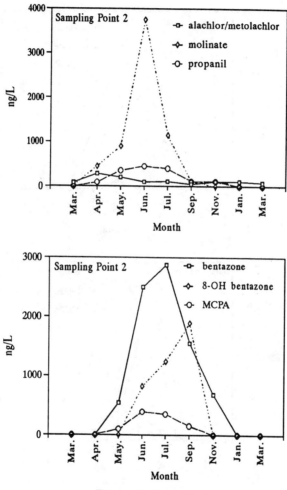

Figure 3. *Continued*

Continued on next page

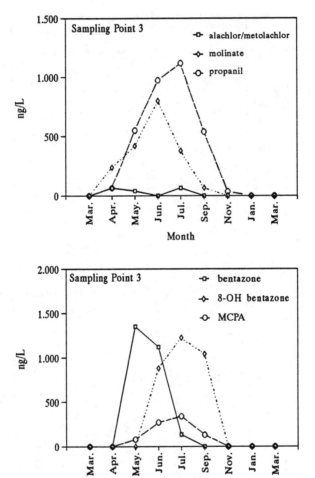

Figure 3. *Continued*

sites, but decreased in sampling points 2 and 3 (triazine and chloroacetanilide herbicides varied between non detectable values up to 0.05 µg/L, with an average levels of 0.0 2 µg/L), in a way similar to that reported in a previous survey (4,5).

Corn herbicides and metabolites The findings shown in Fig. 3, sampling point 1, are significant for several reasons. First, the values of DIA are important throughout the entire year and after application (May, June) are higher than those reported for DEA and atrazine. The main source of DIA is the degradation of atrazine, simazine and cyanazine (3) and as reported in Table I, the Ebro river basin uses important amounts of these three herbicides and also terbutylazine, which also degrades to DIA. This shows a remarkable difference when compared to the Midwest Corn Belt (3) where practically no simazine and terbutylazine are used. Second, the presence of DIA and not of cyanazine can be explained by the instability of cyanazine as compared to other corn herbicides (probably due to the oxidation of the cyano group) (7). In addition, DIA has a much higher aqueous solubility as compared to cyanazine (670 versus 170 mg/L) and higher expected half life, therefore DIA can be transported along the river basin more easily than cyanazine, which was not detected in the estuary. Third, degradation of DIA is fast after May and concentration values of DIA reach non-detectable levels in September. This is not surprising since continued dealkylation of DIA is fast as compared to DEA . The removal of an ethyl chain is preferential over an isopropyl chain, so DEA is more stable (21). Chlortoluron and isoproturon were observed at detectable concentrations (0.03-0.05 µg/L) only in January and only at sampling point 1 (results not reported).

Rice herbicides Concentrations of these herbicides in the dissolved phase are shown in Figure 3 , sampling points 2 and 3. Rice is grown under flooded conditions. Molinate and propanil are applied before or immediately after rice seeding to control water grasses and broadleaf plants, while bentazone and MCPA are post-emergence herbicides and are often detected in water after treatment (amounts used, see Table I). The major inputs occurred in early summer (June-July) just after application with concentration levels up to 3-4 µg/L and 0.5-1 µg/L, for bentazone/molinate and MCPA/propanil, respectively. Observed concentrations decreased quickly through the summer with non-detectable levels in October and November. The maximum contamination levels were observed in the drainage ditches (sampling point 2) receiving waters from rice fields. The lower concentrations in sampling point 3 can be attributed to the dilution effect.Bentazone is moderately stable in natural waters with regard to hydrolysis and photolysis. Laboratory studies using river waters irradiated by a xenon arc lamp reported half lives of 2hr 45min versus 1hr 25 min for alachlor (22). Under abiotic transformation processes bentazone leads to the formation of two hydroxy derivates, 6 and 8-hydroxy bentazone. Trace levels of 8-hydroxybentazone were identified (11). The maximum metabolite contribution appeared two months after the application of parent compound. Molinate is the most heavily used pesticide in the Ebro River Estuary (Table I) and was encountered at the highest concentrations , up to 4 µg/L, at sampling point 2 in June 1992, following which was a sharp decrease in concentration. Photodegradation was reported to be the major route of dissipation of molinate in water to give 2-ketomolinate among other photoproducts (23). Photooxidation of molinate probably takes place in rice fields rather than in the river itself and occurs quickly in the summer. Propanil is the least hazardous compound in the surface water, since the drainage ditches were remarkably polluted only immediately after treatment, at concentration levels up to 1 µg/L. In a similar manner as molinate, propanil concentrations decreased very rapidly through-out the summer. Finally MCPA, (1.5 tons per year, see Table I), has a relevant contribution to the total pollution with a maximum contamination peak of 0.4 µg/L in June and July, which demonstrates its stability under the environmental conditions of this study area.

Ground Water   The global pollution behavior is similar in the four wells despite a great variation in the hydrologic conditions through-out the aquifer (see Table II). The carbamate insecticides, methiocarb, methomyl and carbofuran, were the main pollutants. The seasonal variation of carbamate insecticide concentration over one year period, March 93 to February 94, are shown in Figure 4.

   Carbamate insecticides. Carbamate insecticide peak pollution was higher with concentration levels   of 0.8 µg/L of methomyl in March 1993 in well 3, but carbamates are not stable in an aqueous environment. Methomyl is usually sprayed in September and October, and appeared in the ground water in February with a maximum contamination peak in March. Carbofuran and methiocarb are less water soluble than methomyl and leached to ground water within 6 months after application, with a maximum contamination peak in June and July. Methiocarb and carbofuran patterns closely match that of irrigation, which is a continuous practice through-out the year except in July and August. However, irrigation rates are higher in spring (twice the usual), during green production leading to a higher loading of pesticides in the underlying aquifers. The persistence of carbamate insecticide pollution was estimated to be in the range of 3-4 months. The fate of carbamate insecticides in ground water is basically linked to hydrolysis pathway. Laboratory stability studies in slightly basic and biologically inhibited distilled waters (pH = 7.8, acetate buffer), maintained in the dark at 10°C have indicated that methomyl, carbofuran and methiocarb remained stable 60, 45 and 30 days, respectively (17). The discrepancy between the observed carbamate values and the laboratory data should be attributed to the particular environmental factors that affect degradation. As we have observed, the soil has low microbial activity and the pesticides are probably kept in quite anaerobic conditions, leading to the increased stability for these compounds. The results obtained here will not be extrapolated to other areas, since the soil type and the climate conditions affect the stability of the compounds.

   Pesticide adsorption on soil particulate was regarded as a minor transformation pathway due to the high water solubility of the compounds studied. The striking feature was that carbamate leaching occurred very quickly in response to winter and spring furrow irrigation (3 months for methomyl and 6 months for carbofuran and methiocarb with the highest peak observed matching the high-flow irrigation period). The soil is predominantly silt-loam. Clay content at the surface is 20% but has a tendency to increase with depth. Organic matter content averages 2% in the 1-50 cm surface layer but drops rapidly with depth to very low level. Thus the Almeria soil has some pesticide adsorptive capacity in the surface, however the hydraulic conductivity is high and consequently pesticide movement can be expected. Soil pH is very similar between samples ranging from 7.5 to 8. Rapid transport of carbamate insecticides through the soil may be explained by a macropore flow through the deeper soil core connected directly to the upper layer of soil (20-30 cm). This vertical channel bypasses the soil matrix and permits rapid leaching to ground water (24), and occurs preferentially in non-tillage soils like those of the Almeria area (25). Finally, carbaryl and butocarboxim were also occasionally found in spring at concentrations up to 0.5 µg/L and 0.2 µg/L respectively, and seemed not to be a significant problem to ground water.

   Carbamate insecticide metabolites Methiocarb sulfone and 3-hydroxycarbofuran, the main metabolites of methiocarb and carbofuran respectively, were detected in real ground waters for the first time and their concentration fluctuations in concentration followed very closely those of the parent insecticides (see Figure 5). The most likely hypothesis is that part of both insecticides were adsorbed in Almeria soil, metabolized in the unsaturated soil layer and transported from the overlying unsaturated zone into the aquifer according to seasonal irrigations. Once deeper layers of soil are reached, lower degradation rates can be expected due to decreased carbon content as observed for aldicarb degradation in soil (26). Carbofuran and methiocarb loading most likely occur during the year following their application. Carbofuran was used as a soil treatment and 3-hydoxycarbofuran resulted from metabolic degradation as described elsewhere (19), but methiocarb was foliage-sprayed and the question was whether methiocarb sulfone appeared as a result of the metabolic activity of soil bacteria and fungi or as a sunlight photo-oxidation process before reaching the upper soil layer. Laboratory studies (experiments not reported) show that photo-degradation processes lead only to the formation of methiocarb sulfoxide. Additionally, foliage residue studies indicate that only methiocarb sulfoxide is formed as the main metabolic breakdown product (experiments not reported). Methiocarb sulfone can be regarded as a result of microbial oxidation in the unsaturated soil layer and not as a result of an abiotic process such as photolysis.

The fact that important levels of methiocarb sulfone and 3-hydroxycarbofuran are found in the aqueous environment is quite surprising in regard to our previous work (17). Up to 100% and 20% losses of methiocarb sulfone and 3-hydroxycarbofuran respectively, have been observed after 20 days in distilled water kept in the dark at pH = 4.8 with ammonium acetate-acetic acid buffer and stored at 4°C. This can be explained by a continuous metabolite loading over at least 3 months. Both metabolites are much more water soluble than their parent compounds and thus are leached more rapidly to the underlying aquifer. The lower stability of the degradation products with respect to the parent compounds can explain why the metabolite concentrations have never exceeded those of the parent compounds during the monitoring program. Other carbofuran degradation products such as 3-hydroxycarbofuranphenol and 3-ketocarbofuranphenol were never detected in well water,   probably because of the lack of sensitivity below 0.1 μg/L under UV (17) and using thermospray (TSP) mass spectrometer detection (20).

Conclusion

The present paper provides a clear example that on-line SPE followed by LC techniques can be used in a routine way for the determination of pesticide pollution in surface waters and ground waters from Spain. Although the more frequently occurring pesticides in European ground waters are herbicides (atrazine, simazine, and alachlor) typically used in large quantities on various crops, this work has demonstrated that specific crops and vulnerable sites such as the Ebro estuary and the Almeria area can prompt the need to develop "tailor-made" monitoring studies adapted to local pesticide usage. Rice herbicides are usually found in the Ebro estuary and the concentrations are related to application, although after application a relatively fast degradation occurs. Triazine herbicides are a valuable indicator of the Ebro river basin contamination and exhibit conservative behavior through-out the year. The degradation products of the triazine herbicides, DEA and DIA, are indicators of herbicide usage (atrazine, simazine, cyanazine and terbutylazine are used along the river basin) and contamination is likely to occur to a higher extend in groundwater as compared to surface runoff.

Carbamate insecticides, methomyl, methiocarb and carbofuran were found in high amounts (above 0.5 μg/L) in a representative Almeria aquifer. Their presence is a source of concern because of the high acute toxicity of carbamate insecticides. Fortunately they are not stable in a slightly basic aqueous environment and their persistence has never exceeded 3 months. Detection of carbamate metabolites, 3-hydroxycarbofuran and methiocarb sulfone will logically drive us in future work to investigate the carbamate degradation pathways in an aquatic environment in order to identify further possible toxic metabolites.

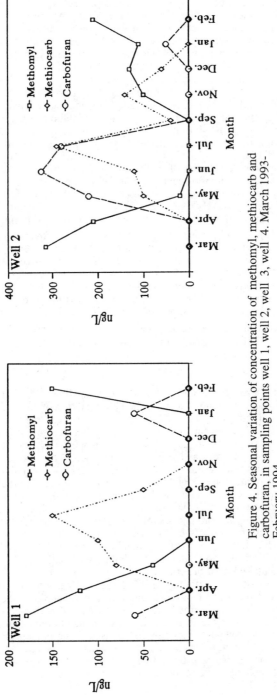

Figure 4. Seasonal variation of concentration of methomyl, methiocarb and carbofuran, in sampling points well 1, well 2, well 3, well 4. March 1993-February 1994.

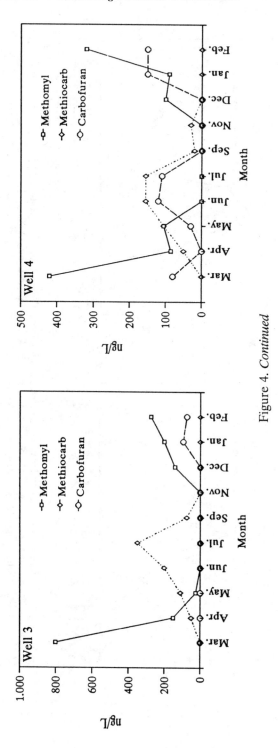

Figure 4. *Continued*

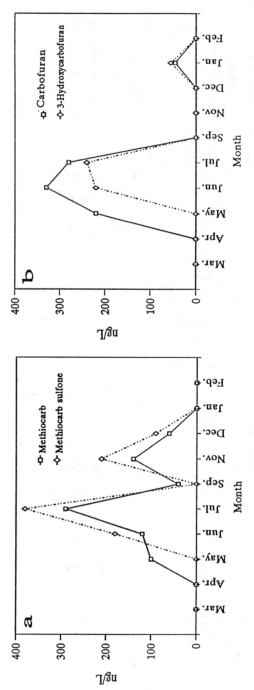

Figure 5. Correlation between (a) concentration of methiocarb and its metabolite methiocarb sulfone, (b) concentration of carbofuran and its metabolite 3-hydroxycarbofuran, March 1993-February 1994.

Acknowledgments. This work was supported by the Environment R&D Program 1991-1994 the Commission of the European Communities (Contract No. EV5V-CT94-0524) and by CICYT (AMB95-0075-C03).

Literature Cited

1.   McGehee Marsh, J., *Arch. Environ. Contam. Toxicol.*,**1993** , *25*, 446-455.
2.   Thurman, E.M.; Goolsby, D.A.; Meyer, M.T.; Kolpin, D.W., *Environ. Sci. Technol.*, **1991**, *25,* 1794-1796.
3   Thurman, E.M.; Meyer, M.T.; Mills, M.S.; Zimmerman, L.R.;Perry, C.A.; Goolsby,D.A. *Environ. Sci. Technol.*, **1994**, *28*, 2267-2277.
4.   Readman, J.W.; Albanis, T.A.; Barceló, D.; Galassi, S.; Tronczynski, J.; Gabrielides, G.P. *Mar. Pollut. Bullet.*, **1993**, *26*, 613-619.
5.   Gascón, J.; Durand, G.;Barceló, D. *Environ. Sci. Technol.*, **1995**, *29*, 1551-1556.
6.   Thurman, E.M.; Goolsby, D.A.; Meyer, M.T.; Kolpin, D.A. *Environ. Sci. Technol.,* **1991**, *25*, 1794-1796.
7.   Thurman, E.M., Goolsby, D.A.; Meyer, M.T.; Mills, M.S.; Pomes, S.; Kolpin, D.A., *Environ. Sci. Technol*. **1992**, *26*, 2440-2447.
8.   Kochany, J.; Maguire, R.J., *J. Agric. Food Chem.*, **1994**, *42*, 406-412.
9.   Johnson, W.G.; Lavy, T.L., *J. Environ. Qual.*, **1994**, *23*, 556-562.
10.   Pereira, W.E.; Rostad, C.E.;*Environ. Sci. Technol.*, **1990**, *24*, 1400-1406
11.   Chiron, S.; Papilloud, S.; Haerdi, W.; Barceló, D. *Anal. Chem.*, **1995**, *67*, 1637- 1643.
12.   Ritter, W.F. *J. Environ.Sci. Health.*, **1990**, *B25(1)*, 1-29.
13.   US Environmental Portection Agency. Another Look: National Pesticide Survey. Phase II report, Washington D.C. **1992**, 1-166.
14.   Verstraeten, I.M.; Cannia, J.C.; Tanner, D.Q., US Geological Survey, Reconnaissance of Gorund Water Quality in the North Platte Natural Resources   District, Western Nebraska, June-July 1991. Lincoln, NE,**1995**, 1-114.
15.   Samanidou, V.; Fytianos, K.; Pfister, G.; Bahadir, M., *Sci. Total Enviton.*, **1988**, *5*, 559-566.
16.   Barceló, D. *J. Chromatogr.*, **1993**,*643*,117-143.
17.   Chiron, S.; Fernandez-Alba, A.; Barceló, D. *Environ. Sci. Technol.*, **1993**, *27*, 2352-2359.
18.   Taberner, A.; Fabregues, C.; Godall, M. Proceedings of the "Congreso de la Sociedad Española de Malherbologia", 1993, Lugo, Spain pp 212-217.
19.   Chiron, S.; Valverde, A.; Fernandez-Alba, A.; Barceló, D. J. *Assoc. Off. Anal. Chem. Intl.,* **1995** , vol. 78. No. 6 (in press)
20.   Chiron, S.; Dupas, S.; Scribe, P.; Barceló, D. J. *Chromatogr., A*, **1994**, 665, 295-305
21.   Mills, M.S.; Thurman, M.E., *Environ. Sci. Technol.*, **1994**, *28* , 600-605.
22.   Chiron, S., Abian, J.; Ferrer, M.; Sanchez-Baeza, F.; Messeguer, A.; Barceló, D. *Environ. Toxicol. Chem.*, **1995**, 14, 1287-1298.
23.   Pereira, W.E.; Hostettler, F.D. *Environ. Sci. Technol.*, **1993**, 27, 1542-1552.
24.   Mills, M.S.; Thurman, E.M. *Environ. Sci. Technol.* **1994**, *28*, 73-79.
25.   Piyush Singh; Rameswar, S. K. *J. Environ. Qual.* **1991**, 20, 295-300.
26.   Ou; Li Tse; Rao, P.S.C.; Edvarsson, K.S.V.; Jessup, R.E.; Hornsby, A.G. *Pest. Sci.* **1988**, 23, 1-12.

Chapter 19

Hydroxylated Atrazine Degradation Products in a Small Missouri Stream

Robert N. Lerch[1], William W. Donald[1], Yong-Xi Li[2],
and Eugene E. Alberts[1]

[1]Cropping Systems and Water Quality Research Unit,
Agricultural Research Service, U.S. Department of Agriculture,
269 Agricultural Engineering Building, University of Missouri,
Columbia, MO 65211
[2]Analytical Bio-Chemistry Laboratories, Inc., 7200 East ABC Lane,
Columbia, MO 65202

This research assessed the occurrence of hydroxylated atrazine degradation products (HADPs) in streamwater from Goodwater Creek watershed in the claypan soil region of northeastern Missouri. Streamwater was sampled weekly from June, 1992 to December, 1994 at a V-notch weir used to measure streamflow for this 7250-ha watershed. Filtered water samples were prepared by cation exchange solid-phase extraction and analyzed for hydroxyatrazine (HA), deethylhydroxyatrazine (DEHA), and deisopropylhydroxyatrazine (DIHA) by high performance liquid chromatography with UV detection. HADPs were confirmed by electrospray HPLC/MS/MS and direct probe MS methods. Frequency of HADP detection was 100% for HA, 25% for DEHA, and 6% for DIHA. Concentrations ranged from 0.18-5.7 µg L^{-1} for HA, <0.12-1.9 µg L^{-1} for DEHA, and <0.12-0.72 µg L^{-1} for DIHA. These results establish that HADPs can contaminate surface water and that HA contamination of surface water is a significant fate pathway for atrazine in this watershed.

Widespread contamination of surface waters by atrazine and its N-dealkylated degradation products in the corn production areas of the USA and Canada has been well established over the last twenty years (1-5). However, surface water contamination by hydroxylated atrazine degradation products (HADPs; hydroxyatrazine, deethylhydroxyatrazine, and deisopropylhydroxyatrazine) has never been thoroughly assessed. HADPs have been shown to contaminate groundwater at only parts per trillion levels (6), and their potential to significantly contaminate groundwater appears to be minimal (7-10). However, HADP contamination of surface waters has received little consideration. In a study of ozonation treatment of atrazine contaminated surface waters, HA was detected at 1-2 µg L^{-1} from two reservoir samples (11). Although these findings were confirmed by alternative methods, the sites were sampled only once. Atrazine degradation in a small Iowa stream was postulated to degrade to HA via photolysis, but direct proof of HA in the stream was not provided (*12*). Using a polyclonal immunoassay, HA was not detected in any environmental water samples (*13*), but long-term monitoring for HA was not conducted.

Hydroxylated atrazine degradation products (HADPs) are chemically distinct from their chlorinated analogues. They have lower solubility in water and organic solvents, and HADPs are stronger bases with dissociation constants near 10^{-5} compared to about 10^{-2} for most chlorinated *s*-triazines (*14-17*). Furthermore, HADPs are tautomeric compounds with the stability of the enol, keto, and ionic forms dependent upon pH (*18-19*). In soils, HA adsorption and persistence are greater than atrazine and chlorinated atrazine degradation products (*7,8*). HADP formation is generally thought to detoxify atrazine because HADPs are not toxic to plants or aquatic photosynthetic microorganisms (*14,20,21*). HA is essentially non-toxic to rats with an acute LD_{50} of >3000 mg kg^{-1}, and it is not mutagenic or teratogenic. However, HA has been found to mechanically damage rat kidneys by crystallization (D. D. Sumner, Ciba-Geigy Corp., personal communication). The overall impact of HADPs in terrestrial or aquatic ecosystems has not been assessed to date.

In general, high performance liquid chromatography (HPLC) methods are best suited for the separation and quantitation of HADPs because of their moderate non-polarity and low volatility (*15,22-25*). Enzyme-linked immunoassays have also been developed for analysis of HA in soil and water (*13,26*). HPLC/mass spectrometry (MS) and tandem mass spectrometry (MS/MS) methods have been successfully used for confirmation of HADPs in environmental samples (*6,25,27*). Thermospray HPLC/MS/MS was used to obtain daughter ion spectra for confirmation of HADPs in laboratory aqueous photolysis experiments (*27*). Recent work by Cai et al. (*6*), confirmed the presence of HA in groundwater by fast atom bombardment (FAB)/MS and FAB/MS/MS.

HA formation in the environment occurs by hydrolysis of atrazine (*7,14,18,27-34*). In water, chemical hydrolysis of atrazine to HA is pH dependent, and the presence of humic and fulvic acids increases the reaction rate (*29,35-37*). Photolytic hydrolysis of atrazine to HA in water and on solid surfaces proceeds indirectly from ·OH radical generating photosensitizers, such as humic acid, NO_3^-, polycyclic aromatic hydrocarbons, and biological reactions (*20,32,38-40*). In soils, degradation of atrazine to HA is generally the major degradation pathway with lesser amounts of the chlorinated degradation products, deethylatrazine (DEA) and deisopropylatrazine (DIA) (*7,18,29-31*) . HA formation in soils has been solely attributed to chemical hydrolysis of adsorbed atrazine (*18,29*), but recent work indicates that biological hydrolysis of atrazine can also occur (*33,34*). Thus, the formation of HA in the environment results from both biotic and abiotic pathways.

The N-dealkylated HADPs, deethylhydroxyatrazine (DEHA) and deisopropyl-hydroxyatrazine (DIHA), are usually minor soil degradation products of atrazine or HA in short- term studies (*41,42*). However, in a long-term field study of atrazine fate in soil, N-dealkylated HADPs comprised a significant proportion of the bound residues in bulk soil and humic acid fractions (*43*). Biological hydrolysis of DEA, DIA, and didealkylatrazine to their respective hydroxy analogues by *Pseudomonas* species has been shown to occur (44). DEHA and DIHA were also shown to be the major degradation products from photolysis of DEA and DIA in water (27).

Based on the behavior of HADPs in soil, annual atrazine application likely results in their accumulation in surface soil. Thus, HADPs could contamination streamwater by transport in surface runoff, either in solution or sorbed to sediment. Furthermore, atrazine contamination of surface waters is well documented, and subsequent degradation to HADPs in streamwater may be possible. The overall environmental fate of atrazine is currently incomplete without direct study of HADPs in surface water of agricultural watersheds. Methods for routine analysis of HADPs in surface water are well established (15,23,25), but HPLC/MS/MS confirmation methods for HADPs in surface water are still in the development phase (25) . Therefore, the objectives of this research were: 1) to conduct long-term monitoring of HADPs in streamwater of Goodwater Creek watershed in the claypan soil region of northeastern Missouri; and 2) to confirm the presence of HADPs using electrospray HPLC/MS/MS and direct probe MS methods.

Experimental Procedures

Chemicals and Standard Materials. Hydroxyatrazine (HA) [2-hydroxy-4-ethylamino-6-isopropylamino-s-triazine], deethylhydroxyatrazine (DEHA) [2-hydroxy-4-amino-6-isopropylamino-s-triazine], deisopropylhydroxyatrazine (DIHA) [2-hydroxy-4-ethylamino-6-amino-s-triazine] were 94-99% pure (Ciba-Geigy Corp. Greensboro, NC). Atrazine [2-chloro-4-ethylamino-6-isopropylamino-s-triazine] was 98% pure (Ultra Scientific, N. Kingstown, RI), and deethylatrazine (DEA) [2-chloro-4-amino-6-iso-propylamino-s-triazine] and deisopropylatrazine (DIA) [2-chloro-4-ethylamino-6-amino-s-triazine] were 97% pure (Crescent Chemical Co., Inc., Hauppauge, NY). All solvents used were HPLC grade. All chemicals used for the routine analysis of the HADPs by cation exchange solid-phase extraction and C_8 reverse-phase HPLC were described by Lerch and Donald (15). Ammonium formate (NH_4CHO_2) was reagent grade (J. T. Baker Chemical Co., Phillipsburg, NJ), and formic acid (CHOOH) was supplied by Sigma Chemical, Co. (St. Louis, MO).

Stream Sampling and Site Description. Goodwater Creek was sampled at a V-notch weir used to measure daily streamflow for the entire 7250-ha Goodwater Creek watershed, located in the claypan soil region of northeastern Missouri (latitude, 92° 03' W; longitude, 39° 18' N). About 70% of the watershed is cropped with approximately 23% of the cropped acreage (990 ha) planted to corn or sorghum in 1992 and 1993 (L. K. Heidenreich, University of Missouri, personal communication). Naturally formed claypan soils (argillic horizon derived from clay loess parent material) of the Mexico-Putnam soil association predominate in the watershed. These soils are somewhat poorly to poorly drained on slopes of 0-3% with silt loam surface horizons and clay loam or silty clay loam subsoils containing 40-60% montmorillonitic clays.

Two to five 1-L grab samples were collected in amber glass jars with Teflon-lined screw caps on a weekly basis or as permitted by streamflow from June, 1992 to December, 1994. All samples were filtered through 0.45 μm nylon filters to remove suspended sediment and then combined and mixed to provide a uniform sample. Field samples were stored at 2-4°C before analysis. Samples were filtered within two days of receipt and usually analyzed within 10 d but always within 40 d.

Routine Analyses

Hydroxylated Atrazine Degradation Products. Analysis of HADPs by propylbenzenesulfonic acid cation exchange (SCX) solid-phase extraction (SPE) and C_8 reverse-phase HPLC with UV detection (C_8-HPLC/UV) was previously described by Lerch and Donald (*15*). For DEHA and DIHA, the limit of detection was 0.12 μg L^{-1} and the limit of quantitation was 0.40 μg L^{-1}. For HA, the limit of detection was 0.04 μg L^{-1} and the limit of quantitation was 0.13 μg L^{-1}. Quality control samples included corresponding sets of field and lab spikes at 0.5, 1.0, or 5.0 μg L^{-1} and blanks. All (untreated) stream samples were analyzed in duplicate or triplicate. Reported HADP concentrations represent the average concentration of replicates corrected for the corresponding field spike recovery which were typically 90 ± 10% (*15*).

Atrazine and Chlorinated Metabolites. Sample clean-up by C_{18} SPE for the determination of atrazine, DEA, and DIA was performed as described by Thurman et al. (*45*). GC analysis was performed on a Varian 3400 equipped with a N-P detector, an 8200 autosampler, a splitless injector, and a Restek XTI-5 fused-silica capillary column (30 m x 0.53 mm) with a 5% diphenyl, 95% dimethyl polysiloxane stationary phase (Restek Corp., Bellafonte, PA). Specific GC conditions were described by Lerch et al. (*25*).

Confirmation Analyses. HADPs were qualitatively confirmed using electrospray reverse-phase HPLC/MS/MS and direct probe MS. All confirmation analyses were conducted by Analytical Bio-Chemistry Laboratories, Inc. (Columbia, MO) on HPLC fractions supplied to them as described below. Details of other confirmation methods for HADPs isolated from streamwater were described by Lerch et al. (*25*).

Electrospray HPLC/MS/MS. The daughter ion mass spectra of HA and DEHA were obtained using fractions collected from samples prepared for routine analysis by SCX SPE and C_8-HPLC/UV. The Beckman HPLC system (San Ramon, CA) used for routine analysis was equipped with an Isco Foxy 200 fraction collector (Lincoln, NE), and fractions of DEHA and HA were collected from 6 different samples, representing about 6% of the samples collected for the entire study. Twenty 1-mL fractions were pooled into glass culture tubes for each sample and analyte. Fractions were also collected from standards of DEHA and HA. The fractions were evaporated to 1 to 2 mL in a Savant Speedvac SS-4 evaporator (Farmingdale, NY) at 60°C and about 670 Pa and further evaporated just to dryness under a stream of N_2(g) at ambient temperature (22-25°C). Samples were reconstituted in 5 mL of CH_3OH, by repeated vortex mixing, and evaporated under a stream of N_2(g) to about 100 μL for DEHA and about 250 μL for HA. Standard samples were analyzed by HPLC, and overall efficiencies for the evaporation procedure were about 80% for both DEHA and HA.

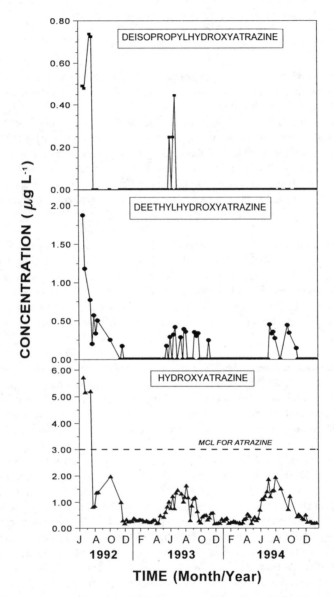

Figure 1. Concentrations of hydroxyatrazine (HA), deethylhydroxyatrazine (DEHA), and deiso-propylhydroxyatrazine (DIHA) in Goodwater Creek from June, 1992 to December, 1994.

DIHA was omitted from these analyses because its final concentrations were below the MS/MS detection limits of 10 to 12 µg L^{-1}. A Finnigan TSQ-7000 MS/MS system (San Jose, CA) and a Thermo Separation Products CM4100 HPLC system (San Jose, CA) with an electrospray ionization interface (electrospray HPLC/MS/MS) were used to obtain daughter ion spectra from the fraction samples of DEHA and HA. HPLC separations were performed using a Hypersil microbore C$_{18}$, 25 cm by 1.0 mm (i.d.) column, sample injection volume of 5 µL, and a flow rate of 75 µL min^{-1}. Mobile phase A was 2% aqueous CH$_3$CN with 0.1% CHOOH/NH$_4$CHO$_2$, pH 3, and mobile phase B was 10% H$_2$O: 80% CH CN: 10% tetrahydrofuran (C$_8$H O) with 0.1% CHOOH/NH$_4$CHO$_2$, pH 3. Mobile phase gradient separation conditions were described by Lerch et al. (*25*). Retention times were 12.9 min for HA and 23.3 min for DEHA. MS/MS conditions for [M+H]$^+$ masses of 198 (HA) and 170 (DEHA) were: electrospray ionization voltage, 4.5 kV; capillary temperature, 225°C; scan time, 1 s; scan range, 50-225 amu; collision energy, 25 eV; collision gas, Ar; multiplier voltage, 1200 V.

Direct Probe MS. Full scan mass spectra were obtained by direct probe MS for DEHA fractions described in the Electrospray HPLC/MS/MS. A HP 5989A MS engine (Palo Alto, CA) was used under the following conditions: N$_2$ head pressure, 0.55 MPa; electron impact source temperature, 250°C; quadrupole temperature, 100°C; scan range, 40-250 amu; repeller voltage, 7 V; electron energy, 70 eV; multiplier voltage, 2906 V; MS mode, positive ion detection. A 2 µL sample was put in a capillary tube and placed on the probe tip. The probe tip was held at 40°C for 1 min; then heated to 120°C at 30°C min^{-1} and held for 1 min.

Results and Discussion

Monitoring of HADPs in Surface Water. HADPs were consistently detected in Goodwater Creek over the 2.5 year monitoring period reported (Figure 1). Concentrations ranged from 0.18-5.7 µg L^{-1} for HA, <0.12-1.9 µg L^{-1} for DEHA, and <0.12-0.72 µg L^{-1} for DIHA. Total runoff to the stream varied greatly during the study with 15 cm in 1992, 75 cm in 1993, and 30 cm in 1994, compared to the twenty year (1971-91) annual average of 29 cm. Despite the extreme variability in runoff, DEHA and HA showed similar seasonal concentration changes from year to year. Their concentrations increased during the spring, reached their highest levels in June or July, slowly dissipated from late summer to fall, and reached very low, or non-detectable, levels in the winter. This same general pattern has been reported for other herbicides and metabolites in surface water (*4,5*) except that the extent of dissipation by fall was greater than that observed for HA and DEHA.

Concentrations of HA and DEHA varied inversely relative to streamflow throughout the study (Figures 1 and 2). For example, a small runoff event in mid-July, 1992 reduced HA concentration from 5.2 to 0.82 µg L^{-1} in a week. Thereafter, HA concentrations steadily increased over the next 70 days to 2.0 µg L^{-1}, during which time there were no significant runoff events (Figure 2). This relationship also was observed in 1993 and 1994 when several major runoff events led to lower concentrations of HA or DEHA followed by increases in concentration under baseflow (i.e. low streamflow)

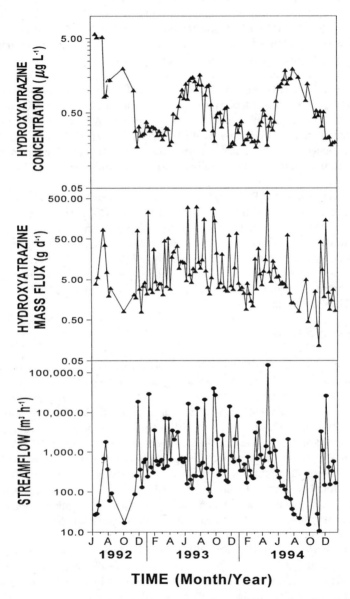

Figure 2. Hydroxyatrazine (HA) concentration and mass flux in relation to Goodwater Creek streamflow from June, 1992 to December, 1994.

conditions. Thus, for this watershed, HA or DEHA concentrations decreased due to dilution by the surface runoff. The rise in HADP concentrations during baseflow conditions indicates that there was a source of HADPs in the creek during these periods. Quarterly samples (June, September, and December, 1994 and March, 1995) of shallow groundwater in the creek's alluvium indicated that no HADPs were present, and tile drainage is not used in the watershed. Thus, soluble HADPs in baseflow originated from other sources such as desorption from sediment and chemical, photochemical, or microbial hydrolysis of atrazine in the stream.

In contrast to the concentration data, the mass flux of HA was directly related to streamflow (Figure 2). During surface runoff events, the total mass of HA increased even though its concentration decreased in the creek. Therefore, the increased mass of HA in the stream was a direct consequence of HA transported in solution during runoff events. This strongly suggests that one of the mechanisms by which HA reached the streamwater was desorption from soils, in the field, and subsequent transport in solution by surface runoff from the field to the stream. Some transported HA may have come from the field soil solution levels maintained by the equilibrium between adsorbed and solution HA. However, the soil solution is unlikely to be a significant source of HA transported during runoff events because this equilibrium strongly favors adsorption (*8*). Photolytic hydrolysis of atrazine on soil and plant surfaces is another potential source of soluble HA transported by surface runoff.

Comparison of HADPs to Chlorinated Analogues. Frequency of HADP detection in surface water samples was 100% for HA, 25% for DEHA, and 6% for DIHA (Figure 3). Atrazine, HA, and DEA were the most frequently detected surface water contaminants resulting from atrazine use in the watershed. As a group, the HADPs were detected less frequently than their chlorinated analogues, and the N-dealkylated HADPs were relatively insignificant surface water contaminants. Atrazine and DEA showed much broader concentration ranges and higher mean concentrations than HA. Because of a few high concentration spikes, mean concentrations of atrazine (4.9 μg L^{-1}) and DEA (1.3 μg L^{-1}) were greater than their respective 75th percentile concentrations. HA concentrations did not greatly fluctuate like those of atrazine and DEA; therefore, its mean (0.74 μg L^{-1}) and median (0.38 μg L^{-1}) concentrations were similar. However, the year-round persistence of HA in streamwater resulted in median HA concentrations slightly greater than DEA and only 0.19 μg L^{-1} lower than atrazine. The N-dealkylated HADPs were infrequently detected with both occurring at levels below 2 μg L^{-1}. While DEHA was generally present from late spring through the fall, DIHA was detected only in late spring to early summer of 1992 and 1993. DEHA levels were not closely associated with DEA. Instead, DEHA concentrations in the stream were more related to HA concentrations showing a similar seasonal pattern and response to streamflow while DIHA was so infrequently detected that comparisons to other metabolites were not valid.

Atrazine and DEA were more readily transported by surface runoff events than HA. This resulted in higher streamwater concentrations of atrazine and DEA early in the growing season compared to HA, but lower levels compared to HA in the late summer and fall of 1992 and 1993 (Figure 4). In 1994, dry conditions from July

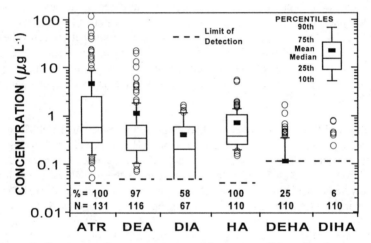

Figure 3. Summary of concentrations and frequency of detection (expressed as percent) of atrazine and its degradation products in Goodwater Creek from June, 1992 to December, 1994 (N = sample size).

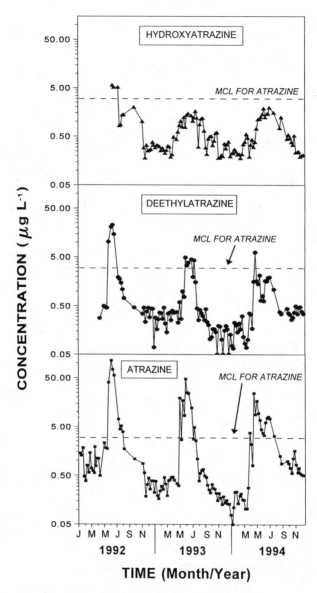

Figure 4. Concentrations of hydroxyatrazine (HA), deethylatrazine (DEA), and atrazine in Goodwater Creek from January, 1992 to December, 1994.

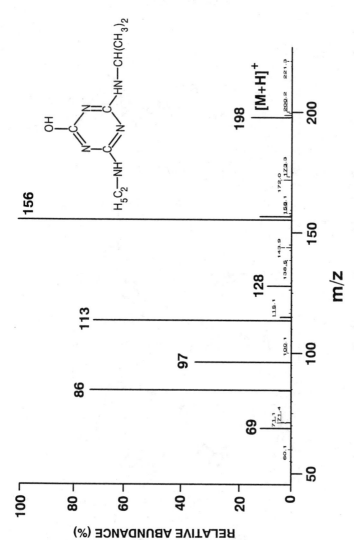

Figure 5. Electrospray HPLC/MS/MS daughter ion mass spectra of the corresponding [M+H]$^+$ ion of hydroxyatrazine (HA) isolated from a streamwater sample (note: the molecular structure depicted represents the tautomeric form fragmented).

through September led to less atrazine and DEA transported from soil, and therefore, a greater mass was available in the surface soil for transport later in the year. This resulted in stream concentrations of atrazine and DEA as high or higher than HA through the fall. The generally greater persistence of HA in the stream during late summer and early fall relative to atrazine and DEA primarily resulted from four processes. First, atrazine and DEA have lower soil adsorption than HA; thus, atrazine and DEA were more easily transported in runoff during the growing season, particularly from late spring to early summer. By fall, less atrazine and DEA remain available in the surface soil for transport. In 1993 and 1994, average mass transport during the 8 week period following atrazine application accounted for 89% of the atrazine, 57% of the DEA, and only 36% of the HA annually transported from the watershed. Second, as adsorption and bound residue formation of atrazine increased with time after application, atrazine hydrolysis would become progressively more significant (*19,35*). Third, atrazine and DEA are less persistent than HA in agricultural soils (*7*). Thus, the greater persistence and adsorption of HA resulted in proportionally more remaining in surface soil later in the growing season. Fourth, atrazine and DEA will both be subject to leaching to a greater extent than HA (*9,10*), further increasing the proportion of HA to atrazine or DEA in the plow layer with time. Both atrazine and DEA leached to depths of 0.9 m in field plots equipped with pan lysimeters, and DEA has frequently been detected in groundwater wells to depths of 15 m within Goodwater Creek watershed (unpublished data). In contrast, no HADPs were detected in groundwater, as mentioned above.

Confirmation of HADPs in Streamwater. Daughter ion mass spectra of HA were obtained by electrospray HPLC/MS/MS via Ar collision of the corresponding [M+H]$^+$ ions (Figure 5). The [M+H]$^+$ parent ion at m/z 198 was observed for all HA samples, and tentative identification of all major daughter ions provided further evidence for identification of HA. The most abundant daughter ion for all HA samples was at m/z 156 formed by loss of the isopropyl group. The most diagnostic daughter ions were at m/z 128, 113, and 97 in which the dealkylated and increasingly deaminated, 2-hydroxy substituted triazine ring was present. Based on our assignment of the daughter ion fragments, the HPLC conditions resulted in HA fragmentation in the enol form (as depicted in Figure 5). The HA mass spectra by electrospray HPLC/MS/MS very closely matched that produced by thermospray HPLC/MS/MS (*27*).

 Daughter ion mass spectra of DEHA by electrospray HPLC/MS/MS consistently showed the presence of the [M+2H]$^+$ parent ion (m/z 171) for all samples (Figure 6). Given the results of the HA mass spectra, this was an unexpected finding, but the formation of a [M+2H] parent ion for several triazines by negative mode thermospray HPLC/MS has been reported (*27*). In addition, the assignment of all major daughter ions could be explained in terms of [M+2H]$^+$ parent ion fragmentation. The most abundant daughter ions were at m/z 57 with a relative abundance of 100% for all samples and at m/z 73 with relative abundances of 36-72%. These ions were not, however, uniquely diagnostic of DEHA since they represented breakage of the triazine ring by removal of keto (C=O) and HCN groups. The most diagnostic daughter ion (i.e. possessing an intact triazine ring with an O at the 2 position) in the DEHA spectra was

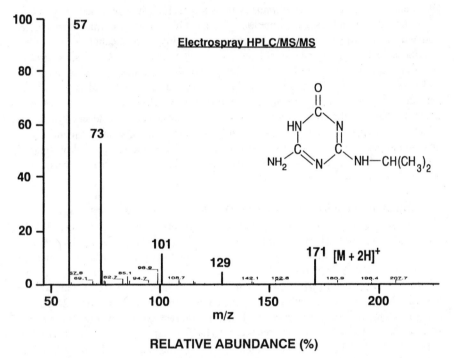

Figure 6. Mass spectra of deethylhydroxyatrazine (DEHA) isolated from a streamwater sample using electrospray HPLC/MS/MS via the corresponding [M+H]⁺ ion and direct probe MS (note: the molecular structures depicted represent the tautomeric forms fragmented).

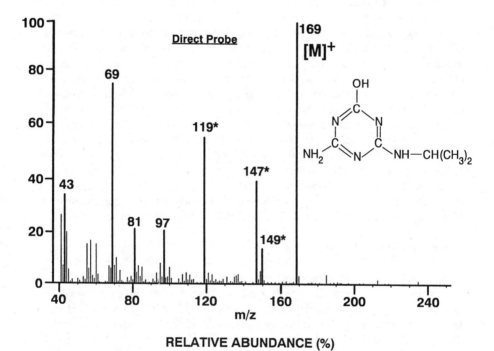

Figure 6. *Continued*

* Peaks from Impurities

of very low relative abundance at m/z 129, formed by loss of the isopropyl group. Unlike HA, the assignment of daughter ion fragments for DEHA indicated loss of a keto group rather than loss of a hydroxyl group at the 2 position of the triazine ring. Therefore, under the same HPLC conditions used for HA, DEHA was in the keto form (Figure 6).

Further work was needed to confirm DEHA and validate the electrospray HPLC/MS/MS spectra because the [M+2H]+ parent ion was observed and abundant diagnostic daughter ions were lacking. Therefore, direct probe MS was used to obtain full scan mass spectra of DEHA (Figure 6). Direct probe MS of all DEHA samples exhibited the expected [M]+ parent ion at m/z 169 with high relative abundance. In addition, abundant diagnostic daughter ions, m/z 97 and m/z 81, were observed. Other higher mass daughter ions of DEHA may have been masked due to interactions with dibutyl phthalate (base peak at m/z 149) and unidentified contaminants which resulted in high relative abundance ions at m/z 119 and 147. Nonetheless, the DEHA spectra obtained by direct probe MS supported the conclusions that DEHA was present in the samples, and that the electrospray HPLC/MS/MS spectra for DEHA was valid. Based on the assignment of fragments in the direct probe spectra of DEHA and its closer correspondence to the electrospray HPLC/MS/MS spectra of HA, DEHA was apparently in the enol form under conditions used for direct probe MS (Figure 6). The enol tautomer of DEHA apparently predominated in methanol for the direct probe samples whereas the pH 3 buffer used in the mobile phase of the electrospray HPLC/MS/MS analyses favored formation of the keto tautomer. Therefore, the unexpected differences in mass spectra between HA and DEHA using electrospray HPLC/MS/MS likely resulted from fragmentation of different tautomers.

Summary and Conclusions

More than two years of monitoring Goodwater Creek established that HADPs are surface water contaminants. HA was the most significant HADP with greater concentrations and frequency of detection than the N-dealkylated HADPs. Changes in HA and DEHA concentrations in the stream were inversely related to streamflow, indicating dilution during surface runoff events and a source of HADPs or formation of HADPs under baseflow conditions. HA mass flux was directly related to streamflow demonstrating that HA was transported in solution by surface runoff events. HA fluctuated less in the stream than atrazine and DEA resulting in lower average concentrations but similar median concentrations. HA persistence in the stream was greater in the late summer and fall than atrazine or DEA. The presence of HA and DEHA in streamwater was confirmed by electrospray HPLC/MS/MS, and in addition, DEHA was confirmed by direct probe MS. Verification of the stream monitoring data represented the first direct proof of HADPs as surface water contaminants and established HA contamination of surface waters as an important fate pathway for atrazine in Goodwater Creek watershed. On the basis of these results, the most likely mechanisms responsible for HADP contamination of surface water were: 1) transport of HADPs in solution by surface runoff events; soluble HADPs most likely occur via desorption of HADPs from soil; 2) desorption of HADPs from stream sediment deposited by surface runoff events; and 3) chemical, microbial, and aqueous photolytic hydrolysis of atrazine and its chlorinated degradation products in the stream. Further studies to establish which of these processes are most important are currently under way.

Acknowledgments

The authors wish to thank Dr. Paul E. Blanchard for his insights on the hydrogeology of the watershed. We also wish to thank Joe Absheer, Kathy Burgan, Jenny Fusselman, Mike Gresham, Kathy Neufeld, and Mark Olson for their contributions. A special thanks to Drs. E. Michael Thurman and Michael T. Meyer for organizing the symposium.

Literature Cited

(1) Richard, J. J.; Junk, G. A.; Avery, M. J.; Nehring, N. L.; Fritz, J. S.; Svec, H. J. *Pestic. Mont. J.* **1975**, *9*, 117-123.

(2) Muir, D. C. G.; Yoo, J. Y.; Baker, B. E. *Arch. Environ. Contam. Toxicol.* **1978**, *7*, 221-235.

(3) Pereira, W. E.; Rostad, C. E.; Leiker, T. J. *Sci. of the Total Environ.* **1990**, *97/98*, 41-53.

(4) Thurman, E. M.; Goolsby, D. A.; Meyer, M. T.; Kolpin, D. W. *Environ. Sci. Technol.* **1991**, *25*, 1794-1796.

(5) Thurman, E. M.; Goolsby, D. A.; Meyer, M. T.; Mills, M. S.; Pomes, M. L.; Kolpin, D. A. *Environ. Sci. Technol.* **1992**, *26*, 2440-2447.

(6) Cai, Z.; V. Ramanujam, M. S.; Gross, M. L.; Monson, S. J.; Cassada, D. A.; Spalding, R. F. *Anal. Chem.* **1994**, *66*, 4202-4209.

(7) Winkelmann, D. A.; Klaine, S. J. *Environ. Toxicol. Chem.* **1991**, *10*, 335-345.

(8) Clay, S. A.; Koskinen, W. C. *Weed Sci.* **1990**, *38*, 262-266.

(9) Schiavon, M. *Ecotoxicol. Environ. Safety* **1988**, *15*, 46- 54.

(10) Kruger, E. L.; Somasundaram, L.; Kanwar, R. S.; Coats, J. R. *Environ. Toxicol. Chem.* **1993**, *12*, 1959-1967.

(11) Adams, C. D.; Randtke, S. J. *Environ. Sci. Technol.* **1992**, *26*, 2218-2227.

(12) Kolpin, D. W.; Kalkhoff, S. J. *Environ. Sci. Technol.* **1993**, *27*, 134-139.

(13) Wittman, C.; Hock, B. *Acta Hydrochim. Hydrobiol.*, **1994**, 22:60-69.

(14) Esser, H. O.; Dupuis, G.; Ebert, E.; Vogel, C.; Marco, G. J. In *Herbicides: Chemistry,Degradation, and Mode of Action*; Kearney, P. C.; Kaufman, D. D., Eds.; Marcel Dekker, Inc.: New York, 1975; Vol 1, Chapter 2.

(15) Lerch, R. N.; Donald, W. W. *J. Agric. Food Chem.* **1994**, *42*, 922-927.

(16) Vermeulen, N. M. J.; Apostolides, Z.; Potgieter, D. J. J.; Nel, P. C.; Smit, N. S. H. *J. Chromatogr.* **1982**, *240*, 247-253.

(17) Shiu, W. Y.; Ma, K. C.; Mackay, D.; Seiber, J. N.; Wauchope, R. D. In *Reviews of Environmental Contamination and Toxicology*; Ware, G. W. Ed.; Springer-Verlag: New York, 1990; Vol. 116, pp. 15-187.

(18) Russell, J. D.; Cruz, M.; White, J. L.; Bailey, G. W.; Payne, W. R.; Pope, J. D.; Teasley, J. I. *Science.* **1968**, *160*, 1340-1345.

(19) Skipper, H. D.; Volk, V. V.; Frech, R. *J. Agric. Food Chem.* **1976**, *24*, 126-129.

(20) Jordan, L. S.; Farmer, W. J.; Goodin, J. R.; Day, B. E. In *Residue Reviews*; Ware, G. W., Ed.; Springer-Verlag: New York, 1970; Vol. 32, pp. 267-286.

(21) Stratton, G. W. *Arch. Environ. Contam. Toxicol.* **1984**, *13 (1), 35-42.*

(22) Steinheimer, T. R.; Ondrus, M. G. *Water Resour. Invest. (U.S. Geol. Surv.)* **1990**, *No.89-4193.*

(23) Steinheimer, T. R. *J. Agric. Food Chem.* **1993**, *41*, 588-595.

(24) Wenheng, Q.; Schultz, N. A.; Stuart, J. D.; Hogan, J. C.; Mason, A. S. *J. Liq. Chromtogr.* **1991**, *14(7)*, 1367-1392.

(25) Lerch, R. N.; Donald, W. W.; Li, Y-X; Alberts, E. E. *Environ Sci. Technol.* **1995**, *29*, (In Press).

(26) Schlaeppi, J.; Fory, W.; Ramsteiner, K. *J. Agric. Food Chem.* **1989**, *37*, 1532-1538.

(27) Abian, J.; Durand, G.; Barcelo, D. *J. Agric. Food Chem.* **1993**, *41*, 1264-1273.

(28) Pelizzetti, E.; Minero, C.; Vincenti, C. M.; Pramauro, E.; Dolci, M. *Chemosphere* **1992**, *24*, 891-910.

(29) Armstrong, D. E.; Chesters, G.; Harris, R. F. *Soil Sci. Soc. Amer. Proc.* **1967**, *31*, 61-66.

(30) Obien, S. R.; Green, R. E. *Weed Sci.* **1969**, *17(4)*, 509-514.

(31) Li, G.; Felbeck, G. T. *Soil Sci.* **1972**, *114*, 201-209.

(32) Minero, C.; Pramauro, E.; Pelizzetti, E.; Dolci, M.; Marchesini, A. *Chemosphere* **1992**, *24*, 1597-1606.

(33) Mandelbaum, R. T.; Wackett, L. P.; Allan, D. L. *Environ. Sci. Technol.* **1993**, *27*, 1943-1946.

(34) Radosevich, M.; Traina, S. J.; Hao, Y-L.; Tuovinen, O. H. *Appl. Environ. Microbiol.* **1995**, *61*, 297-302.

(35) Gamble, D. S.; Khan, S. U. *J. Agric. Food Chem.* **1990**, 38:297-308.

(36) Hance, R. J. In *Progress in Pesticide Biochemistry and Toxicology*; John Wiley and Sons: Chichester, W. Sussex, 1987; Vol. 6, pp. 223-247.

(37) Khan, S. U. *Pestic. Sci.* **1978**, 9:39-43.

(38) Goldberg, M. C.; Cunningham, K. M.; Squillace, P. J. *Water Resour. Invest. (U.S. Geol. Surv.)* **1991**, *No. 91-4034*, pp. 232-238.

(39) Choudhry, G. G.; Webster, G. R. B. In *Residue Reviews*; Ware, G. W., Ed.; Springer-Verlag: New York, 1985; Vol. 96, pp. 79-136.

(40) Hessler, D. P.; Gorenflo, V.; Frimmel, F. H. *J. Water SRT - Aqua.* **1993**, 42:8-12.

(41) Best, J. A.; Weber, J. B. *Weed Sci.* **1974**, *22*, 364-373.

(42) Kruger, E. L.; Somasundaram, L.; Kanwar, R. S.; Coats, J. R. *Environ. Toxicol. Chem.* **1993**, *12*, 1969-1975.

(43) Capriel, P.; Haisch, A.; Khan, S. U. *J. Agric. Food Chem.* **1985**, *33*, 567-569.

(44) Behki, R. M.; Khan, S. U. *J. Agric. Food Chem.* **1986**, *34*, 746-749.

(45) Thurman, E. M.; Meyer, M.; Pomes, M.; Perry, C. A.; Schwab, A. P. *Anal. Chem.* **1990**, *62*, 2043-2048.

Chapter 20

Assessment of Herbicide Transport and Persistence in Groundwater: A Review

S. K. Widmer[1] and Roy F. Spalding

Water Sciences Laboratory, University of Nebraska–Lincoln,
103 Natural Resources Hall, Lincoln, NE 68583–0844

The mobility and persistence of commonly-detected herbicides and herbicide degrades in groundwater were measured in a natural gradient field transport study. Retardation factors (Rs) were calculated by nonlinear regression of experimental breakthrough curves using two physical-chemical models. Butachlor was the most highly retained herbicide (R=1.7-1.9); all other compounds showed retardation <30% (R<1.3). Alachlor and butachlor exhibited losses of 40% and 70%, respectively, in 50-60 days; all other compounds showed no detectable loss. The results of the field transport study were compared with those obtained from equilibrium sorption estimates. Batch equilibration procedures assuming a linear model tended to overpredict retention; the Freundlich model provided a more accurate prediction of herbicide transport at low concentrations.

Pesticide detections at trace concentrations in groundwater are widespread, and contamination is typically correlated to leaching vulnerability. Areas with permeable soils, short distances to groundwater, and heavy application of herbicides are especially susceptible to nonpoint source contamination, as found throughout the Corn Belt of the United States (1,2).

The transport of organic solutes in groundwater has been measured in both laboratory and field transport studies. The transport of reactive solutes is retarded relative to the rate of groundwater flow. A retardation factor (R) is a measure of a solute's retention in the system of interest, and indicates the extent of interaction between a solute and the matrix. Several methods of calculating R from transport data have been reported, including using the center of mass of breakthrough curves (3-5), the time required to attain a relative concentration of

[1]Current address: Department of Marine Chemistry and Geochemistry, Woods Hole Oceanographic Institution, Woods Hole, MA 02543–1543

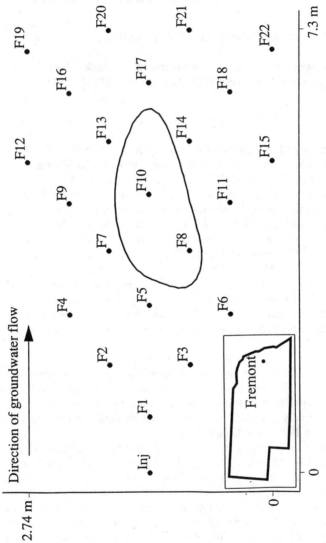

Fig. 1. Plan view of study site with bromide plume ($C/C_0=0.20$) observed in the first injection experiment 1 month after injection. Inset shows the location of the study site.

0.5 *(6)*, the position of the peak maximum *(4)*, or estimating R as a parameter in nonlinear regression of breakthrough curves using the advection-dispersion equation *(7)*, the model often used to describe solute transport in a porous medium.

Equilibrium sorption estimates are often used in groundwater transport models to predict field behavior *(8-10)*. Sorption isotherms often follow a linear model, where the sorbed concentration is directly proportional to the solution concentration (C); the slope of the isotherm is the partition coefficient (K_d). Nonlinear isotherms are often described by the Freundlich or Langmuir model. A retardation factor may be calculated from sorption isotherm parameters:

$$R = 1 + \frac{\rho_B K_d}{n} \qquad \textit{Linear isotherm} \tag{1}$$

$$R = 1 + \frac{\rho_B}{n} K_f b C^{b-1} \qquad \textit{Freundlich isotherm} \tag{2}$$

$$R = 1 + \frac{\rho_B \alpha \beta}{n (1 + \alpha C)^2} \qquad \textit{Langmuir isotherm} \tag{3}$$

where ρ_B is the bulk density and n the porosity of the aquifer matrix, K_f is the Freundlich partition coefficient and b is the Freundlich isotherm power (the value of b may be either <1 or >1, resulting in a concave upward or downward isotherm), α indicates the binding strength and β is the sorption maximum. Retardation factors for Freundlich and Langmuir isotherms are concentration-dependent, complicating their use in analytical models.

This study was conducted to: (i) determine the sequential mobility of several commonly-detected herbicides and herbicide degradates under natural-gradient conditions in a sand and gravel aquifer; (ii) determine the persistence of the compounds under aquifer conditions; (iii) calculate retardation factors for each compound using several methods and two nonlinear models, and (iv) compare retardation estimates as determined by batch equilibration with those obtained in the field study.

Materials and Methods

Field Transport Study. The study was conducted in a shallow fluvial Quaternary Age sand and gravel aquifer near Fremont, Nebraska (Figure 1). Shallow aquifer hydraulic conductivities (K) range from about 30-60 m d^{-1} *(11)*. The measured average rate of groundwater flow was 0.15 m d^{-1} based on the transport of the bromide tracer. The water table was at a depth of approximately 2-3 m and the hydraulic gradient was 0.0013 m m^{-1} S 60° E.

A system of multilevel samplers (MLSs) was used to delineate the solute plumes. Fences of MLSs were arranged in arcs, such that 8 arcs were longitudinally located within the 7.3 m monitored (Figure 1). The construction and installation of MLSs are described in Widmer and Spalding *(12)*. The injection well was a 0.10 m diameter PVC pipe screened from 3.7 to 4.3 m. This well was located 0.9 m upgradient from the nearest MLS (Figure 1).

Injection and Monitoring of Solutes. Two injection studies were conducted, the first including 15 mg L^{-1} bromide as a conservative tracer, 3 µg L^{-1} atrazine, 2 µg L^{-1} alachlor, 10 µg L^{-1} cyanazine, and 10 µg L^{-1} metolachlor. These are the four most heavily-applied herbicides in Nebraska and are detected in groundwater at similar concentrations *(2)*. The second injection included 10 mg L^{-1} bromide, 1 µg L^{-1} butachlor and 3 µg L^{-1} atrazine, deethylatrazine (DEA), and deisopropylatrazine (DIA). Background concentrations of all compounds were less than the laboratory's quantitation limit. Herbicides and NaBr were dissolved in 1300 L of native groundwater; 950 L were injected in 7 hours (first injection, 1991) or 10 hours (second injection, 1992). The resulting plumes were monitored for 2 months (first injection) or 3 months (second injection).

The injected mass remained within the area monitored by the MLSs throughout the experiment (Figure 1). The plumes ranged less than 2 m in the vertical and transverse directions. Several million liters were removed from the aquifer upon completion of each experiment to remove the introduced herbicides.

Chemical Analyses. Bromide analysis was conducted using an ion-selective electrode coupled with a single-junction reference electrode. Herbicides were analyzed using solid-phase extraction and isotope dilution GC/MS *(13)*.

Data Analyses. Breakthrough curves (plots of concentration as a function of time) were constructed for each solute at each monitoring point. The advection-dispersion equation with finite pulse input (derived from *14*) was fitted to the experimental breakthrough curves.

$$A(t) = \frac{C_0}{2} \left[erfc \left(\frac{x - vt}{\sqrt{4Dt}} \right) + exp \left(\frac{vx}{2D} \right) erfc \left(\frac{x + vt}{\sqrt{4Dt}} \right) \right] \qquad (4)$$

C = A(x,t) for t$\leq$t$_0$
C = A(x,t) - A(x,t-t$_0$) for t$>$t$_0$
where C is the solute concentration (µg L^{-1}), C$_0$ is the mean injected concentration (µg L^{-1}), x is the distance from the injection well (m), v is the transport velocity (m d^{-1}), t is time (d), t$_0$ is the pulse width (d), and D is the hydrodynamic dispersion coefficient (m^2 d^{-1}).

Nonlinear regression of bromide data using equation 4 provided estimates of v, t$_0$, and D at each sampling point. For retarded solutes, v and D are interpreted as the apparent values v* and D* (D*=D/R). Nonlinear regression of herbicide data gives an estimate of the apparent solute velocity (v*=v/R). Little transverse dispersion was observed at this site, and no transverse dispersion term was included in equation 4. Only breakthrough curves obtained from MLSs along the center line of flow, where the theory holds, were analyzed using the advection-dispersion equation.

Experimental breakthrough curves were also analyzed using an exponentially-modified Gaussian (EMG) equation. This equation is the standard model used to fit asymmetric chromatographic peaks *(15-17)*.

$$C(t) = \frac{a}{2\tau} \, exp \left[\frac{\sigma^2}{2\tau^2} + \frac{t_r - t}{\tau} \right] \left[erf \left(\frac{t - t_r}{\sqrt{2}\,\sigma} - \frac{\sigma}{\sqrt{2}\,\tau} \right) + 1 \right] \tag{5}$$

where C(t) is the concentration of solute as a function of time, a is the peak area, t_r is the retention time (defined at the peak maximum), σ is the width of the Gaussian peak, and τ is the exponential time constant, which describes the extent of tailing *(18)*. Breakthrough curves from all MLSs were analyzed using equation 5.

Retardation. Comparison of the breakthrough curves for the conservative tracer (bromide) and each herbicide at a sampling point provides a measure of herbicide retention in the aquifer. An R may be calculated as the ratio of the time required for a solute to attain half-maximum concentration as compared to that for a conservative tracer:

$$R = \frac{t_{half-maximum\ concentration,\ herbicide} - t_i}{t_{half-maximum\ concentration,\ bromide} - t_i} \tag{6}$$

where t_i is the time of the start of the injection and is identical for both herbicide and tracer, since the injections were simultaneous. This method of calculating R provides an estimate of the difference in breakthrough times, and is equivalent to considering a relative concentration of 0.5 if the maximum concentration is 1.0, as used by Agertved et al. *(6)*.

The centroid (center of mass) of the fitted curves can also be used to calculate a retardation factor, where R is the ratio of time required for the centroid of a herbicide plume (in the x-dimension) to pass a monitoring point compared to that for bromide. In chromatographic systems, the retention time may be based on the peak centroid or on the peak maximum *(16)*. When estimates of solute velocities are available, as in the case of nonlinear regression using the advection-dispersion equation, the retardation factor may be calculated as the ratio of the bromide to herbicide velocity; the bromide velocity indicates the groundwater flow velocity. A measure of retardation may also be calculated from first moment of the data *(19)* which requires no curve fitting; the ratio of moments (herbicide to bromide) gives R.

Peak tailing is taken into account in measurements of the center of mass (peak centroid, velocity, and first moment); estimates of R calculated from the center of mass may be useful in estimating the pumping time required to effectively remove a contaminant plume from an aquifer (as in a pump-and-treat remediation system).

Persistence. Persistence of injected compounds is indicated by conservation of mass. In the case of nonlinear regression, persistence is demonstrated by constant peak areas with time. Peak area can be determined from the raw data by the zeroth moment *(3)*. The relative area, normalized to the conservative tracer, is 1 for a persistent solute. Relative areas <1 indicate removal from solution by degradation and/or irreversible sorption. An estimate of loss is given by the reduction in relative area.

Alternatively, a degradation term may be included in the advection-dispersion equation to provide an estimate of persistence *(20)*

$$C = B(x,t) \text{ for } t \le t_0$$

$$B(x,t) = \frac{C_0}{2} \left[exp\left(\frac{vx(1-m)}{2D}\right) erfc\left(\frac{Rx-mvt}{\sqrt{4DRt}}\right) + \right.$$

$$\left. exp\left(\frac{vx(1+m)}{2D}\right) erfc\left(\frac{Rx+mvt}{\sqrt{4DRt}}\right) \right] \qquad (7)$$

$$C = B(x,t) - B(x,t-t_0) \text{ for } t > t_0$$

where R is the retardation factor and m is defined as

$$m = \sqrt{1 + \frac{4kD}{v^2}}$$

where k is the first-order degradation rate constant. Nonlinear regression using this model estimates two variables: R and k, from which a half-life $(t_{1/2})$ may be calculated, assuming first-order kinetics.

Laboratory Sorption Study. Procedures summarized in Roy et al. *(21)* were used to quantify the sorption of the aforementioned herbicides and degradates to the aquifer matrix and particle size separates. Both variable and constant soil:solution ratio methods were used; solution concentrations (approximately 1-20 $\mu g\ L^{-1}$) were consistent with concentrations used in the field transport study.

Sediment Sample Collection and Analysis. Sediment samples were collected from the field site using a hollow-stem auger. Samples were collected from depths 3 to 4 m below the ground surface. Sediments were fractionated using wet-sieving to separate the sands (63 μm-2 mm) from gravels and fines. The silt fraction (2-63 μm) was separated from the clay-sized fraction (<2 μm) by timed centrifugation *(22)*. The Oceanographic International ampule system for sediments was used to measure the total organic carbon (TOC) content of the aquifer matrix and the particle size separates. X-ray diffraction was used to determine the mineralogy of the clay fraction.

Batch Equilibrations. For variable soil:solution ratio tests, ratios of 1:1, 1:2, and 1:3 were used for whole sediments and sand fraction; 1:2, 1:4, and 1:6 for the silt fraction; and 1:20, 1:30, and 1:40 for the clay fraction. Triplicate samples were used for all but the clay fraction, where the small amount of material provided only one sample per ratio. The solution contained approximately 10 µg L^{-1} of each compound in distilled, deionized water. Samples were equilibrated on a thermostated water bath shaker (25 °C) for 24 h (48 h for clay fraction). Herbicide concentrations in the supernatant were determined using solid-phase extraction and GC/MS *(13)*. Sorption was determined from the decrease in solution concentration between samples and blanks (no sediment added).

For constant soil:solution ratio tests, a 1:4 ratio was used (1:20 for clay fraction); solution concentrations were approximately 1, 5, 10, and 20 µg L^{-1}. Duplicate samples were used, except for the clay fraction, for which a single sample for two solution concentrations (~10 and 20 µg L^{-1}) was used. All samples were equilibrated for 24 h.

Data Analysis. Linear, Freundlich, and Langmuir isotherm models were fitted to the sorption data for each compound. A one-dimensional finite-difference model was used to determine the effect of different isotherm equations on predicted breakthrough curves. The model includes degradation and sorption and requires inputs of transport parameters *(23)*. Velocity, dispersion, pulse width, and degradation rates determined from nonlinear regression of the field data (equation 4 or 7) were used as model inputs.

Results and Discussion

Field Transport Study. Breakthrough curves were constructed for each compound at each sampling point *(12)*. Typical breakthrough curves for the 1992 injection are shown in Figure 2. Initial breakthrough curves near the point of injection showed little resolution between solutes; resolution and peak broadening increased as travel time and distance increased. For a conservative (non-sorbed, non-degraded) solute, peak broadening is due to dispersion, and peak area remains constant. For a non-conservative solute, broadening may be due to dispersion and chemical interactions, and peak area (mass) may be reduced if the solute undergoes transformation and/or irreversible sorption.

Retardation. Retardation factors were calculated for each compound using each of the seven approaches discussed above (half-maximum for both models, centroid of each model, peak maximum for EMG model, velocity ratio, and first moment). Comparison of retardation factors shows small differences due to the method of calculation both within and between models *(24)*. For the 1991 study, differences between the herbicide classes ($\alpha=0.05$) were observed

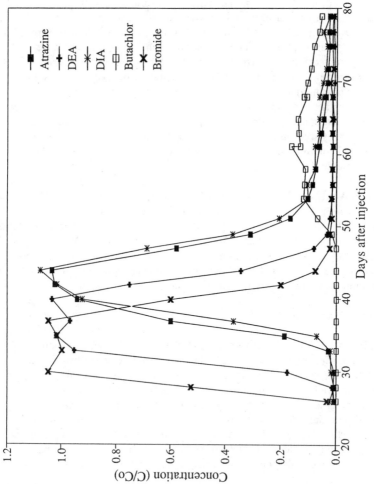

Fig. 2. Set of breakthrough curves for sampler F10 (4.6 m from the injection well) at a depth of 3.65 m; 1992 injection.

for all methods of calculating R except the centroid of the advection-dispersion equation: The triazines (atrazine and cyanazine) were retained to a greater extent than the acetanilides (alachlor and metolachlor) (Table I). All herbicide Rs were equal when using the advection-dispersion equation and the centroid approach. Metolachlor and cyanazine Rs were not significantly different as given by first moment or the centroid of the EMG model. No significant differences in herbicide Rs were detected using the peak-maximum of the EMG model *(24).*

All herbicide Rs were significantly different in the 1992 study using the first moment approach. Retardation factors for atrazine and DIA were not significantly different using the centroid approach with either the advection-dispersion or EMG model; all other herbicides were significantly different *(24).*

The observed retardation factor of 1.2 for atrazine is consistent with other studies *(6,25).* Although R values need not be constant in different aquifer matrices, the similarity of atrazine Rs may be due to similar reported organic carbon contents, resulting in little retention of atrazine.

Table I. Retardation factors[†] for injected herbicides calculated using seven methods discussed in the text

Injected Compound	Mean ± standard error	Range
Atrazine (1991)	1.17 ± 0.05	1.12 - 1.25
Atrazine (1992)	1.19 ± 0.05	1.13 - 1.24
DEA	1.08 ± 0.02	1.05 - 1.10
DIA	1.22 ± 0.05	1.15 - 1.26
Cyanazine	1.16 ± 0.06	1.07 - 1.25
Alachlor	1.12 ± 0.05	1.08 - 1.19
Metolachlor	1.16 ± 0.04	1.09 - 1.23
Butachlor	1.74 ± 0.09	1.65 - 1.91

[†]Values reported are least-square means (± standard error) of seven Rs, each calculated from the least-square mean of 5-26 monitoring points (as reported in ref. *24).*

Comparison of the fitted curves shows that both the EMG and the advection-dispersion model fit the data well *(24).* The EMG model better describes the asymmetry demonstrated in the breakthrough curves. Considering the utility of the advection-dispersion equation in estimating other transport parameters, such as velocity and dispersion, this model will generally be preferred. Care must be taken in choosing the analytical solution to the advection-dispersion to be used in modeling efforts. Considerations include the source geometry and temporal characteristics, the nature of the flow field, the extent of the medium, properties of the porous medium, and whether equilibrium prevails during transport.

Persistence. Atrazine, DEA, DIA, cyanazine, and metolachlor showed no detectable loss from solution *(12)*. The resistance of atrazine to biotic degradation is verified by non-increasing levels of the metabolites DEA and DIA above background (for the 1991 injection) or above that injected (in the 1992 study). Based on the decrease in relative area, observed losses of alachlor and butachlor were approximately 40% and 70%, respectively, in 50-60 d *(12)*. Loss from solution due to irreversible sorption were not distinguished from those due to degradation; no metabolites of alachlor or butachlor were analyzed.

Losses approximated by the relative area (the zeroth moment) agreed with estimates of the decrease in area contained under the EMG model. The area under the advection-dispersion model was less than that given by the zeroth moment, since the advection-dispersion model did not describe the tailing portion of the curves well. Estimates of herbicide half-lives from regression using equation 7 predict more loss of alachlor and butachlor than was observed *(24)*.

Dispersion. Estimates of longitudinal hydrodynamic dispersion coefficients (Ds) were given by nonlinear regression of bromide data, and ranged from 0.0007 to 0.04 m^2 d^{-1} *(24)*. No correlation between D and depth, distance from the injection well, particle size, or velocity were observed ($r \leq 0.6$) *(24)*.

Laboratory Sorption Study. Aquifer material characterization. The organic carbon content decreased as particle size increased: The mean TOC content of the sand, silt, and clay fraction were 43, 280, and 7200 μg C (g soil)$^{-1}$, respectively. The TOC content of the whole sediment was approximately 62 μg C (g soil)$^{-1}$. While the clay fraction comprises a very small portion of the sample weight (<0.2%), its organic carbon content is very high relative to the larger size fractions. Illite is the predominant clay mineral.

Batch Equilibrations. Equilibrium sorption estimates using variable soil:solution ratios indicated that the clay fraction was about 10 times more sorptive than the silt fraction for all compounds but DIA. The sand fraction and whole sediments showed minimal sorption capacity, and sorption was not significantly greater than zero for most compounds. The sorptive characteristics of the whole sediment were dominated by those of the sand fraction, since sand comprises about 95% of the total sample weight. The silt fraction showed significant (α=0.05) sorption of all compounds. Sorption was weakly correlated to organic carbon contents (r=0.44) (Widmer, S.K.; Spalding, R.F. *J. Contam. Hydrol.*, submitted).

For the whole sediments and silt fractions, the triazines atrazine and cyanazine were retained to a lesser extent than the acetanilides alachlor and metolachlor. This contradicts the results of the 1991 field study, where the triazines exhibited greater retardation than the acetanilides. These discrepancies may be due to the lack of equilibrium in advectively flowing groundwater, aquifer heterogeneity, and a complex pore-space structure in field situations. The accessibility of transient solute particles to small pores may be limited in flowing groundwater, reducing sorption and retention as compared to a batch equilibration system, where all particles are accessible through destruction of the sediment structure and continuous agitation.

Batch equilibrations using a constant soil:solution ratio allowed for the construction of sorption isotherms (Widmer, S.K.; Spalding, R.F. *J. Contam. Hydrol.*, submitted). Sorption isotherms for alachlor on the whole sediment and each particle size separate are shown in Figure 3. For herbicides exhibiting appreciable sorption, isotherms appeared linear, with r^2s ≥ 0.7 for whole sediments and ≥ 0.85 for the silt fraction. Few significant differences between herbicides were observed for the whole sediment and sand fraction, and only butachlor exhibited significant sorption ($\alpha=0.05$). The silt fraction demonstrated higher sorptivity; sorption of butachlor, alachlor, cyanazine, DIA, and metolachlor was significantly greater than zero.

The validity of the local equilibrium assumption (LEA) was tested for alachlor and butachlor transport, using the criteria given by Bahr *(26)*. Based on the time of transport, the LEA should hold, since the travel time was about two orders of magnitude greater than the equilibration time. The LEA may be violated in this study on the basis of the "dispersion time" (dispersivity divided by velocity), which should be at least five times greater than the equilibration time for the LEA to hold *(26)*.

Sorption coefficients could not be normalized to the organic carbon content of the particle size separates to produce a constant K_{oc} for any herbicide showing measurable sorption. Estimates of K_{oc}s for herbicides sorbing to the whole sediment and silt fraction were an order of magnitude greater than those reported for surface soils *(27)*. This is probably due to their very low TOC content and low sorption. For the clay fraction, K_{oc}s tended to be much lower than those reported for surface soils *(27)*. Since the TOC content of the clay-sized particles (approaching 1%) is above the threshold proposed for the validity of normalized sorption *(28)*, this result indicates that the organic carbon associated with the clay fraction of the aquifer material is a less effective sorbent than the organic carbon of surface soils.

Introducing nonlinearity by fitting the Freundlich model improved the goodness-of-fit (as measured by the root mean squared error, RMSE) for cyanazine and butachlor isotherms, but showed a decrease in RMSE for all other compounds sorbing to the whole sediment. For the silt fraction, the linear model was as good or better at describing all sorption isotherms except those of cyanazine. Only two points were available for the clay fraction, prohibiting investigation of isotherm nonlinearity. The Langmuir model did not describe the data as well as either the linear or Freundlich models.

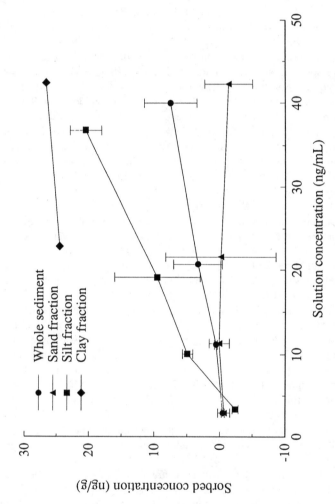

Fig. 3. Sorption isotherms for alachlor determined by batch equilibration. Error bars indicate standard deviation of duplicate samples.

Comparison to Results of the Field Transport Study. Retardation factors were calculated for each solute based on isotherm parameters (Table II). Bulk density was taken as 1.6 g cm^{-3} and porosity as 0.39 *(11)*. The retardation of three herbicides was overpredicted by batch equilibration using a linear model compared to the results of the field study: The minimum R value (mean minus the standard error) indicates that metolachlor retention is overpredicted by 65%, cyanazine by 73%, and butachlor by 340%. Two of the more mobile compounds, atrazine and DEA, showed no significant sorption in the laboratory study, and their retention is underpredicted by batch equilibration.

For nonlinear isotherms, where R is a function of concentration, sorption can be characterized at a single solution concentration *(29)*. Retardation factors presented in Table II were calculated using the mean injected concentration used in the field study. For a pulse input of solute, however, the solution

Table II. Retardation factors determined from whole sediment sorption isotherms ± standard error. Standard errors were computed from the uncertainty associated with each parameter, assuming 25% error in ρ_B and n.

Herbicide	Linear R intercept≠0	Linear R intercept=0[†]	Freundlich R[‡]	First Moment R[§]
DIA	1.7 ± 1.6	1.4 ± 1.3	1.4 ± 10.2	1.23 ± 0.004
DEA	0.8 ± 1.9	1.0 ± 5.6	1.1 ± 15.8	1.07 ± 0.005
Atrazine	1.1 ± 5.1	1.0 ± 130	1.0 ± 25.7	1.16 ± 0.04
Alachlor	1.9 ± 0.9	1.7 ± 0.7	1.3 ± 3.5	1.04 ± 0.004
Cyanazine	3.3 ± 1.4	2.1 ± 2.3	1.6 ± 53.9	1.10 ± 0.004
Metolachlor	3.0 ± 1.2	2.4 ± 1.0	3.3 ± 6.0	1.09 ± 0.004
Butachlor	11.7 ± 4.4	8.5 ± 3.2	4.6 ± 3.6	1.65 ± 0.07

† Langmuir model produced the same fit as linear model with intercept forced through zero except for DIA (R=1.6±33.9), DEA (R=1.3±34.5), and atrazine (R=1.1±54.8).
‡ Retardation factors calculated using solution concentration injected in the field study.
§ Retardation factors from field study as reported in ref. *12*. Standard errors indicate deviation in multiple measures of R (at multiple sampling points).
SOURCE: Adapted from Widmer, S.K.; Spalding, R.F. (J. Contam. Hydrol., submitted).

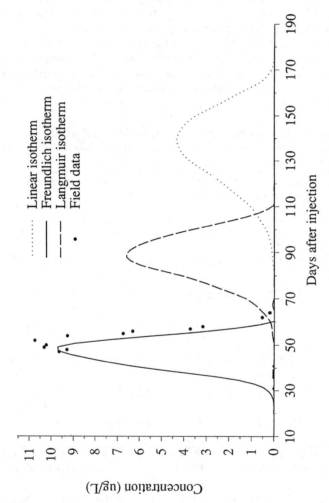

Fig. 4. Observed (in field study) and predicted (from isotherm parameters) breakthrough curves for cyanazine at a monitoring point 4.6 m from the injection well.

concentration is always changing at a single monitoring point. To investigate the effect of isotherm nonlinearity on predicted retention, a one-dimensional finite-difference transport model (BIO1D, GeoTrans, Inc.) was used to predict breakthrough curves from isotherm parameters (Widmer, S.K.; Spalding, R.F. *J. Contam. Hydrol.*, submitted). Observed (field data) and predicted breakthrough curves for butachlor for a monitoring point 4.56 m from the injection well show that the linear model overpredicts the estimated transport time by several hundred days. The Langmuir model is slightly more accurate, but overpredicts retention by <100 d. The Freundlich model provides a relatively accurate prediction of butachlor transport. The same trend was observed for other compounds (Figure 4). The Freundlich model provided a better description of solute transport even when the linear model appeared to give the best description of the sorption isotherm, indicating the importance of the concentration-dependence of R in estimating retardation at low solution concentrations.

Conclusions

This research demonstrates the utility of carefully-controlled field transport studies in obtaining accurate estimates of contaminant mobility and persistence *in situ.* The transport of commonly-detected herbicides and herbicide degradates was only minimally retarded relative to a conservative tracer. Structurally-similar compounds demonstrated differing transport velocities: DEA was more mobile than atrazine, while DIA was less mobile. These results are consistent with the observed DEA/atrazine ratios reported in groundwater.

Breakthrough curves obtained from a field transport study were analyzed using nonlinear regression and two physical-chemical models. The advection-dispersion equation provides meaningful estimates of groundwater flow velocity and dispersion, and its use is generally preferred. An empirical model, such as one often used in chromatographic systems, may be useful in analyzing data from an experiment including a controlled, finite input, relatively long travel distances, and slow flow velocities. Small differences were observed in retardation factors calculated from various peak parameters. The method of determining transport parameters may become more important as retardation or tailing increase.

Batch equilibration tended to overestimate retention as compared to that observed in the field study. Extrapolation beyond the conditions of the sorption experiment may produce invalid estimates of retention in low organic carbon matrices. The linear model is simple to employ in transport models and is therefore often used to predict solute transport; however, the concentration-dependence of sorption may become important in determining the transport of solutes at low solution concentrations, as are often found in groundwater contaminated by nonpoint sources.

Acknowledgments

This research was funded through the USDA-Cooperative State Research Service, the Nebraska Research Initiative, and the University of Nebraska Medical Center's Environmental Toxicology and Carcinogenesis Graduate Program. The authors wish to thank the staff of the Water Sciences Laboratory for assistance with well installation, groundwater sampling, and chemical analyses. Dr. Joseph Skopp assisted with nonlinear regression. Dr. Steven Comfort, Dr. Joseph Skopp, and Dr. Patrick Shea provided editorial review. The authors acknowledge the City of Fremont Department of Utilities, under the direction of John McCafferty, for their cooperation.

Literature Cited

1. Kolpin, D. W.; Goolsby, D. A.; Aga, D. S.; Iverson, J. L.; Thurman, E. M. In *Selected paper on agricultural chemicals in water resources of the Midcontinental United States;* Goolsby, D. A.; Boyer, L. L.; Mallard, G. E., Eds.; USGS Open-File Report 93-418; USGS: Denver, CO, 1983; pp 64-73.

2. Spalding, R. F.; Burbach, M. E.; Exner, M. E. *Ground Water Monit. Rev.* **1989**, *9,* 126-133.

3. Roberts, P. V.; Goltz, M. N.; Mackay, D. M. *Water Resour. Res.* **1986**, *22,* 2147-2158.

4. Winters, S. L.; Lee, D. R. *Environ. Sci. Technol.* **1987**, *21,* 1182-1186.

5. Bianchi-Mosquera, G. C. *In situ determination of transport parameters for organic contaminants in ground water;* Ph.D. dissertation, University of California, Los Angeles, CA, 1993.

6. Agertved, J.; Rügge, K.; Barker, J. F. *Ground Water.* **1992**, *30,* 500-506.

7. Kookana, R. S.; Schuller, R. D.; Aylmore, L. A. G. *J. Contam. Hydrol.* **1993**, *14,* 93-115.

8. Voss, C. I. *A finite-element simulation model for saturated-unsaturated, fluid-density-dependent ground-water flow with energy transport or chemically-reactive single-species transport;* USGS Water-Resources Investigations Report 84-4369; USGS: Reston, VA, 1984.

9. Goode, D. J.; Konikow, L. F. *Modification of a method-of-characteristics solute-transport model to incorporate decay and equilibrium-controlled sorption or ion-exchange;* USGS Water-Resources Investigations Report 89-4030; USGS: Reston, VA, 1989.

10. S. S. Papadopulos and Associates, Inc. *MT3D: A modular three-dimensional transport model;* S. S. Papadopulos and Associates, Inc.: Rockville, MD, 1991.

11. Ferlin, M. A. *Slug tests in highly permeable aquifers;* M. S. Thesis, University of Nebraska, Lincoln, NE, 1993.

12. Widmer, S. K.; Spalding, R. F. *J. Environ. Qual.* **1995**, *24,* 445-453.

13. Cassada, D. A.; Spalding, R. F.; Cai, Z.; Gross, M. L. *Anal. Chim. Acta.* **1994**, *287,* 7-15.

14. Lapidus, L.; Amundson, N. R. *J. Phys. Chem.* **1952**, *56*, 984-988.
15. Barber, W. E.; Carr, P. W. *Anal. Chem.* **1981**, *53*, 1939-1942.
16. Foley, J. P.; Dorsey, J. G. *Anal. Chem.* **1983**, *55*, 730-737.
17. Olivé, J.; Grimalt, J. O. *J. Chrom. Sci.* **1991**, *29*, 70-77.
18. Rundel, R. *Peakfit Technical Guide version 3.0;* Jandel Scientific: San Rafael, CA, 1991.
19. Levenspiel, O. *The chemical reactor omnibook;* O. Levenspiel: Corvallis, OR, 1979.
20. Van Genuchten, M. Th.; Alves, W. J. *Analytical solutions of the one-dimensional convective-dispersive solute transport equation;* USDA Technical Bulletin 1661; USDA: Washington, DC, 1982.
21. Roy, W. R.; Krapac, I. G.; Chou, S. F. J.; Griffin, R. A. *Batch-type procedures for estimating soil adsorption of chemicals;* EPA Technical Resource Document PB92-188515; EPA: Washington, DC, 1991.
22. Jackson, M. L. *Soil chemical analysis -- Advanced course;* 2nd ed.; M. L. Jackson: Madison, WI, 1975.
23. Srinivasan, P.; Mercer, J. W. *BIO1D, One-dimensional model for comparison of biodegradation and adsorption processes in contaminant transport;* GeoTrans, Inc.: Washington, DC, 1987.
24. Widmer, S. K.; Spalding, R. F.; Skopp, J. *J. Environ. Qual.* **1995**, *24*, 439-444.
25. Schneider, A. D.; Wiese, A. F.; Jones, O. R. *Agron. J.* **1977**, *69*, 432-436.
26. Bahr, J. M. *Transport and mass exchange processes in sand and gravel aquifers: Field and modelling studies, Int. Conf. and Worksh.;* Atomic Energy Canada: Ottawa, ON, 1990.
27. Wauchope, R. D.; Butler, T. M.; Hornsby, A. G.; Augustijn-Beckers, P. W. M.; Hurt, J. P. *Rev. Environ. Contam. Toxicol.* **1992**, *123*, 1-164.
28. Schwarzenbach, R. P.; Westall, J. *Environ. Sci. Technol.* **1981**, *15*, 1360-1367.
29. Sanchez-Martin, M. J.; Sanchez-Camazano, M. *J. Agric. Food Chem.* **1984**, *32*, 720-725.

Chapter 21

Cyanazine, Atrazine, and Their Metabolites as Geochemical Indicators of Contaminant Transport in the Mississippi River

M. T. Meyer[1,3], E. M. Thurman[1], and D. A. Goolsby[2]

[1]U.S. Geological Survey, 4821 Quail Crest Place, Lawrence, KS 66049
[2]U.S. Geological Survey, Denver Federal Center, Building 25, Denver, CO 80225

The geochemical transport of cyanazine and its metabolite cyanazine amide (CAM) was compared to atrazine and its metabolite deethylatrazine (DEA) at three sites in the Mississippi River basin during 1992 and six sites during 1993. The floods of 1993 caused an uninterrupted exponential decline in herbicide concentrations; whereas, in 1992 herbicide concentrations varied mostly in response to two discrete discharge pulses in the spring and midsummer and were stable during an extended period of summer low-flow. Concentration half-lives calculated from the 1993 data for atrazine were approximately twice those of cyanazine at all sites. The half-life for atrazine and cyanazine was shortest, 22 and 14 days, respectively at the Mississippi River at Clinton, Ill. -- the farthest upstream site -- and longest, 42 and 22 days, respectively, at the Baton Rouge, La. site -- the farthest downstream site. The concentration of CAM exceeded the concentration of DEA through September at all sites where the mean ratio of atrazine-to-cyanazine (ACR) was less than 4.0. The ratio of CAM-to-cyanazine (CAMCR) increased from 0.2 to more than 1.0 and the ratio of DEA-to-atrazine (DAR) increased from less than 0.1 to 0.3 from application in May through early to mid-July. Temporal changes in the CAMCR were used to identify pre- and post-application "slugs" of water transported along the reaches of the Mississippi River.

Cyanazine and atrazine (Fig. 1) are widely applied pre-emergent herbicides in the midcontinental United States (1). Since the late 1980's the usage of cyanazine has increased from approximately one-fourth to one-half that of atrazine (1-3). Because cyanazine is less persistent than atrazine, it degrades to a suite of compounds that are readily transported to surface water (4-6). The study of herbicide metabolites in

[3]Current address: U.S. Geological Survey, 3916 Sunset Ridge Road, Raleigh, NC 27607

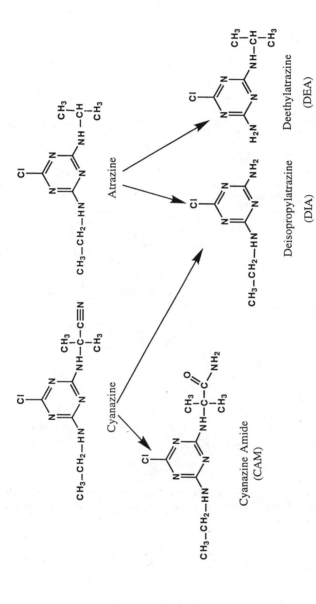

Figure 1: Structure of cyanazine, atrazine, and three of their metabolites: cyanazine amide (CAM), deethylatrazine (DEA), and deisopropylatrazine (DIA).

Figure 2. Map showing the sampling sites for the 1992 and 1993 Mississippi River basin study. Sites with open triangles were sampled in 1992 and all sites were sampled in 1993.

surface water and ground water has been heavily weighted towards two metabolites, deethylatrazine (DEA) and deisopropylatrazine (DIA), ascribed primarily to the degradation of atrazine (Fig. 1). Recent studies have shown that metabolites of other widely used herbicides are frequently detected in surface water (*4-8*) and ground water (*9-10*). Furthermore, herbicide metabolites can account for a significant amount of the total herbicide pool. For example, a regional herbicide study of reservoirs showed that herbicide metabolites accounted for approximately 50 percent of the total herbicide load (*7*).

The Mississippi River is the main conduit for the surface-water drainage of two-thirds of the midwestern United States (*4*). Therefore, regional application and degradation patterns of cyanazine and atrazine should be reflected in their aggregate drainage into the Mississippi River. Cyanazine, atrazine, and two of their metabolites, cyanazine amide (CAM) and deethylatrazine (DEA), respectively, were studied at three sites in the Mississippi River in 1992 and six sites in 1993 (Figure 2). The purpose of this study was to compare the temporal and spatial variation of cyanazine, atrazine, CAM, and DEA and assess their use as geochemical indicators of contaminant transport processes in the Mississippi River.

Experimental Methods

Study Design. In 1992 two sites, the Mississippi River at Thebes, Illinois, and the Ohio River at Grand Chain, Illinois, were sampled approximately every 2 weeks from late-April through July. The Mississippi River at Baton Rouge, Louisiana, was sampled from mid-April through November. These same three sites were sampled in 1993 along with the Mississippi River at Clinton, Illinois, the Illinois River at Valley Falls, Illinois, and the Missouri River at Hermann, Missouri. Water samples were collected from July through September in 1993 at all but the Mississippi River at Baton Rouge site where samples were collected from July 1993 to March 1994.

Analytical Methods. Herbicides and selected metabolites were isolated from water samples and analyzed for triazine herbicides and metabolites using solid-phase extraction (SPE) after the methods of Thurman and others (*11*) and Meyer and others (*12*). Water samples were spiked with 100 µL of a 1.23 ng/µL solution of terbuthylazine as a surrogate standard. A millilab 1A workstation (Waters, Milford, MA) washed the SPE cartridges with 2 mL of methanol, 6 mL of ethylacetate, 2 mL of methanol, followed by 2 mL of distilled water. Then 100 mL of sample was pumped through the cartridge at a rate of 20 mL/min. The SPE cartridge then was eluted with 2.5 mL of ethylacetate. The eluate then was spiked with 500 µL of internal standard, d10-phenanthrene, at a concentration of 0.2 ng/µL. The sample eluate was subsequently evaporated to a volume of approximately 100 µL using nitrogen evaporation, transferred to a 200 µL glass-lined polystyrene crimp-top vial, and refrigerated at -10 °C until analyzed by GC/MS.

GC/MS analyses for a suite of eleven triazine compounds including atrazine, cyanazine, DEA, and DIA, and two chloro-acetanilide compounds were done on a Hewlett-Packard 5890A GC (Palo Alto, CA) and a 5970A mass selective detector (MSD) using selected-ion monitoring (SIM) after the method of Thurman and others (11). Operating conditions were ionization voltage, 70 eV; ion source temperature 280 °C; electron multiplier, 400 V above autotune; direct capillary interface at 280 °C; initial GC oven temperature, 50 °C; ramp rate 6 °C/min, daily tuned with perflurotributylamine, and 25 ms dwell per ion. The compounds were separated

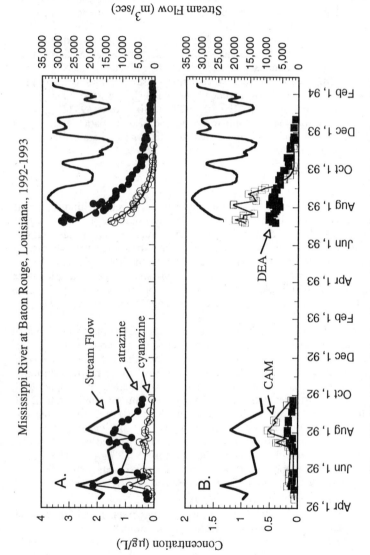

Figure 3A. Variation of stream flow and the concentration of cyanazine and atrazine and B. variation of the concentration of cyanazine amide (CAM) and deethylatrazine (DEA) with time for the Mississippi River at Baton Rouge, Louisiana in 1992 and 1993.

using a Hewlett Packard, 12 m x 0.2 mm i.d., HP-1 or Ultra-1 capillary column with a 0.33-μ methylsilicone film (Palo Alto, CA), with a helium carrier gas at a flow rate of 1 mL/min and a head pressure of 35 kPa. Quantification of the base peak of each compound was based on the response of the 188 ion of the internal standard, phenanthrene-d$_{10}$. Confirmation of the compound was based on the presence of the molecular ion (if present) or the base peak and one to two confirming ions with a retention-time match of ± 0.2 percent relative to phenanthrene-d$_{10}$. The quantitation limit for all of the compounds analyzed by this method was 0.05 μg/L.

For the analysis of the cyanazine amide a Hewlett Packard 12-m x 0.2 mm, HP-2 or Ultra-2 capillary GC column was used (5). The GC/MS conditions for these analyses were the same as above except for the following: direct capillary interface at 210 °C; initial GC oven temperature, 140 °C; ramp rate 15 °C/min to 250 °C. Atrazine, cyanazine, and DEA were also analyzed to compare against the initial GC/MS analysis. The quantitation limit for atrazine, cyanazine, CAM, and DEA was 0.05 μg/L.

Results and Discussion

Temporal Variation of Herbicide Concentrations. Figure 3A shows the variation in the concentration of atrazine and cyanazine for samples from the Mississippi River at Baton Rouge, Louisiana for 1992 and 1993. Stream flow was characterized by two flushing events that occurred in the spring and mid-summer, 1992; a period of low-flow occurred in between the two discharge events (late May through July). However, in 1993 the Baton Rouge site was characterized by fairly continuous high stream-flow from early spring into the winter due to extensive flooding in the Upper Mississippi River Basin. The contrast in flow conditions between the 1992 and 1993 studies provided an opportunity to examine the effects of stream flow on the transport of herbicides and their metabolites.

In 1992 atrazine and cyanazine were transported along the Mississippi River to Baton Rouge, Louisiana as pulses in response to increased discharge in the spring and again in the midsummer (Fig. 3A). The rapid increase in herbicide concentrations during the "spring flush" demonstrates that much of the herbicide application occurred between late April and early May. The peak concentration of cyanazine was 40 percent that of atrazine during the spring flush, thus indicating that the usage of atrazine in the Mississippi River basin was only approximately twice that of cyanazine. The concentration of atrazine and cyanazine declined only slightly during the period of low-flow from late May through July of 1992. In addition, the concentration of cyanazine was 32 percent that of atrazine both at the beginning and end of the period of low-flow. The concentration of atrazine and cyanazine increased at the beginning of the midsummer flush and declined thereafter, from late July through September. The concentration of cyanazine was only 22 to 17 percent that of atrazine during this period.

The decline in the proportion of cyanazine relative to atrazine, from 40 to 20 percent, throughout the 1992 growing season was due to cyanazine degrading more rapidly in the soil than atrazine (13-16). Thus, during the period of low-flow, the unchanging proportion of cyanazine to atrazine for more than a month indicates that the compounds were discharged into the Mississippi River from an environment with very slow degradation rates. The best explanation for this phenomenon is that herbicide laden runoff was stored in reservoirs and river banks and then metered into the Mississippi River during low-flow. Finally, the early increase and steady decline

Table I: Disappearance half-lives of atrazine and cyanazine at three Mississippi River main-stem sites and the Ohio and Illinois River sampling sites during the 1993 Floods

Site	[1]K (slope)	r^2	[2]n	[3]Half-Life days	Significance
Mississippi R., Clinton, IL					
atrazine	0.032	0.93	11	22	0.001
cyanazine	0.045	0.97	10	14	0.001
Mississippi R., Thebes, IL					
atrazine	0.024	0.95	15	30	0.001
cyanazine	0.034	0.81	14	20	0.001
Mississippi R., Baton Rouge, LA					
atrazine	0.017	0.98	46	42	0.001
cyanazine	0.031	0.96	30	22	0.001
Ohio R., Grand Chain, IL					
atrazine	0.023	0.70	10	30	0.05
cyanazine	0.037	0.75	10	19	0.01
Illinois R., Valley Falls, IL					
atrazine	0.032	0.96	11	22	0.001
cyanazine	0.045	0.98	10	16	0.001

1. K= Linear regressed slopes for selected sites from the 1993 Mississippi River study.
2. n= number of samples.
3. half-life = 0.693/K

in herbicide concentrations during the midsummer flush shows that the majority of herbicides had been flushed from the fields and cycled through the reservoirs by early fall of 1992.

Figure 3A shows that the temporal concentration pattern of atrazine and cyanazine was much different in 1993 than in 1992 for the Mississippi River at Baton Rouge. Atrazine and cyanazine concentrations fluctuated as pulses in response to changes in discharge in 1992, but in 1993 the concentration of herbicides declined exponentially with no spikes. Also, the concentration of atrazine and cyanazine was approximately two times higher in July of 1993, than during the spring flush in 1992. This result indicates that herbicides were being scavenged from large tracts of farmland during the 1993 floods. In addition, the large and numerous discharge pulses along with the smooth decline in the concentration of atrazine and cyanazine indicate that in 1993 herbicides were almost continuously flushed from the fields by frequent rain storms and also transported directly from flood inundated fields adjacent to streams and rivers.

The systematic decline in herbicide concentrations allowed for the calculation of the disappearance or concentration half-lives of atrazine and cyanazine at five of the 1993 sites. Table I shows that the shortest half-life for atrazine and cyanazine was 22 and 14 days, respectively -- at the site farthest upstream (Clinton, Ill.) -- and longest 42 and 22 days, respectively -- the farthest downstream site(Baton Rouge, La.) (Fig. 1). The concentration half-life was approximately two times longer for atrazine than for cyanazine at each site. These results along with the data from the 1992 study is consistent with the interpretation that cyanazine is more labile than atrazine (*13-16*). These data also suggest that the concentration half-life increases as the drainage scale increases. This is most easily explained by the combination of retarded herbicide transport due to herbicide storage in regulated streams (reservoirs) and the increase in the degradation half-lives of most herbicides and their metabolites in water relative to soil (*17*).

The concentration of cyanazine varied from 45 to 49 percent that of atrazine in early July, when the first samples were collected. However, additional atrazine data from Goolsby and others (18) shows that the first herbicide flush occurred in May and that the peak concentration of herbicides occurred in late June. Because sampling started well after herbicide application and cyanazine degrades more rapidly than atrazine (*13-16*), the proportion of cyanazine to atrazine may have been considerably higher earlier in the season. Thus, the increase in the proportion of cyanazine to atrazine from 40 percent in 1992 to more than 45 percent in 1993, may represent a real increase in cyanazine usage between the two years. Figure 3A shows that cyanazine was not detected after early November while atrazine was detected into February. The shorter half-life and lower use of cyanazine relative to atrazine explain why atrazine was detected for a much longer time after application than cyanazine. The ubiquitous presence of atrazine and cyanazine in the Mississippi River throughout the growing season indicates that their metabolites may also be present.

Temporal Variation of Metabolite Concentrations.   DEA, a metabolite of atrazine, has been found in surface water and ground water throughout the United States (*4, 7, 10, 18-24*). Recent studies have found that CAM is also a frequent contaminant in surface water (*4-6*) and its presence has also been reported in ground water (*9-10*). In 1992 and 1993 CAM and DEA were detected throughout the growing season in the Mississippi River (Fig. 3B). Trace levels of CAM were detected prior to the spring flush in 1992. DEA and CAM were present at concentrations of less than 0.2 µg/L through the spring flush and the period of low-flow. The highest concentration of CAM and DEA occurred with the peak discharge in the midsummer flush; their concentrations declined thereafter. In addition, CAM declined more rapidly in concentration than DEA.

In 1993 the highest concentrations of DEA and CAM occurred between early July, when sampling first started, and early August. The concentration of both DEA and CAM declined thereafter. As in the previous year CAM declined more rapidly in concentration than DEA.

Figure 3B also shows that the concentration of CAM exceeded that of DEA through the midsummer in both sampling years (Fig. 3B); in July and August the concentration of CAM was approximately twice that of DEA. However, the concentration of DEA exceeded the concentration of CAM by early September of both sampling years. Similar temporal patterns were observed between the concentration of DEA and CAM at the other sampling sites in 1992 and 1993. For example, the concentration of CAM exceeded the concentration of DEA through the spring and midsummer, at all but the Missouri River site, but by early September the concentration of DEA exceeded the concentration of CAM. Research has shown that cyanazine rapidly degrades to CAM; however, CAM is also rapidly degraded (25-26). In addition, while DEA is not as readily formed as CAM, it degrades much more slowly than CAM (5). Thus, in basins where atrazine usage is only about two to four times more than cyanazine more CAM than DEA is flushed into surface water from spring through the early to midsummer; however, because cyanazine is preferentially degraded relative to atrazine, by mid to late summer the amount of DEA flushed into surface water exceeds that of CAM.

Figure 3 shows that the highest concentration of atrazine and cyanazine in 1992 occurred with the spring flush, shortly after application, but the highest concentration of DEA and CAM occurred in the midsummer flush. This may also have been the case in 1993, as it appears that the concentration of DEA and CAM was peaking when sampling first started in early July (Fig. 3B). A later season metabolite pulse has been recently reported by Thurman and others (27). Field study indicates that the maximum level of DEA in the soil occurs approximately three to four weeks after application (5, 27). Also if the rainfall patterns are favorable this buildup of metabolites in the soil is expressed in the temporal concentration pattern of a stream (27). These data and previous research demonstrate that findings from local scale studies of herbicide degradation in the soil can be expressed on a regional scale. The herbicide and metabolite data that have been presented show some interesting patterns that may be further explored using herbicide and metabolite ratios.

Metabolite Ratios. Atrazine, cyanazine, and DEA are commonly analyzed in studies of water quality. This and other studies (4-6, 18, 23) have shown that atrazine, cyanazine, and DEA are detected in stream water throughout the Mississippi River Basin during the growing season, with atrazine also being detected well into the winter. CAM is also a common contaminant in stream water during the growing season, but is not yet an analyte that is routinely analyzed in studies of water quality. Because the transport and occurrence of atrazine and its metabolite DEA is well studied, it would be useful to find a simple method to evaluate the transport of CAM relative to DEA for studies in which CAM has not been analyzed.

For example, Table II shows the mean ratio of the concentration of atrazine-to-cyanazine (ACR) and CAM-to-DEA (CAMDEAR) for each of the 1992 and 1993 sampling sites. At each site with a mean ACR less than or equal to 4.0, the mean CAMDEAR was greater than or equal to 1.0. In addition, the mean CAMDEAR was less than 1.0 at the Missouri River site; the only site with a mean ACR greater than 4.0. These data show that streams where the mean ACR is less than 4.0 more CAM is transported than DEA. Similar results have been obtained for streams throughout the midwest (*5*). In addition, the mean CAMDEAR exceeded unity at the Baton Rouge site in both sampling years. This result demonstrates that more CAM than DEA was transported in the Mississippi River and discharged into the Gulf of Mexico during a substantial portion of the growing season. Thus, the mean ACR can be used as a general predictor of the transport of CAM relative to DEA on a regional scale. Recent studies have shown that herbicide metabolite ratios are useful indicators geochemical transport processes (*6, 23, 27-29*).

The ratio of DEA-to-atrazine (DAR) has been used as a geochemical indicator of surface water and ground water interaction (*23, 27*), to distinguish point from nonpoint source contamination (*28*), and to determine seasonal differences in atrazine degradation rates (*29*). In addition, seasonal variation of the ratio of CAM-to-cyanazine (CAMCR) indicate that it may also be a useful indicator of contaminant transport processes in surface water (*5*).

Figure 4A shows the variation of the DAR and CAMCR for the Ohio River, and the Mississippi River at Thebes and Baton Rouge sampling sites for 1992. The lowest DAR (less than 0.1) and CAMCR ($\approx$ 0.2) was obtained in late April at the Ohio River and Mississippi River at Thebes sites and in early May at the Mississippi River at Baton Rouge site (Fig. 4A). Studies have shown that a low DAR of less than 0.1 is indicative of herbicides flushed into surface water shortly after herbicide application (*21,23*). Low CAMCR's of approximately 0.2 were also associated with the flushing of recently applied herbicides from the fields in a regional reconnaissance of herbicides in stream water (*5*). Thus, the low DAR and CAMCR obtained from samples at the Mississippi River at Thebes and the Ohio River sites demonstrate that herbicides had been applied throughout the midwest by late April. This figure also shows that the DAR and CAMCR increased throughout the spring and midsummer.

For example, the DAR increased from less than 0.1 to approximately 0.2 and the CAMCR from approximately 0.2 to 0.6 from late April into late May (Fig. 4A). The DAR and CAMCR remained fairly constant through the period of low-flow from late May through June and then increased during the midsummer flush from early July into August. By midsummer the DAR had increased to approximately 0.3 and the CAMCR to more than 1.0. These data also show that while the DAR and CAMCR increased with time, the CAMCR increased much more rapidly than the DAR.

Table III shows that there was a systematic increase in the DAR and CAMCR with time during the spring and midsummer flushing events in 1992 and 1993; the CAMCR increased approximately ten times more rapidly than the DAR. For example, the CAMCR increased at a rate that ranged from 0.015 to 0.030 per day and the DAR at a rate that ranged from 0.002 to 0.007 per day. Similar results have been obtained from several midwestern streams (*5*). Because the increase in the CAMCR is much more pronounced than the DAR it may be useful as an indicator of early to mid to late season transport of herbicides.

Table II: Concentration ratio of atrazine-to-cyanazine (ACR) and cyanazine amide- to-cyanazine (CAMDEAR) for the Mississippi River study sites

Site	[1]ACR mean	[2]n	[1]CAMDEAR mean	n
Mississippi R., Clinton, IL	2.1	10	1.9	8
Illinois R., Valley Falls, IL	2.5	11	1.5	9
Ohio R., Grand Chain, IL	34.0, 3.8	10, 10	1.0, 1.1	6, 7
Mississippi R., Thebes, IL	2.8, 2.9	9, 14	1.5, 1.4	8, 11
Missouri R., Hermann, MO	7.4	6	0.51	5
Mississippi R., Baton Rouge, LA	3.7, 4.0	17, 30	1.3, 1.5	12, 24

1. Mean ratio of water samples for which both compounds occurred above the reporting level (0.05 μg/L).
2. n= number of samples.
3. Data for 1992 and 1993, respectively.

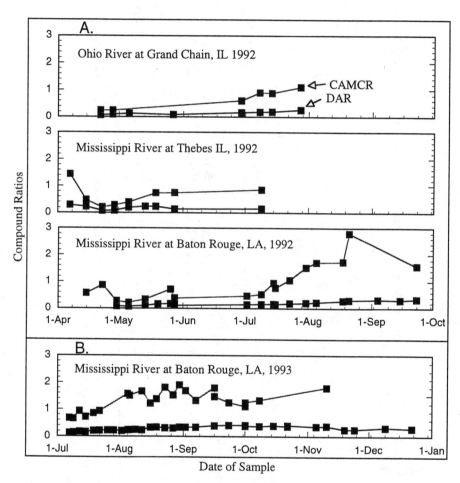

Figure 4A. Variation of stream flow and the concentration ratio of cyanazine amide-to-cyanazine (CAMCR) and deethylatrazine-to-atrazine (DAR) for the 1992 sampling sites and B. for the Mississippi River, Baton Rouge sampling site in 1993.

Table III: Regressed relationship and slopes of the ratio of cyanazine amide-to-cyanazine (CAMCR) and deethylatrazine-to-atrazine (DAR) with time

Site	Slope	r^2	n[2]	Significance
Ohio R., Grand Chain, IL; 1992				
CAMCR	**0.016**	0.89	4	0.05
DAR	**0.003**	0.96	5	0.01
Mississippi R., Thebes, IL; 1992				
CAMCR	*0.015*[1]	1.00	3	0.001
DAR	*0.007*	0.90	6	0.01
Mississippi R., Baton Rouge, LA; 1992				
CAMCR	***0.024, 0.026***	0.99, 0.88	3, 8	0.01, 0.01
DAR	***0.005, 0.002***	0.91, 0.86	4,12	0.05, 0.001
Mississippi R., Baton Rouge, LA; 1993				
CAMCR	**0.030**	0.93	9	0.001
DAR	**0.004**	0.93	21	0.001

1. Regressed slopes of the CAMCR and DAR for 1992 samples collected prior to summer low-flow are in italics and after summer low-flow are in normal type.
2. n = number of samples.

Samples collected in early and mid April at the Mississippi River at Thebes were characterized by CAMCR's that ranged from 0.5 to 1.5 (Fig. 4A). The high CAMCR and elevated preapplication DAR obtained at the Thebes site indicate that some of previous years herbicides were flushed into the Mississippi River. Increased flow did occur during this time supporting this hypothesis (*18*). This figure also shows that the CAMCR was greater than 1.0 at the Baton Rouge site in late April when the first post-application herbicide flush was identified at the Ohio River and Mississippi River at Thebes sampling sites. The upper midwest herbicide flush was received at the Baton Rouge site one to two weeks later. Thus, the CAMCR provided a way to identify "slugs" of water transported down the Mississippi River. Additional knowledge on the seasonal variation of the DAR and CAMCR was provided by the 1993 study of the Mississippi River at Baton Rouge.

Figure 4B shows that the CAMCR increased from 0.6 to more than 1.0 and the DAR from approximately 0.1 to 0.2 from early July through early August. The initial CAMCR of approximately 0.6 obtained at the Baton Rouge site in 1993 indicates that the majority of herbicide application had occurred at least three to four weeks before sampling had occurred. This determination would have been difficult to discern using the DAR because of its slow increase relative to the CAMCR with time. Thus, elevated CAMCR suggests that the highest concentration of herbicides had been flushed through before the 1993 sampling was started. This is supported by atrazine concentration data form Goolsby and others (*18*) in June of 1993.

In addition, Figure 4B shows that the CAMCR varied between 1.0 and 2.0 from mid August into November. Thus, through the early and mid season, when the majority of herbicides are transported in surface water (*23*), the CAMCR increases systematically with time. However, in mid to late season the CAMCR varies in an uncorrelated manner. For example, data from the 1992 study of the Mississippi River at Baton Rouge shows that the CAMCR did not increase in a systematic way after early August. This pattern of temporal variation in the CAMCR has been observed in other midwestern streams (*5*). Thus, CAMCR can be used an indicator of early mid and late season transport of herbicides on a local and regional scale. These data demonstrate that the CAMCR used in conjunction with the DAR is an important geochemical indicator of the seasonal transport of herbicides in the Mississippi River.

Conclusions

This study shows that the temporal pattern of herbicide transport is highly dependent on seasonal rain patterns. Also herbicide laden water stored in reservoirs meter out a steady source of herbicides during low-flow. The amount of cyanazine relative to atrazine that is transported in the Mississippi River decreased with time and the concentration half-life of atrazine and cyanazine increased with increased drainage area. Also the proportion of cyanazine to atrazine during the spring flush indicated that only about two times more atrazine than cyanazine was applied in the Midwest in 1992 and 1993.

CAM was shown to be a frequent contaminant in the Mississippi River throughout the growing season. A midsummer metabolite pulse indicated that the majority CAM and DEA was transported more than a month after application. During the spring and early to midsummer the concentration of CAM exceeded the concentration of DEA along the main stem of the Mississippi River. The mean ACR and CAMDEAR provided a simple means to assess the amount of CAM relative to

DEA transported in surface water. At sites where the CAMDEAR exceeded unity the mean ACR was less than 4.0 and more CAM than DEA was transported. More CAM than DEA was discharged into the Gulf of Mexico in 1992 and 1993.

Finally, the CAMCR and DAR are useful geochemical indicators of contaminant transport processes in surface water. A low CAMCR and DAR are indicative of herbicide transport shortly after application. The CAMCR and DAR increase linearly from spring after application though much of the growing season. Pre- and post-application "slugs" of water were identified and their transport traced down the Mississippi River using the CAMCR.

Acknowledgments: The authors wish to thank Gail Mallard of the U. S. Geological Survey's Toxic Substances Program for funding this research. The use of brand names in this paper is for identification purposes only and does not imply endorsement by the U. S. Geological Survey.

Literature Cited

1. Gianessi, L.P.; Anderson, J.E. *Pesticides in U. S. Crop Production: National Data Report,* National Center for Food and Agriculture Policy, Washington, D.C., **1995**, pp. 587.
2. Gianessi, L.P.; Puffer, C. M. *Herbicide use in the United States: Resources for the Future*, National Summary Report, Washington, D.C., **1991**, pp. 128.
3. Gianessi, L. P.; and Puffer, C. M. *Herbicide use in the United States; Resources for the Future*, National Summary Report, Washington, D.C., **1988**, pp. 490.
4. Pereira, W.E.; Hostettler, F.D. *Environ. Sci. Technol.*, **1993**, *27*, pp. 1542-1552.
5. Meyer, M. T. PhD. Thesis, University of Kansas, Dec. **1994**, pp. 362.
6. Meyer, M. T., and Thurman, E. M. *Cyanazine metabolites in surface water: The transport and degradation of labile herbicides*: Morganwalp, D.W., and Aronson, D.A., Eds., U.S. Geological Survey Toxic Substances Hydrology Program--Proceedings of theTechnical Meeting, Colorado Springs, Colorado, September 20-24, 1993: **1995**, U.S. Geological Survey Water-Resources Investigations Report *94-401*.
7. Goolsby, D.A.; Thurman, E.M., Koplin, D.W.; Meyer, M.T. *Preprints of Papers Presented at the 209th ACS National Meeting*, Anaheim, CA, American Chemical Society, Environmental Division, **1995**, *35*, No.1, pp.278-281.
8. Thurman, E. M.; Goolsby, D.A.; Aga, D.S.; Pomes, M.L.; Meyer, M.T.. *Environ. Sci. Technol.*, **1995**, *30*, pp. 569-574.
9. Baker, D.B.; Bushway, R.J.; Adams, S.A.; Macomber, C.S. *Environ. Sci. Technol.*, **1993**, *27*, pp. 562-564.
10. Koplin, D.W.; Burkart, M.R.; Thurman, E. M., *Hydrogeologic, Water Quality, and land-use for the reconnaissance of herbicides and nitrate in near-surface aquifers of the midcontinental United States, 1991*: **1993**, U. S. Geological Survey Open-File Report, *93-114*, pp. 61.
11. Thurman, E.M.; Meyer, M.T.; Pomes, Michael; Perry, C.A.; Schwab, A.P. *Anal. Chem.* **1990**, *62*, pp. 2043-2048.
12. Meyer, M.T.; Mills, M.S.; Thurman, E.M. *J. Chromatogr.* **1993** , 629, 55-59.
13. Sirons, G. J.; Frank, R.; Sawyer, T. *J. Agric. Food Chem.* **1973**, 21, pp.1016.
14. Muir, D.C.; Baker, B.E. *Weed Res.* **1978**, *18*, p. 111.

15. Wauchope, R.D.; Butler, T.M.; Hornsby, A.G.; Augustijn-Beckers, P.W.M.; Burt, J.P. In *Reviews of Environmental Contamination*; Hare, G.W., Ed.; Springer-Verlag, New York, **1991**; *123*, pp. 164.

16. Blumhorst, M.R.; Weber, J.B. *J. Agric. Food Chem.* **1992**, *40*, pp.894

17. Grover, R.; and Cessna, A.J. eds. *Environmental Chemistry of Herbicides, v. II*: CRC Press Inc., Boca Raton, Florida, 1991, 302 pp.

18. Goolsby, D. A.; Battaglin, W. A.; and Thurman, E. M. *Occurrence and transport of agricultural chemicals in the Mississippi River Basin, July through August 1993*: U. S. Geological Survey, Circular, **1993**,1120-C, pp.22.

19. Glotfelty, D. E.; Taylor, A. W.; Inensee, A. R.; Jersey, J.; and Glen, S. *J Environ. Qual.*, **1984**,*13*, pp. 115-121.

20. Leonard, R.A. In *Environmental chemistry of Herbicides*; Grover, R., Ed., CRC Press: Boca Raton, FL,**1988**, pp. 45-88.

21. Pereira, W.E.; Rostad, C.E. *Environ. Sci. Technol.*, **1990**, *24*, pp. 1400-1406.

22. Thurman, E. M.; Goolsby, D. A.; Meyer, M. T.; and Koplin, D. W. *Environ. Sci. Technol.*, **1991**, *25*, pp. 1794-1796.

23. Squillace, P.J.; Thurman, E. M. *Environ. Sci. Technol.*, **1992**, *29*, pp. 1719-1729.

24. Thurman, E. M.; Goolsby, D.A.; Meyer, M.T.; Mills, M. S.; Pomes, M.L.; Koplin, D.W. *Environ. Sci. Technol.*, **1992**, *26*, pp. 2440-2447.

25. Benyon, K. I.; Stoydin, G.; and Wright, A. N. *Pest. Sci.*, **1972**, 3, pp. 293-305.

26. Sirons, G. J.; Frank, R.; and Sawyer, T., *J Agricul. Food Chem.*, **1973**, *21*, pp. 1016-1020.

27. Thurman, E. M.; Meyer, M.T.; Mills, M. S.; Zimmerman, L.R.; Perry, C.A. *Environ. Sci. Technol.*, **1994**, *28*, pp. 2267-2277.

28. Adams, C. D.; Thurman, E. M. *J Environ. Qual.*; **1991**, *20*, pp. 540-547.

29. Schottler; S.P.; Eisenreich, S.J.; Capel, P.D. *Environ. Sci. Technol.*, **1994**, *28*, pp. 1079-1089.

Author Index

Alberts, Eugene E., 254
Alpendurada, M. F., 237
Barceló, D., 237
Barrett, Michael R., 200
Beckert, Werner F., 63
Benedicto, Janet, 63
Best, K. B., 151
Bridges, D. C., 165
Capel, Paul D., 34
Cessna, A. J., 151
Chiba, Misako, 77
Chiron, S., 237
Clay, D. E., 117
Clay, S. A., 117
Coats, Joel R., 140
Conn, J. S., 125
Donald, William W., 254
Eckhardt, David A. V., 101
Elliott, J. A., 151
Fernandez-Alba, A., 237
Fleeker, James R., 43
Goolsby, D. A., 288
Grover, R., 151
Hansen, P.-D., 53
Harper, S. S., 117
Herzog, David P., 43
Hock, Bertold, 53
Hottenstein, Charles S., 43
Kawata, Mitsuyasu, 77
Koskinen, W. C., 125
Krotzky, A., 53
Kruger, Ellen L., 140
Laird, David A., 86
Lawruk, Timothy S., 43

Lerch, Robert N., 254
Lewis, D. T., 178
Li, Yong-Xi, 254
Liu, Z., 117
Lopez-Avila, Viorica, 63
Ma, Li, 226
McCallister, Dennis L., 178
Meitzler, L., 53
Meulenberg, E., 53
Meyer, M. T., 1,288
Müller, G., 53
Nicholaichuk, W., 151
Obst, U., 53
Parkhurst, Anne, 178
Rubio, Fernando M., 43
Schroyer, Blaine R., 34
Smith, A. E., 165
Sorenson, B. A., 125
Spalding, Roy F., 226,271
Spener, F., 53
Stearman, G. Kim, 18
Strotmann, U., 53
Takagi, Yasushi, 77
Thurman, E. M., 1,178,288
Valverde, A., 237
Verstraeten, Ingrid M., 178
Wagenet, R. J., 101
Weil, L., 53
Wells, Martha J. M., 18
Widmer, S. K., 271
Wittmann, C., 53
Yagi, Katsura, 77
Yanase, Daisuke, 77

Affiliation Index

Agricultural Research Service, 86,125,254
Agriculture and Agri-Food Canada, 151
Almeria Campus (Spain), 237

Analytical Bio-Chemistry Laboratories,
   Inc., 254
CID-CSIC, 237

Cornell University, 101
Deutsches Institut für Normung, 53
Iowa State University, 140
Midwest Research Institute, 63
National Hydrology Research Institute, 151
North Dakota State University, 43
Ohmicron Environmental Diagnostics, 43
Otsuka Chemical Company, Limited, 77
South Dakota State University, 117
Technical University of Munich, 53
Tennessee Technological University, 18

Tennessee Valley Authority, 117
U.S. Department of Agriculture,
 86,125,254
U.S. Environmental Protection Agency,
 63,200
U.S. Geological Survey, 1,34,101,178,288
University of Georgia, 165
University of Minnesota, 125
University of Nebraska—Lincoln,
 178,226,271
University of Porto, 237

Subject Index

A

Acetanilide herbicides, leaching potential
 evaluation, 212–216
Adsorptivity of photosynthesis-inhibiting
 herbicides, determination using in vivo
 fluorometry, 77–83
Advection–dispersion equation, 274–275
Agricultural chemicals, transport, 179
Agricultural practices, impact on water
 quality, 151
Agricultural runoff
 herbicide mobility and concentration,
 226–234
 sources, 151,153
Alachlor
 analysis without sample preparation, 19
 analytical methods, 34–35
 degradation pathways, 179
 environmental concern, 2
 HPLC-based screening method for
 analysis, 34–41
 monitoring in surface water and
 groundwater, 237–249
 presence in groundwater and surface
 water, 178–179
 role in herbicide and degradate
 concentration in regolith in
 northeastern Nebraska, 178–194
 transport and persistence in
 groundwater, 271–285
 usage changes, 2

Alachlor oxanilic acid, HPLC-based
 screening method for analysis, 34–41
Alachlor oxoethanesulfonic acid, HPLC-
 based screening method for analysis,
 34–41
Alachlor sulfonic acid degradate,
 groundwater leaching, 212–213,215–216
4-Amino-6-(1,1-dimethylethyl)-3-
 (methylthio)-1,2,4-triazine-5(4H)-one,
 See Metribuzin
Ammonia-based fertilizers, interactions
 with herbicides, 118
Ammonia, effect on atrazine sorption and
 transport
 atrazine adsorption, 120–122
 atrazine desorption, 122
 atrazine movement, 123
 experimental procedure, 118–120
 hydroxyatrazine adsorption, 123
 solution effects on atrazine, 122–123
Analytical strategies, coordinating sample
 preparation with final determination
 of herbicides, 18–32
Anion-exchange packing, use in Empore
 disk for herbicide metabolite isolation
 from water, 9
Aqueous agricultural runoff samples,
 analysis without sample preparation, 19
Aqueous dilution, sample matrix
 control, 28
Aqueous solubility, role in fate and
 transport of herbicide metabolites, 11–13

Asymmetric triazine herbicide, fate in
   silt–loam soils, *See* Triazine
   herbicide, fate in silt–loam soils
Atrazine
   ammonia, effect on sorption and
      transport, 117–123
   analysis without sample preparation, 19
   analytical methods, 34–35
   application to soils, 101
   degradation mechanisms, 227,228*f*
   degradation pathways, 179
   degradation to metabolites, 288,291
   degradation vs. water solubility, 6,9
   factors affecting fate, 101–102
   geochemical indicator of contaminant
      transport in Mississippi River, 288–301
   HPLC-based screening method for
      analysis, 34–41
   mobility and concentration in
      agricultural runoff, 226–234
   monitoring in surface water and
      groundwater, 237–249
   presence in groundwater and surface
      water, 178–179
   role of clay on adsorption, 86–87
Atrazine
   standardization of immunoassays for
      water and soil analysis, 53–61
   structure, 288,289*f*
   transport and persistence in groundwater,
      271–285
   usage changes, 2
Atrazine and degradates in soils in Iowa
   fate
      atrazine, 141–143,144*f*
      deethylatrazine, 141–143,144*f*
      deisopropylatrazine, 141–143,144*f*
      hydroxyatrazine, 141–143,144*f*
      influencing factors, 141,149
   mobilities
      atrazine, 145–149
      deethylatrazine, 145–149
      deisopropylatrazine, 147–149
      didealkylatrazine, 147–148
      hydroxyatrazine, 147–148
      pesticide history effect, 143,145

Atrazine and degradates in soils in Iowa—
   Continued
   soil characteristics, 141,142*t*
   soil moisture effect, 143,146*f*
Atrazine degradation products,
   hydroxylated, *See* Hydroxylated
   atrazine degradation products
Atrazine immunoassay
   2-position substitution, 47–48,49*t*
   6-position substitution, 48–49
Atrazine–smectite surface interactions
   hydrolysis of atrazine, 95
   model of reaction, 96–98
   smectites
      hydration of surfaces, 89–91
      structure, 87–89
      surface acidity, 91,93
      sorption of atrazine, 92*f*,93–94
Atrazine transport estimation in silt–loam
   soil
   experimental description, 102
   modeling study
      derivation of frequency distributions
         for 22-year precipitation record,
         106*t*,107,109
      experimental procedure, 102–105
      model calibration and evaluation, 107
   results
      field study, 109
      model calibration and evaluation
      effects of uncertainty in model
         parameters, 112–114
      predictions of atrazine concentrations
         in drainage, 110*f*,111
      results, 108*f*,109
   study problems, 102

B

Beaver Creek watershed, Nebraska,
   herbicide mobility and concentration
   in agricultural runoff, 226–234
Benefin, movement after application to
   golf courses, 165–175
Bentazone, monitoring in surface water and
   groundwater, 237–249

Bromoxynil, transport in runoff from
corrugation irrigation of wheat, 151–163
Butachlor, transport and persistence in
groundwater, 271–285
N-Butyl-*N*-ethyl-2,6-dinitro-4-(trifluoromethyl)benzenamine, *See* Benefin

C

Carbamate insecticides
contamination of groundwater, 239
monitoring in surface water and
groundwater, 237–249
Carbofuran, monitoring in surface water
and groundwater, 237–249
Chemical analysis of metabolites of
herbicides
degradation vs. water solubility, 6,9
identification methods, 9–11
isolation methods, 6,9
Chemical derivatization, detectability, 23,28
Chlorinated atrazine analogues, comparison
to hydroxylated atrazine degradation
products, 261–263,265
2-Chloro-4-amino-6-(isopropylamino)-*s*-
triazine, *See* Deethylatrazine
2-Chloro-4,6-diamino-*s*-triazine, *See*
Didealkylatrazine
2-Chloro-2',6'-diethyl-*N*-(methoxymethyl)-
acetanilide, *See* Alachlor
2-Chloro-4-(ethylamino)-6-amino-1,3,5-
triazine, *See* Atrazine
2-Chloro-4-(ethylamino)-6-amino-*s*-
triazine, *See* Deisopropylatrazine
(4-Chloro-2-methylphenoxy)acetic acid,
See MCPA
(±)-2-(4-Chloro-2-methylphenoxy)
propanoic acid, *See* Mecoprop DMA
Chlorophenoxy acid herbicides
conventional analytical methods, 63
in situ derivatization–supercritical
fluid extraction method for
chlorophenoxy acid in soil, 63–75
3-(4-Chlorophenyl)-1,1-dimethylurea,
See Monuron
Chlorophyll *a* fluorescence, bioprobe for
yield, 77

Chlorothalonil, movement following
application to golf courses, 165–175
Chlorpyrifos, movement following
application to golf courses, 165–175
Chromatographic method, identification
of herbicide metabolites, 9–10
Chromophores, detectability, 23,26–27*f*
Clays
role in adsorption of nonionic organic
compounds, 86–87
See also Smectites
Concentration of herbicides in agricultural
runoff, *See* Herbicide mobility and
concentration in agricultural runoff
Contaminant transport in Mississippi River
analytical procedure, 291,293
experimental description, 288
metabolite ratios, 296–301
site, 290*f*,291
study design, 291
time of year vs. concentration
herbicide, 292*f*,293–295
metabolite, 295–296
Corrugation irrigation of wheat, transport
of nutrients and postemergence-applied
herbicides in runoff, 151–163
Cross-reactivity, sample matrix control,
31–32
Cyanazine
degradation pathway, 3–4,5*f*
degradation to metabolites, 288,291
degradation vs. water solubility, 6,9
fate and transport studies, 11
geochemical indicator of contaminant
transport in Mississippi River,
288–301
geographic information system map, 4,7*f*
mobility and concentration in
agricultural runoff, 226–234
structure, 288,289*f*
transport and persistence in groundwater,
271–285
usage changes, 2
Cyanazine immunoassay, 47
N-Cyclohexyl-5-hydroxy-3-methyl-2-
oxoimidazolidine-1-carboxamide, *See*
OK–9201

D

2,4-D
analysis without sample preparation, 19
transport in runoff from corrugation
irrigation of wheat, 151–163
2,4-D diethylamine salt formulation,
movement following application to golf
courses, 165–175
Deethylatrazine
degradation product of atrazine, 227
degradation vs. water solubility, 6,9
environmental concern, 2
fate in soils of Iowa, 140–149
HPLC-based screening method for
analysis, 34–41
mobility and concentration in
agricultural runoff, 226–234
monitoring in surface water and
groundwater, 237–249
role in herbicide and degradate
concentration in regolith in
northeastern Nebraska, 178–194
transport and persistence in
groundwater, 271–285
Deethylhydroxyatrazine
HPLC-based screening method for
analysis, 34–41
occurrence in Missouri stream,
254–269
Degradates of pesticides, environmental
impact in groundwater, 200–221
Degradation
latitude effect, 125–126
water solubility effect, 6,9
Deisopropylatrazine
degradation product of atrazine, 227
degradation vs. water solubility, 6,9
environmental concern, 2
fate in soils of Iowa, 140–149
HPLC-based screening method for
analysis, 34–41
mobility and concentration in
agricultural runoff, 226–234
monitoring in surface water and
groundwater, 237–249

Deisopropylatrazine—*Continued*
role in herbicide and degradate
concentration in regolith in
northeastern Nebraska, 178–194
transport and persistence in
groundwater, 271–285
Deisopropylhydroxyatrazine
HPLC-based screening
method for analysis, 34–41
occurrence in Missouri stream, 254–269
Demethoxymethylalachlor, HPLC-based
screening method for analysis, 34–41
Depth, role in herbicide and degradate
concentration in regolith in
northeastern Nebraska, 178–194
Derivatization–supercritical fluid
extraction method for chlorophenoxy
acid herbicide determination in soil,
See In situ derivatization–
supercritical fluid extraction method
for chlorophenoxy acid herbicide
determination in soil
Detectability
chemical derivatization, 23,28
chromophores, 23,26–27f
electrophores, 23
Determination of herbicides, coordination
with sample preparation, 18–32
3,5-Dibromo-4-hydroxybenzonitrile,
See Bromoxynil
Dicamba diethylamine salt formulation,
movement following application to golf
courses, 165–175
3,6-Dichloro-2-methoxybenzoic acid,
165–175
2,4-(Dichlorophenoxy)acetic acid,
See 2,4-D
(±)-2-[(2,4-Dichlorophenoxy)phenoxy]-
propionic acid, *See* Diclofop
Diclofop, transport in runoff from
corrugation irrigation of wheat,
151–163
Didealkylatrazine
fate in soils of Iowa, 140–149
HPLC-based screening
method for analysis, 34–41

O,O-Diethyl *O*-(3,5,6-trichloro-2-
pyridinyl)phosphorothioate, *See*
Chlorpyrifos
2,6-Diethylaniline, HPLC-based screening
method for analysis, 34–41
2-{[(2,6-Diethylphenyl)methoxymethyl]
amino}-2-oxoethanesulfonate, *See*
Alachlor
Dilution, aqueous, sample matrix
control, 28
S,S-Dimethyl 2-(difluoromethyl)-4-(2-
methyl-propyl)-6-(trifluoromethyl)-3,5-
pyridinedicarbothioate, *See* Dithiopyr
DIN approach
categories of immunoassays, 55,57,58*t*
draft standard, 57
interlaboratory tests, 59,60*f*
limitations, 61
procedure, 55
round robin tests, 57–59
Direct-probe MS
analysis of hydroxylated atrazine
degradation products, 254–269
procedure, 259
Dithiopyr, movement after application to
golf courses, 165–175
Diuron, microplate fluorometry, 79–80

E

Electrophores, detectability, 23
Electrospray HPLC–MS–MS
analysis of hydroxylated atrazine
degradation products, 254–269
procedure, 257,259
Empore disk
herbicide metabolite isolation from
water, 9
packing materials, 9
Environmental control, factors affecting
efficiency, 53–54
Environmental impact of pesticide
degradates in groundwater
environmental concerns
groundwater residues, 201–203
toxicological concerns, 201

Environmental impact of pesticide
degradates in groundwater—*Continued*
experimental description, 201
leaching potential evaluation
acetanilide herbicides
groundwater residue studies,
212–213,215–216
soil residue studies, 212–213,214*t*
compounds, 203–206
sulfonylurea herbicides
application rates, 216–217
degradates, 217–218
mobility, 216
nomenclature, 217,218*t*
soil residue studies, 218–220
structures, 217
s-triazine herbicides
groundwater residual studies,
207,208*t*,211–212
soil residue studies, 207,209–211
Enzyme-linked immunosorbent assay
identification of herbicide metabolites,
10–11,18
triazine herbicides, 43–44
Equilibrium sorption estimates,
calculation of retardation factor, 273
N-(1-Ethylpropyl)-3,4-dimethyl-2,6-
dinitrobenzenamine, *See*
Pendimethalin

F

Fate and transport of herbicide
metabolites, 11–13
2'-Fluoro-5-hydroxy-3-methyl-2-oxo-
imidazoline-1-carboxanilide, *See*
OK–8901
Fluorometry for soil mobility and
adsorptivity determination of
photosynthesis-inhibiting herbicides,
See In vivo fluorometry for soil mobility
and adsorptivity determination of
photosynthesis-inhibiting herbicides
Freundlich isotherm, calculation of
retardation factor, 273

G

Gas chromatography, use for herbicide analysis, 18
Gas chromatography–MS, identification of herbicide metabolites, 9–10
Geochemical indicators, contaminant transport in Mississippi River, 288–301
Geographic information system maps, usage of herbicides, 4–6,7–8*f*
Golf courses
  movement of pesticides following application, 165–175
  turfgrass acreage, 166
Groundwater
  contamination by carbamate insecticides, 239
  environmental impact of pesticide degradates, 200–221
  herbicide metabolites, 1–13
  herbicide transport and persistence, 271–285
  monitoring of pesticides and metabolites, 237–249
Groundwater contamination by pesticides, 200–201
Groundwater Loading Effects of Agricultural Management Systems model, prediction of 2,4-D transport, 174–175

H

Half-life, role in fate and transport of herbicide metabolites, 11–13
Herbicide(s)
  acetanilide, *See* Acetanilide herbicides
  environmental concern, 2
  photosynthesis inhibiting, in vivo fluorometry for soil mobility and adsorptivity determination, 77–83
  postemergence applied, transport in runoff from corrugation irrigation of wheat, 151–163
  sulfonylurea, *See* Sulfonylurea herbicides
  triazine, *See* 1,3,5-Triazine herbicides

Herbicide and degradate concentration in regolith in northeastern Nebraska
  chemical and physical analytical methods, 181–182
  experimental objective, 179
  herbicides and degradates
    batch equilibration experiments, 192*f*,193
    concentrations, 187
    correlations of concentration and properties
      canonical discriminant analyses, 189–191,193
      multivariate and univariate analyses, 189
    difference by landscape position and depth
      canonical discriminant analyses, 185–186
      chemical and physical properties of upper regolith, 182–183
      multivariate and univariate analyses, 183–185
    groundwater detection, 193–194
  site selection, 180–181
  statistical analytical methods, 182
Herbicide degradation, latitude effect, 125–126
Herbicide leaching, *See* Leaching of herbicides
Herbicide metabolites, *See* Metabolites of herbicides
Herbicide mobility, relationship to sorption, 117
Herbicide mobility and concentration in agricultural runoff
  analytical procedure, 230
  application conditions, 227,230
  concentrations
    first postapplication runoff, 231–232
    seasonal runoff events, 232–234
  experimental objectives, 227
  sampling procedure, 230
  site description, 227,229*f*,230
Herbicide persistence, role of soil temperatures, 126
Herbicide residues, dispersion into environment by surface runoff, 226–227

Herbicide sorption to soil, measurement by laboratory batch equilibration techniques, 117–118

Herbicide transformation in soil environments, factors affecting rates, 227

Herbicide transport and persistence in groundwater
comparison of laboratory sorption study to field transport study, 283
experimental objectives, 273
field transport study
procedure
chemical analyses, 274
data analyses, 274
injection and monitoring of solutes, 274
persistence, 276
retardation, 275–276
site description, 272f,273–274
results
breakthrough curves, 277,278f
dispersion, 280
persistence, 280
retardation, 277,279–280
laboratory sorption study
procedure
batch equilibrations, 277
data analysis, 277
sediment sample collection and analysis, 276–277
results
batch equilibrations, 280–282
organic carbon content vs. particle size, 280–285

Herbicide usage, See Usage of herbicides

High-performance liquid chromatography, use for herbicide analysis, 10,18

High-performance liquid chromatography-based screening method for pesticide analysis
advantages, 40–41
analytical procedure, 35–36,39f
experimental materials, 35
methodology, 37–38
requirements, 35
soil analysis, 38,39f,40t,41f
soil extraction procedure, 36

High-performance liquid chromatography-based screening method for pesticide analysis—Continued
water analysis, 38,40,41t
water analysis procedure, 36

High-performance liquid chromatography–mass spectrometry, identification of herbicide metabolites, 10

History of pesticide, role in atrazine degradation, 143,145

Hydrolysis, atrazine, 95

Hydrophobicity–polarity–ionogenicity, solute property analysis, 19–21

Hydroxyatrazine
ammonia effect on adsorption, 123
fate in soils of Iowa, 140–149
HPLC-based screening method for analysis, 34–41
occurrence in Missouri stream, 254–269

8-Hydroxybentazone, monitoring in surface water and groundwater, 237–249

3-Hydroxycarbofuran, monitoring in surface water and groundwater, 237–249

2-Hydroxy-4-(ethylamino)-6-(isopropylamino)-s-triazine, See Hydroxyatrazine

Hydroxylated atrazine degradation products
analytical techniques, 255
chemical properties, 255
comparison to chlorinated analogues, 261–263,265
concentrations in surface water, 258f,259–261
confirmation analytical procedure, 257,259
confirmation in stream water, 264f,265–268
experimental description, 256
formation, 255
mechanisms, 268–269
occurrence, 255–256
routine analytical procedure, 257
site description, 256
stream sampling procedure, 256–257

Hyphenated techniques for sample
preparation and final determination
detectability, 23,26–28
examples, 19,20*f*
sample matrix control, 28–32
solute properties, 19–25

I

Immunoassays
degradate screening, 221
standardization of water and soil
analysis, 53–61
use as screening tools, 54,56*f*
In situ derivatization–supercritical
fluid extraction method for
chlorophenoxy acid herbicide
determination in soil
accuracy and precision, 74–75
analytical procedure, 67,69*t*
percent recoveries, 69,71–74
reagents, 65
rejection of outliers, 67,69
soil samples, 65–66
soil spiking procedure, 66–67
standards, 64–65
supercritical fluid extraction procedure,
67,68*t*
In vivo fluorometry for soil mobility and
adsorptivity determination of
photosynthesis-inhibiting herbicides
compound properties, 77,78*t*
fluorometric techniques
fluorometry with cucumber cotyledon
disks, 78–79
microplate fluorometry, 79–80
yield of in vivo chlorophyll *a*
fluorescence as bioprobe, 77
soil leaching parameters
soil adsorptivity, 81–82
soil mobility, 80–81
Ionogenicity–hydrophobicity–polarity,
solute property analysis, 19–21
Iowa, atrazine and degradate fate in
soils, 140–149

L

Laboratory batch equilibration techniques,
measurement of herbicide sorption to
soil, 117–118
Langmuir isotherm, calculation of
retardation factor, 273
Latitude, role in herbicide degradation,
125–126
Leaching of herbicides
influencing factors, 126
relationship to herbicide persistence, 126
Leaching potential evaluation
acetanilide herbicides, 212–216
sulfonylurea herbicides, 216–220
s-triazine herbicides, 207–212
Linear isotherm, calculation of
retardation factor, 273
Liquid chromatographic–solid-phase
extraction techniques, monitoring of
pesticides and metabolites in surface
water and groundwater, 237–249
Liquid–liquid extraction, pesticide
monitoring techniques, 239
Liquid–solid extraction, use for herbicide
analysis, 18
Loam–silt soil
atrazine transport estimation, 101–114
triazine herbicide fate, 125–138

M

Magnetic particle based solid-phase
immunoassay method, triazine analysis,
43–51
Mass spectrometry–high-performance
liquid chromatography, identification
of herbicide metabolites, 10
Maximum contaminant levels, pesticides in
potable water, 166
MCPA, monitoring in surface water and
groundwater, 237–249
Mecoprop dimethylamine, movement
after application to golf courses,
165–175

Metabolites
  carbamate insecticides, monitoring in
    surface water and groundwater, 237–249
  herbicides
    chemical analysis, 6–11
    contribution to total herbicide pool in
      surface water and groundwater, 291
    environmental concern, 2
    isolation from water, 9
    toxicity, 2–3
  pesticides, monitoring in surface water and
    groundwater, 237–249
Methiocarb, monitoring in surface water
  and groundwater, 237–249
Methiocarb sulfone, monitoring in surface
  water and groundwater, 237–249
Methomyl, monitoring in surface water and
  groundwater, 237–249
2-Methyl-1,3,5-trinitrobenzene, *See* TNT
Metolachlor
  mobility and concentration in
    agricultural runoff, 226–234
  monitoring in surface water and
    groundwater, 237–249
  transport and persistence in groundwater,
    271–285
Metribuzin, role of soil temperatures in
  persistence, 126
Mississippi River, geochemical indicators
  of contaminant transport, 288–301
Missouri stream, hydroxylated atrazine
  degradation products, 254–269
Mobility
  in agricultural runoff, *See* Herbicide
    mobility and concentration in
    agricultural runoff
  photosynthesis-inhibiting herbicides,
    determination using in vivo
    fluorometry, 77–83
  relationship to sorption, 117
Moisture of soils, role in atrazine
  degradation, 143,146*f*
Molinate
  geographic information system map,
    5–6,8*f*
  monitoring in surface water and
    groundwater, 237–249

Monitoring of pesticides and metabolites
  in surface water and groundwater
  carbamate insecticide(s), 248,250–251*f*
  carbamate insecticide metabolites,
    248–249,252*f*
  experimental objective, 239,241
  pesticide properties and usage, 241,242*t*
  sample collection procedure, 240*f*,241–242
  sample handling and analysis
    procedure, 243
  study areas, 241
  surface water, 243–247
Monuron, photosynthesis II inhibition, 78
Movement of pesticides following
  application to golf courses
  influencing factors, 174
  pesticide extraction and analysis
    procedure, 169–171
  pesticide movement
    measurement procedure, 167–169
    simulated fairways, 173–174
    simulated green, 171
Multistage solid-phase extraction
  procedure, solute property analysis,
    22–23,24–25*f*

N

Nebraska
  herbicide and degradate concentration in
    regolith, 178–194
  herbicide mobility and concentration in
    agricultural runoff, 226–234
Nitrogen, transport in runoff from
  corrugation irrigation of wheat, 151–163
Nonpoint source contamination, influencing
  factors, 271
Nonpoint sources of pollution from
  pesticides, importance of assessment, 237
Nutrients, transport in runoff from
  corrugation irrigation of wheat, 151–163

O

OK–8901 and OK–9201, in vivo
  fluorometry for soil mobility and
  adsorptivity determination, 77–83

Organic contaminants, environmental concern, 1

Organic matter, role in herbicide and degradate concentration in regolith in northeastern Nebraska, 178–194

P

Partition coefficient, role in fate and transport of herbicide metabolites, 11–13

Pendimethalin, movement after application to golf courses, 165–175

Persistence
calculation, 276
in groundwater, *See* Herbicide transport and persistence in groundwater
role of soil temperatures, 126

Pesticide(s)
drinking water risk, 166
environmental concern, 165
groundwater contamination, 200–201
monitoring in surface water and groundwater, 237–249
monitoring study techniques in water samples, 239
movement following application to golf courses, 165–175
transport mechanisms, 237

Pesticide degradates, environmental impact in groundwater, 200–221

Pesticide metabolites, *See* Metabolites of pesticides

Pesticide usage, *See* Usage of pesticides

pH, role in herbicide and degradate concentration in regolith in northeastern Nebraska, 178–194

Phosphorus, transport in runoff from corrugation irrigation of wheat, 151–163

Photosynthesis-inhibiting herbicides, in vivo fluorometry for soil mobility and adsorptivity determination, 77–83

Polarity–hydrophobicity–ionogenicity, solute property analysis, 19–21

Postemergence-applied herbicides, transport in runoff from corrugation irrigation of wheat, 151–163

Propanil
fate and transport studies, 12–13
geographic information system map, 5–6,8*f*
metabolites, 12
monitoring in surface water and groundwater, 237–249

R

Rain pattern, role in contaminant transport in Mississippi River, 288–301

Regolith in northeastern Nebraska, herbicide and degradate concentration, 178–194

Retardation factor, calculation, 271,273,275–276

Rice production, pesticide usage, 239

Root zone mixture, composition for greens, 166–167

Runoff
for corrugation irrigation of wheat, transport of nutrients and postemergence-applied herbicides, 151–163
from agricultural areas, herbicide mobility and concentration, 226–234

S

Sample matrix control
aqueous dilution, 28
cross-reactivity, 31–32
soil moisture, 29
soil texture, 29–31
solvent exchange, 28–29
surface tension, 29

Sample preparation, coordination with final determination of herbicides, 18–32

Sand content, role in herbicide and degradate concentration in regolith in northeastern Nebraska, 178–194

Screening tools, use of immunoassays, 54,56*f*

Sensitivity of triazine immunoassays, influencing factors, 43–51

Silt–loam soil
  atrazine transport estimation, 101–114
  triazine herbicide fate, 125–138
Simazine
  analysis without sample preparation, 19
  monitoring in surface water and
    groundwater, 237–249
Simazine immunoassay
  2-position enzyme conjugation and
    6-position immunogen substitution,
    50–51
  6-position substitution, 50,51*t*
Smectite(s)
  hydration of surfaces, 89–91
  structure, 87–89
  surface acidity, 91,93
Smectite surface–atrazine interactions,
    See Atrazine–smectite surface
    interactions
Soil
  atrazine and degradate fate in Iowa,
    140–149
  atrazine transport estimation, 101–114
  triazine herbicide fate, 125–138
Soil analysis, standardization of
    immunoassays, 53–61
Soil and sediment extracts, analysis
    without sample preparation, 19
Soil-applied pesticides, environmental
    concern, 77
Soil mobility and adsorptivity of
    photosynthesis-inhibiting herbicides,
    determination using in vivo fluorometry,
    77–83
Soil moisture
  role in atrazine degradation, 143,146*f*
  sample matrix control, 29
Soil residence times, indicators, 227
Soil samples, in situ derivatization–
    supercritical fluid extraction method
    for chlorophenoxy acid herbicide
    determination, 63–75
Soil temperatures, role in herbicide
    persistence, 126
Soil texture, sample matrix control,
    29–31

Solid-phase extraction
  herbicide metabolite extraction from
    water, 6,9
  pesticide monitoring techniques, 238*f*,239
  use for herbicide analysis, 18
Solid-phase extraction–liquid
    chromatographic techniques, monitoring
    of pesticides and metabolites in surface
    water and groundwater, 237–249
Solute property techniques
  hydrophobicity–polarity–ionogenicity,
    19–21
  multistage solid-phase extraction
    procedure, 22–23,24–25*f*
  supercritical fluid extraction, 21–22
Solution, role in atrazine sorption and
    transport, 122–123
Solvent exchange, sample matrix control,
    28–29
Sorption
  ammonia effect for atrazine, 117–123
  atrazine, 92*f*,93–94
Sorption capacity, role in fate and
    transport of herbicide metabolites, 11–13
Sorption to soil by herbicides,
    measurement by laboratory batch
    equilibration techniques, 117–118
Spain, monitoring of pesticides and
    metabolites in surface water and
    groundwater, 237–249
Specificity of triazine immunoassays,
    influencing factors, 43–51
Standardization of immunoassays for water
    and soil analysis
  DIN approach
  categories of immunoassays,
    55,57,58*t*
  draft standard, 57
  interlaboratory tests, 59,60*f*
  limitations, 61
  procedure, 55
  round robin tests, 57–59
  role, 54–55,56*f*
Styrene divinyl benzene packing, use in
    Empore disk for herbicide metabolite
    isolation from water, 9

Sulfonylurea herbicides, leaching
   potential evaluation, 216–220
Supercritical fluid extraction
solute property analysis, 21–22
use for herbicide analysis, 18
Supercritical fluid extraction–in situ
   derivatization method for
   chlorophenoxy acid herbicide
   determination in soil, *See* In situ
   derivatization–supercritical fluid
   extraction method for chlorophenoxy
   acid herbicide determination in soil
Surface tension, sample matrix control, 29
Surface water
contamination by atrazine and N-
   dealkylated degradation products, 254
herbicide metabolites, 1–13
monitoring of pesticides and
   metabolites, 237–249
Symmetric triazine herbicide fate in
   silt–loam soils, *See* Triazine
   herbicide fate in silt–loam soils

T

Terbutylazine, HPLC-based
   screening method for analysis, 34–41
2,4,5,6-Tetrachloro-1,3-benzenedicarbo-
   nitrile, *See* Chlorothalonil
TNT, standardization of immunoassays for
   water and soil analysis, 53–61
Toxicity, metabolites of herbicides, 2–3
Transport, ammonia effect for atrazine,
   117–123
Transport estimation in silt–loam soil,
   atrazine, 101–114
Transport
agricultural chemicals, 179
contaminants in Mississippi River, *See*
   Contaminant transport in Mississippi
   River
herbicide(s) in groundwater, *See*
   Herbicide transport and persistence
   in groundwater
herbicide metabolites, *See* Fate and
   transport of herbicide metabolites

Transport—*Continued*
nutrients and postemergence-applied
   herbicides in runoff from corrugation
   irrigation of wheat
experimental description, 153
field operation procedure, 153–154
herbicide analysis procedure, 155
irrigation efficiency, 155
irrigation–rainfall procedure, 154,156*t*
nutrient analysis procedure, 154–155
nutrient runoff, 156–158,162*f*
study site, 152*f*,153
surface runoff water sampling
   procedure, 154
transport of herbicides, 158–163
organic solutes in groundwater,
   measurement, 271
Triallate, latitude effect on degradation,
   125–126
Triazine(s), monitoring in surface water and
   groundwater, 237–249
1,3,5-Triazine herbicide(s)
leaching potential evaluation, 207–212
usage, 43
Triazine herbicide, fate in silt–loam soils
atrazine
   ^{14}C dissipation, 133–134
   ^{14}C leaching, 134
   degradation, 134–138
chemical extraction and analysis
   procedure, 129–130
experimental description
   Alaska, 127–128
   Minnesota, 128–129
metribuzin
   ^{14}C dissipation, 130–131
   ^{14}C leaching, 131–132
   degradation, 132–133
study sites, 126–127
Triazine immunoassays
antibody production procedure, 44
atrazine immunoassay
   2-position substitution, 47–48,49*t*
   6-position substitution, 48–49
cross-reactivity determination, 45
cyanazine immunoassay, 47

Triazine immunoassays—*Continued*
  immunoassay format, 44–45
  simazine immunoassay
    2-position enzyme conjugation and
      6-position immunogen substitution,
      50–51
    6-position substitution, 50,51*t*
  structures of compounds, 45–46
  triazine–enzyme hapten conjugation
    procedure, 44
S-(2,3,3-Trichloro-2-propenyl)bis(1-methyl-
  ethyl)carbamothioate, *See* Triallate
Trifluralin
  analysis without sample preparation, 19
  geographic information system map, 4,7*f*
Turfgrass
  factors affecting quality, 166
  intensively managed biotic system, 166

U

Usage
  herbicides
    amounts, 3,4*f*
    analytical costs, 3–4,5*f*
    geographic information system maps,
      4–6,7–8*f*

Usage—*Continued*
  pesticides
    changes resulting from environmental
      studies, 1–2
    environmental concern, 34
    public awareness, 165
    rice production, 239

V

Vapor pressure, role in fate and transport
  of herbicide metabolites, 11–12

W

Water analysis, standardization of
  immunoassays, 53–61
Water quality
  concern in midwestern United States,
    226–227
  role in agricultural practices, 151
Water solubility, relationship to
  degradation, 6,9
Wheat, transport of nutrients and
  postemergence-applied herbicides in
  runoff from corrugation irrigation,
  151–163

Highlights from ACS Books

Good Laboratory Practice Standards: Applications for Field and Laboratory Studies
Edited by Willa Y. Garner, Maureen S. Barge, and James P. Ussary
ACS Professional Reference Book; 572 pp; clothbound ISBN 0–8412–2192–8

Silent Spring Revisited
Edited by Gino J. Marco, Robert M. Hollingworth, and William Durham
214 pp; clothbound ISBN 0–8412–0980–4; paperback ISBN 0–8412–0981–2

The Microkinetics of Heterogeneous Catalysis
By James A. Dumesic, Dale F. Rudd, Luis M. Aparicio, James E. Rekoske,
and Andrés A. Treviño
ACS Professional Reference Book; 316 pp; clothbound ISBN 0–8412–2214–2

Helping Your Child Learn Science
By Nancy Paulu with Margery Martin; Illustrated by Margaret Scott
58 pp; paperback ISBN 0–8412–2626–1

Handbook of Chemical Property Estimation Methods
By Warren J. Lyman, William F. Reehl, and David H. Rosenblatt
960 pp; clothbound ISBN 0–8412–1761–0

Understanding Chemical Patents: A Guide for the Inventor
By John T. Maynard and Howard M. Peters
184 pp; clothbound ISBN 0–8412–1997–4; paperback ISBN 0–8412–1998–2

Spectroscopy of Polymers
By Jack L. Koenig
ACS Professional Reference Book; 328 pp;
clothbound ISBN 0–8412–1904–4; paperback ISBN 0–8412–1924–9

Harnessing Biotechnology for the 21st Century
Edited by Michael R. Ladisch and Arindam Bose
Conference Proceedings Series; 612 pp;
clothbound ISBN 0–8412–2477–3

From Caveman to Chemist: Circumstances and Achievements
By Hugh W. Salzberg
300 pp; clothbound ISBN 0–8412–1786–6; paperback ISBN 0–8412–1787–4

The Green Flame: Surviving Government Secrecy
By Andrew Dequasie
300 pp; clothbound ISBN 0–8412–1857–9

For further information and a free catalog of ACS books, contact:
American Chemical Society
Customer Service & Sales
1155 16th Street, NW, Washington, DC 20036
Telephone 800–227–5558

Bestsellers from ACS Books

The ACS Style Guide: A Manual for Authors and Editors
Edited by Janet S. Dodd
264 pp; clothbound ISBN 0–8412–0917–0; paperback ISBN 0–8412–0943–X

Understanding Chemical Patents: A Guide for the Inventor
By John T. Maynard and Howard M. Peters
184 pp; clothbound ISBN 0–8412–1997–4; paperback ISBN 0–8412–1998–2

Chemical Activities (student and teacher editions)
By Christie L. Borgford and Lee R. Summerlin
330 pp; spiralbound ISBN 0–8412–1417–4; teacher ed. ISBN 0–8412–1416–6

Chemical Demonstrations: A Sourcebook for Teachers,
Volumes 1 and 2, Second Edition
Volume 1 by Lee R. Summerlin and James L. Ealy, Jr.;
Vol. 1, 198 pp; spiralbound ISBN 0–8412–1481–6;
Volume 2 by Lee R. Summerlin, Christie L. Borgford, and Julie B. Ealy
Vol. 2, 234 pp; spiralbound ISBN 0–8412–1535–9

Chemistry and Crime: From Sherlock Holmes to Today's Courtroom
Edited by Samuel M. Gerber
135 pp; clothbound ISBN 0–8412–0784–4; paperback ISBN 0–8412–0785–2

Writing the Laboratory Notebook
By Howard M. Kanare
145 pp; clothbound ISBN 0–8412–0906–5; paperback ISBN 0–8412–0933–2

Developing a Chemical Hygiene Plan
By Jay A. Young, Warren K. Kingsley, and George H. Wahl, Jr.
paperback ISBN 0–8412–1876–5

Introduction to Microwave Sample Preparation: Theory and Practice
Edited by H. M. Kingston and Lois B. Jassie
263 pp; clothbound ISBN 0–8412–1450–6

Principles of Environmental Sampling
Edited by Lawrence H. Keith
ACS Professional Reference Book; 458 pp;
clothbound ISBN 0–8412–1173–6; paperback ISBN 0–8412–1437–9

Biotechnology and Materials Science: Chemistry for the Future
Edited by Mary L. Good (Jacqueline K. Barton, Associate Editor)
135 pp; clothbound ISBN 0–8412–1472–7; paperback ISBN 0–8412–1473–5

For further information and a free catalog of ACS books, contact:
American Chemical Society
Customer Service & Sales
1155 16th Street, NW, Washington, DC 20036